Handbook Timing Belts

Raimund Perneder · Ian Osborne

Handbook Timing Belts

Principles, Calculations, Applications

Springer

Raimund Perneder
Milinowskistr. 37a
14169 Berlin
Germany
e-mail: raimund.perneder@t-online.de

Ian Osborne
Transmission Developments
Dawkins Road
BH15 4HF Poole
UK
e-mail: iosborne@transdev.co.uk

ISBN 978-3-642-44586-6
ISBN 978-3-642-17755-2 (eBook)
DOI 10.1007/978-3-642-17755-2
Springer Heidelberg Dordrecht London New York

Softcover reprint of the hardcover 1st edition 2012

Cover design: eStudio Calamar S.L.

Printed on acid-free paper

Springer is part of Springer Science+Business Media (www.springer.com)

Foreword

Today, timing belts are used in a wide range of innovative drive solutions for use in mechatronic systems that combine mechanics, sensors, control systems and servo technology with freely programmable and distributed drive solutions. The timing belt has become widely used in this framework and has contributed to many industrial drive innovations. For example, in automation and handling equipment using low-backlash robotic timing belt drives working under high dynamic loads, particularly during start-up. In continuous operation, timing belts offer low maintenance and guarantee accurate positioning at high speed.

This handbook is intended for application engineers in both design and development departments and is also suitable as a guide for students at universities, institutes of higher learning and technical colleges. When it comes to drive design, the target should always be an elegant solution, incorporating the use of simple and robust mechanical concepts that can be implemented both cost-effectively and also satisfy "innovative solution" criteria. Timing belts offer a wide variety of approaches to these problems. The information within this book has been derived from years of experience and enables the user to dimension timing belt drives utilising proven generic examples from the power transmission, conveying and linear drive sectors. In addition, the investigation of non-optimum operating conditions and sources of timing belt damage provides production engineers with the ability to optimize underperforming drives. This work also contains guidelines for the design of auxiliary components and the surrounding support structure. Having a good knowledge of these mechanisms will aid functional drive dimensioning.

The handbook is based on the author's 30 years of professional employment in the mechanical power transmission sector. During this period, the timing belt, as a newly introduced drive component, has gradually established a prominent position in the marketplace. Concurrently with this progress and market acceptance, timing belt manufacturers have refined the manufacturing processes for production as the industrial belt user has always looked for functional and economic solutions. From the author's own experience and due to numerous personal contacts within industrial companies, he has been able to document many examples of concepts and drives, which were initially compiled as a loose-leaf collection. These applications comprise the main component of this handbook.

The units in the equations are defined in the SI system. Derivations have been omitted. The numerical equations are presented so that the physical relationships are clear.

Thanks

For all the ideas for content and editorial design, I would like to thank all who have participated in the origination of this book.

My special thanks go to Prof. Dr.-Ing. habil. Dr. hc. Werner Krause from Dresden Technical University for providing the impetus for this book project. Without his inspiration and encouragement this work would not have begun. He supported my efforts from the beginning and supported the technical content for structure and linguistic implementation. Further special thanks go to Prof. Dr. Henning Meyer of Berlin Technical University, in my hometown, who provided continued scientific support. He also established contacts with the publisher which developed into a most pleasant cooperation between a dedicated team of staff from Springer Verlag in Heidelberg and Berlin Technical University.

The origin of this work was significantly influenced by my long affiliation with Wilhelm Herm. Müller GmbH & Co. KG and the Mulco Group companies. I thank all my colleagues and associates on behalf of my book project. In a professional environment I offer my special thanks to Mr. Rudi Kölling of Breco Antriebstechnik GmbH for permission to reproduce the contents of their documentation.

The Future

This work is intended for use as a handbook. From this the reader can easily deduce that it is an exhaustive work about timing belt technology and essentially contains reliable information on all main and related topics. Unfortunately a 100%-accurate work is not feasible. Therefore any technical errors are attributable to the author and translator. Identification of errors as well as suggestions and constructive criticism, can be sent to the author at: raimund.perneder@t-online.de.

The potential for any revised edition of this handbook would be to enhance current knowledge. The timing belt is a relatively new drive component and further developments are ongoing. This handbook is a balanced mix of knowledge/ scientific theory and reflects current operational practice. It aims to promote technical discussion combined with practical solutions. Interested readers and timing belt users are welcome to contact the author for confidential consultation. Any eventual release of information for the enrichment of the handbook would be left to the discretion of the client.

I hope this handbook delivers as much benefit to the reader, as I have had the joy of writing it.

Berlin, Germany, March 2011 Raimund Perneder

Translator's Note

I would like to thank Raimund Perneder for the opportunity of collaborating with him on the translation of this most important work in the field of timing belt technology. I would have not taken on this task if I had not have believed that the subject matter was important to an English speaking audience. Translating from one language to another is always fraught with problems and assumptions. It could be thought that the translation of technical subjects should be straightforward, but the problem is that those technical subjects are held together by words and concepts that are not directly comparable from one language to another. This being so, I have tried to pitch the translation at an international audience where English may not be the first language of the reader. To the native English reader some of the text may seem to be florid or over-descriptive but that is for the convenience of the whole audience.

I would like to thank my colleagues at Transmission Developments for their support and help over the timeline of this project and my partner Vivienne for the proofreading, suggestions and innumerable cups of tea.

Poole, England, March 2011 Ian Osborne

Contents

Chapter 1
The Right Drive in One Mouse Click

"Please enter your drive parameters including speed, power and the desired drive ratio in the program. Drive calculation is being processed. Please wait ..." then with one mouse click the output could read "... and therefore a toothed belt drive is the right solution!" This may seem to be the future for designers at work! In this or in a similar manner, as predicted by experts, the selection of parts and components of any future machinery will be solved by computer-aided design. Obviously, a price quotation would also be desirable! This may seem plausible and such a program is easy to imagine, but in reality it is not feasible as the scope of drive design is just too diverse. If universal drive selection by a mouse click were possible then this handbook of timing belt technology would not be necessary. See Fig. 1.1.

Therefore, this book both introduces the topic and shows which timing belt drive solutions are preferable. Additionally many examples are presented as

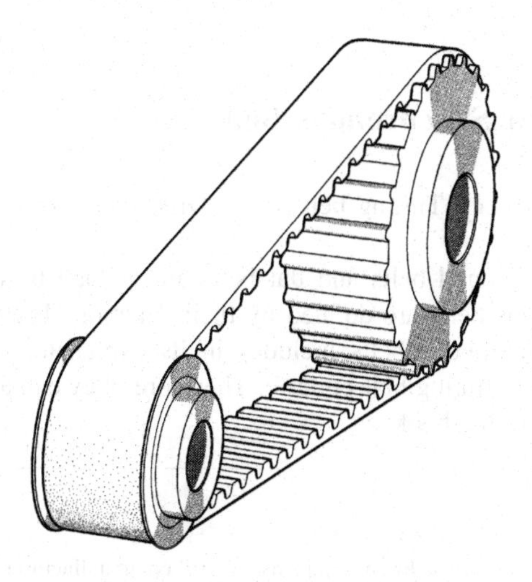

Fig. 1.1 A complete timing belt drive. The basic drive comprises of one timing belt and two matching pulleys

R. Perneder and I. Osborne, *Handbook Timing Belts*,
DOI: 10.1007/978-3-642-17755-2_1,

applications that describe their mission-critical functional advantages. Chapters 3–5 deal with illustrated application examples and provide a wealth of information and solutions within normal drive/space constraints. The illustrated drives can also be regarded as model solutions to be applied to similar but different tasks, and through the combination of two or more known applications, and then innovative new solutions may be developed. Existing designs that can be borrowed from other sources can also considerably reduce the workload.

Also useful are those solutions that are able to access elements from across diverse disciplines, so that the required dynamic properties can be incorporated [72]. The thought processes needed for innovative technical designs call for a creativity that is (still) not available from a mouse click.

The reader will be aware that the main focus and main application areas are timing belt drives. Applications with similar requirements will facilitate the selection process and instructions and recommendations for detailed design are provided in this book. The machine chassis surrounding the drive should be chosen to support the drive components. Timing belts have many positive benefits and the best practice is to use them for their maintenance-free and quiet operation.

Timing belts and their associated pulleys must first be properly dimensioned. For primary drive layout, two sizing steps are required. First, the geometric design of the centre distances, pulley sizes and belt must be considered in the light of the available installation space. Secondly, it must be considered whether the performance requirements and chosen drive geometry can safely transfer the torque and drive forces.

PC-based programs[1] for the design of timing belt drives are readily available and thus some elements of design are available at a mouse click. However, the choice of the drive components and the co-ordination with the surrounding structure will always remain the responsibility of the designer.

Timing Belt or Synchronous Belt?

Which term is correct: Timing belt or the DIN recommended synchronous belt?
Whilst Vee belts, round belts and flat belts are defined by their geometric shape, synchronous belts are defined by their function. The DIN ISO 5296 standard for synchronous belts includes in its explanatory text the terms "timing belt" and "timing belt pulley". Therefore, they can also be found in widespread use in this book.

[1] This book does not contain a PC-program as every belt manufacturer has product-specific programs. Most manufacturers provide a computer-assisted calculation program in connection with their sales catalogues [84].

Available types of timing belt differ in their profile geometry, their pitch, the tension member construction and the elastomers used. Depending on the manufacturer, the belts are available in chloroprene rubber (oil resistant) or made of cast or thermoplastic polyurethane (usually oil-proof). The standard tooth profiles and associated dimension tables are listed in Chapter 2.3.1–2.3.16.

Chapter 2
Foundations

Abstract A sound basis of drive component selection for structures in motion requires an extensive knowledge of the operating characteristics of the components. In the following chapter the reader will discover the properties of commercially available timing belts manufactured from differing elastomers and with differing tension members. In describing their manufacture, their properties can be derived and their main application areas understood. Leading manufacturers' products are referred to by name. The operating characteristics of all timing belt profiles have basically the same force transfer mechanisms. The user will find herein many references to the required pre-tension, achievable transmission accuracy, potential noise and construction design details. The detailed design calculations provide the required technical parameters and optimization methods.

2.1 Timing Belt Drives

Timing belt drives are constructed from components that convert and transfer speeds, directions, torques and forces. Their basic functions embody rotational motion in the pulleys and linear motion in the belt spans. They are used for transporting mechanical forces or used to drive bodies on a predetermined path. They convert the given input speed to the required output speed. Common applications are based on synchronous trains of parts and components. A synchronous timing belt drive consists of the connected members: *pulley-belt-pulley* and additionally there is the support *structure* (design environment) to allow the transmission of the forces produced. The timing belt is classified in the group of *transmission drives*, see Fig. 2.1. Other common names are variable transmissions and positive drives. With this type of drive system, larger centre distances can be bridged without problem and power can be split cost-effectively over several driven pulleys.

The thing that all timing belt transmissions have in common is that the movements of its individual members must follow the principles imposed on them.

R. Perneder and I. Osborne, *Handbook Timing Belts*,
DOI: 10.1007/978-3-642-17755-2_2, © Springer-Verlag Berlin Heidelberg 2012

This practice is referred to as constrained or synchronous motion. In this sort of transmission, positive drive prevails, in any point of any transmission link, so the position of the other drive components are clearly related to that of any other component. Due to the fixed structure of this kind of drive the transmission will always have a uniform ratio.

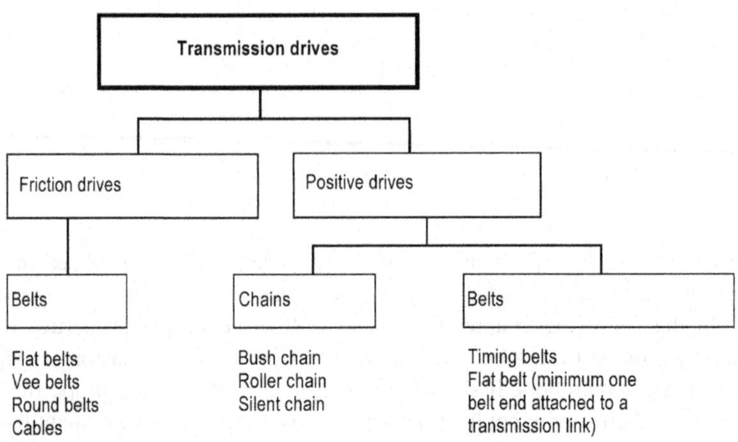

Fig. 2.1 Transmission drives diagram

Why Timing Belts?

> If drive systems have high levels of acceleration and braking, or if they are required to precisely position at high speed, then timing belts must be of primary interest to the designer.

This chapter describes the core properties of a timing belt. Timing belts work simultaneously at both high and low speeds and are particularly suited in acceleration and braking applications aided by their low mass. Timing belts rank as synchronous driving elements within the group of transmission drives and they achieve their high power capacity through low losses, in combination with the associated toothed pulleys. Pulleys are usually made of light metal alloy or sometimes from plastic. The arc of belt wrap over the pulleys always distributes the load across many successive teeth and therefore the greater the number of teeth in mesh, the lower unit load per tooth and thus the greater the torque able to be transmitted by the drive.

Their excellent performance in start-up or braking operations is a result of the interaction between the timing belt and the pulley. Each tooth of flexible elastomer operates between the rigid flanks of the pulley teeth. With every drive reversal from starting to braking and vice versa, the torque reversal will also change the loaded flanks of the belt teeth in mesh with the pulley. This load change occurs gently and without shock loads, thanks to the elasticity of the belt teeth. Timing belts have very benign running characteristics with positive benefits for both upstream and downstream drive components simultaneously. They experience little or no permanent or intermittent fretting during rapid changes in direction of rotation. Under these operating conditions, timing belts demonstrate their superior durability over all other drive elements.

Modern stepper motors and servo technology are often used in production processes to solve point-to-point motion tasks. These processes involve mechanical handling tasks such as gripping, moving and depositing and such tasks are characterized by a limited range of movement, requiring constant starting, braking and positioning. A further complication is that the changes of direction forces in the drive train are concentrated mostly at the same bending points. These are ideal application areas where the use of timing belts will noticeably increase, as they can be perfectly integrated into production processes with many different operating conditions.

The timing belt combines all the advantages of conventional belts (flat, wedge and Vee-ribbed belts) such as high permissible speeds and low-noise operation and as with chain drives, provide slip-free, synchronous motion transfer. The main difference, compared with chain drives, is that the timing belt is of continuous and non-segmented construction and as the timing belt moves from straight to curved running, it experiences no wear or elongation. Moreover, the reduced polygon effect of the belt means that drive noise is correspondingly lower.

The integration of timing belts into technical drive solutions for mechanical and precision engineering is particularly facilitated by the fact that the belt has a large range of applications and, in heavy and continuous use, requires no lubrication. However, it is important to understand that if the immediate environment of the belt drive is subject to lubrication (grease, oil, oil-mist), then an oil-proof belt must be chosen.

2.2 Major Geometric Dimensions

Figure 2.2 shows dimensions of a typical timing belt drive.
Table 2.1 contains the corresponding names and descriptions of each character.

Other variables are co-dependent on the main geometric dimensions and the following useful drive design correlations are listed. For example, the belt length can be expressed as a product of the belt pitch and the number of teeth in the belt:

$$l_{\mathrm{B}} = p \cdot z_{\mathrm{B}}. \tag{2.1}$$

The drive ratio can be calculated from the number of teeth in the pulleys. The ratio is the quotient of whole numbers:

$$i = \frac{z_2}{z_1}. \tag{2.2}$$

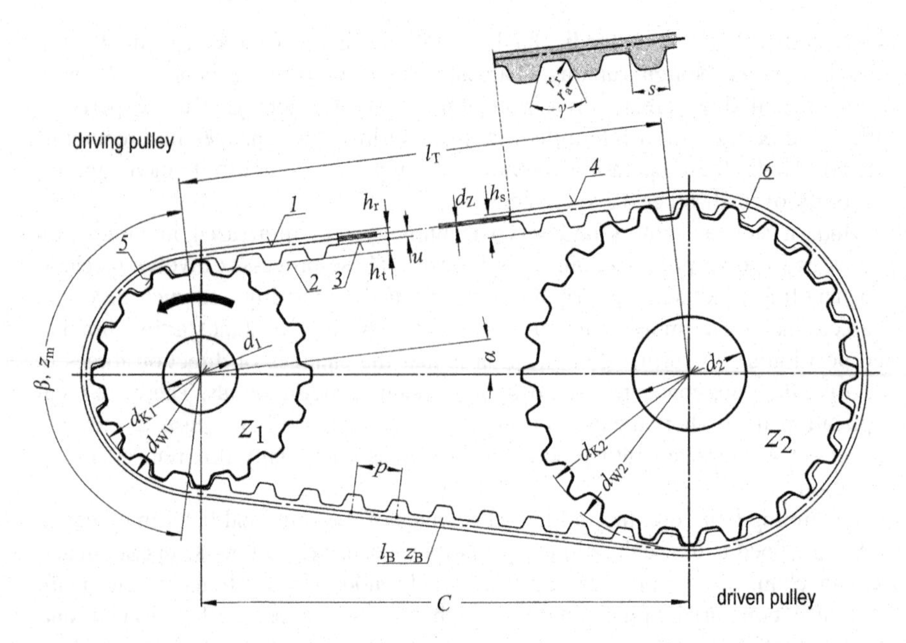

Fig. 2.2 Main geometric dimensions

The centre distance C is calculated from the number of teeth in the pulleys and the belt according to the following relationship:

$$C \approx \frac{p}{4}\left[\left(z_B - \frac{z_2+z_1}{2}\right) + \sqrt{\left(z_B - \frac{z_2+z_1}{2}\right)^2 - \frac{2}{\pi^2}(z_2-z_1)^2}\right]. \tag{2.3}$$

In the event that the only available drive parameters are the centre distance, the number of teeth in the pulleys and the belt pitch, then the length of the belt can be found from the relationship:

$$l_B = \frac{p}{2}(z_2+z_1) + \frac{p\cdot\alpha}{\pi}(z_2-z_1) + 2C\cdot\cos\alpha, \tag{2.4}$$

with

$$\alpha = \arcsin\frac{p(z_2-z_1)}{2\pi\cdot C}, \tag{2.5}$$

or roughly

$$l_B \approx \frac{\pi}{2}(d_{W2}+d_{W1}) + 2C + \frac{(d_{W2}-d_{W1})^2}{4C}. \tag{2.6}$$

For simple drives consisting of pulleys with an equal number of teeth (ratio $i = 1$) the centre distance can be found from:

$$C = \frac{p}{2}(z_B - z) = \frac{1}{2}(l_B - p\cdot z). \tag{2.7}$$

Table 2.1 Designation and description of the main geometric dimensions

Character	Designation (Units)	Definition
–	Timing belt drive	Drive system comprising a timing belt and two or more timing pulleys.
1	Pitch line *also Neutral line*	Circumferential line in the belt that stays the same length when the belt is bent perpendicularly to its base. The pitch line is located in the centre of the tension member.
2	Tip line	The tip line is that line joining the tips of the belt teeth.
3	Root line	The root line is that line joining the roots of the belt teeth.
4	Belt back, back line	The back of the belt or the back line is the outside boundary of the belt.
5	Driving pulley working flank	The working flank of the driving pulley transmits the motion or force from the pulley into the belt.
6	Driven pulley working flank	The working flank of the driven pulley transmits the motion or force from the belt into the pulley.
C	Centre distance (mm)	The centre distance is the shortest distance between two pulley centres subject to the pre-tension load of the belt.
n n_1 n_2	Number of revolutions (min^{-1})	The speed of the small pulley is denoted by n_1 and the large pulley with n_2 (if the large pulley is the driving pulley then the designations should be reversed).
l_B	Belt length (mm)	The belt length is based on the pitch length when under the pre-tension load.
l_t	Span length *also strand length* (mm)	The span length is the distance between neighbouring pulleys where the belt leaves and enters the pulleys at a tangent.
l_1	Loaded belt length (mm) see Fig. 2.14	The belt length under load consists of the loaded belt section plus half of the respective pulley's wrap angle length.
l_2	Unloaded belt length (mm) see Fig. 2.14	The unloaded belt length consists of the unloaded belt section plus half of the respective pulley's wrap angle length.
z_B	Number of teeth in the belt	The total number of teeth on the toothed side of the belt.
z_m	Total number of teeth in mesh	Total number of belt teeth fully meshed with pulley teeth on a pulley.
z_e	Number of meshing teeth for belt calculations	To calculate the peripheral force the value for meshing teeth is rounded down to a whole number. Depending on the belt type the number of meshing teeth is limited to a maximum value e.g. $z_{e\ max} = 12$.
z z_1 z_2	Number of teeth in the pulley	Pulley teeth are used to mesh with the teeth on the belt and also provide radial support for the belt teeth. The number of teeth in the small pulley-usually the driving pulley-is z_1. The number of teeth in the large pulley-is z_2 (if the large pulley is the driving pulley then the designations should be reversed).

(continued)

Table 2.1 (continued)

Character	Designation (Units)	Definition
p p_b p_p	Tooth pitch (mm)	The tooth pitch or nominal pitch equals the distance between two adjacent teeth at the pitch line under pre-tension. To differentiate between belt and pulley pitch use the following nomenclature: p_b = Belt pitch, p_p = Pulley pitch.
d_W d_{W1} d_{W2}	Pitch circle diameter (mm)	The pitch circle diameter lies in the middle of the tension member and through that line of arc formed around the timing pulley centre where the belt pitch p_b and the pulley pitch p_p are the same. The pitch circle diameter is a tolerance-free nominal value where the measurement for the small pulley is designated d_{W1} and the large pulley d_{W2} (if the large pulley is the driving pulley then the designations should be reversed).
d_K d_{K1} d_{K2}	Pulley outside diameter (mm)	The outside diameter is the measurement of the outer surface of the timing pulley around which the timing belt wraps. The outside diameter of the smaller pulley is designated d_{K1} and the larger d_{K2} (if the large pulley is the driving pulley then the designations should be reversed).
d d_1 d_2	Bore diameter (mm)	The bore in the pulley is concentric to the outside diameter of the teeth and it usually serves to accommodate the drive shaft. The bore in the small pulley is designated d_1 and the large pulley d_2 (if the large pulley is the driving pulley then the designations should be reversed).
β	Angle of wrap (°)	The angle of the arc on which the belt wraps around the smaller toothed pulley.
α	Span inclination angle (°)	The angle of which the belt leaves the small pulley taken from the centre line between the small and large pulleys
h_s	Belt total thickness (mm)	The total thickness (total height of the belt) is the measurement from the tip line to the back line of the belt.
h_d	Double-sided belt thickness (mm)	The total thickness of the double-sided belt is the distance between the tip line of one side to the tip line of the other side.
h_t	Belt tooth height (mm)	The belt tooth height is the distance between the base of the tooth to the tip line.
h_r	Belt back height (mm)	The belt back height is the distance between the back line and the root line of the belt.
d_Z	Tension member diameter (mm)	The tension member diameter is the diameter measurement of the tension member.
s	Tooth root width (mm)	The tooth root width is the linear distance between the opposing flanks of a tooth on the root line of a belt under tension.
γ	Flank angle (°)	The belt tooth angle 2γ is the total angle between the two flanks. The half-angle is the flank angle.
r_a	Tip radius (mm)	The tip radius connects the tooth flank and the tip line of the belt.

(continued)

Table 2.1 (continued)

Character	Designation (Units)	Definition
r_r	Root radius (mm)	The root radius connects the tooth flank and the root line of the belt.
i	Drive ratio	The drive ratio is a quotient of the number of teeth in the pulleys z_2/z_1 or the speeds n_1/n_2.
b	Belt width (mm)	The belt width is measured transversely across the belt from the right to the left flank of the timing belt.
B	Toothed width of the pulley (mm) see Fig. 2.28	The toothed width relates to the distance between the two adjacent faces of the teeth. If the pulley has flanges then it relates to the outside distance between the flanges instead.
u	U-value (mm)	The distance between tension member centre and the root line of the belt is denoted as the U-value.

Names and descriptions taken from Krause, W., Metzner, D.: Timing belt transmission [77] and ISO 5288 [55]. Other definitions for pulleys and toothform profiles can be found in Chapter 2.15, Fig. 2.28 and Table 2.6

The pitch circle diameter of a timing pulley is calculated by:

$$d_W = \frac{z \cdot p}{\pi}. \tag{2.8}$$

The pitch circle diameter is a tolerance-free nominal reference variable from which all other important pulley measurements are taken, such as outside diameter, root diameter and back line of the belt.

The pitch circle diameter cannot be directly measured (see pulley quality control in Chapter 2.15) as it lies outside of the pulley body and is the diameter of the circular line around the pulley centre which is formed from the pitch line of the tension member as it passes around the pulley. Timing belts are generally supported on the outside diameter of the pulley and this influences other dimensions of the pulley as well as the actual belt pitch length. This, therefore, depends on the proper calculation of the outside diameter of the pulley which is calculated from the relationship:

$$d_K = d_W - 2(u - v_K) = \frac{z \cdot p}{\pi} - 2(u - v_K). \tag{2.9a}$$

The actual pitch circle that extends around the pulley has a variable diameter that fluctuates with the tooth pitch for the value of d_W. It has a larger effective diameter over the tooth and a slightly smaller effective diameter over the tooth gap.

The value v_K in mm acts as a radial tooth profile correction to the outside diameter in order to reduce the actual belt wrap length to the theoretical ideal length. This reduction results from the deformation forces in the elastomer where the belt is supported on the pulley's outside diameter. Through these support forces, other lateral loads arise on the tension member and, as a consequence of this, flattening occurs in the stranded cable, resulting in a change of the U-value u. Another reduction in the wrap length is caused by the polygon effect on the belt.

The actual pitch line extending around the pulley has a fluctuating value for d_W in rhythm with the tooth profile. Thus, in Eq. 2.9a, the required profile correction v_K has the task of compensating for the sum of all these variations. Equality of pitch between belt and pulley is at the outside or tip diameter of the pulley, where both the theoretical and actual belt wrap length coincides.

The absolute values of v_K for each type of belt are from each manufacturer's empirically determined values and range between 0 and 0.15 mm. The correction values for each belt profile are dependent on the pitch size, the material characteristics of the timing belt components, the pulley tooth gap geometry and, in particular, on whether the tooth root of the belt rests partly or entirely in the pulley tooth gap.

The toothform meshing, described above, often depends on the manufacturer's correctly and precisely maintained values of only a few hundredths of a mm to achieve equality of pitch between the belt and pulley. Only pitche equality of both belt and pulley leads to low-friction tooth meshing and desirable smooth running characteristics. Correspondingly, the profile correction value v_K has a great importance to the quality of pulley manufacture. However, these very small correction parameters are not published by the belt manufacturers.

In real life, the timing belt user will find that the correction values do not explicitly affect the surrounding design. Suffice to say, that the wrap length of the belt around the pulleys is considered to be a perfect circle. Thus, simplifying Eq. 2.9a, we get the relationship:

$$d_K \approx d_W - 2u \approx \frac{z \cdot p}{\pi} - 2u. \tag{2.9b}$$

This simplified relationship between the pitch circle diameter and the outside diameter is also consistent with other references such as DIN 7721 [20].

For some types of belts (e.g. high-power AT types, see Chapter 2.3.3), the belt is supported solely by the tooth root in the pulley. In this case, the important measurement, relating to the arc of contact of the belt, is the tooth root diameter d_F. It is calculated from the relationship:

$$d_F = d_W - 2(h_t + u - v_K) = \frac{z \cdot p}{\pi} - 2(h_t + u - v_K). \tag{2.10a}$$

Equation 2.10a is related to Eq. 2.9b above and applying the simplified calculation the relationship is:

$$d_F \approx d_W - 2(h_t + u) \approx \frac{z \cdot p}{\pi} - 2(h_t + u). \tag{2.10b}$$

2.3 Belt Profiles

The large variety of available belt profiles is inextricably linked with the evolution of the timing belt. The first successful prototype applications led to the realization of the obvious practical benefits and led to rapid market introduction.

The first timing belts capable of being used as drive belts with matching pulleys were developed by US. Rubber Corporation in the 1940's (latterly known as Uniroyal, and today known as the Gates Corporation, Denver, USA) [36]. Their first use was in textile machines and industrial sewing machines. The inventor, Richard Case, [12] decisively improved the synchronization between the needle and bobbin in the Singer Sewing Machine, see Fig. 2.3. He was the first to understand the relationship between the neutral tension member line of the timing belt and the pitch circle diameter of the pulley and defined the basic terms of timing belt technology, which are still in use today. The engineers from the then US Rubber Corporation referred to the group of products as special *toothed flat belts*. The product consisted of a composite structure made of rubber and a special tension member with cotton tooth-facing, later with a nylon (polyamide, PE) fabric tooth-facing. Such endless toothed flat belts were produced by vulcanization in mould-forms. The principles of the positive drive belt soon proved to be such a success, that it was decided to transfer the technology to other applications in the field of mechanical engineering. Due to the obvious benefits, imperial pitch timing belts were introduced into the US market in 1946 and are still in worldwide use today. Eventually, six different standard tooth profiles were successfully introduced into the marketplace. These were standardized from 1977 in DIN ISO 5294 [22] for belts and in DIN ISO 5296 [20] for pulleys.

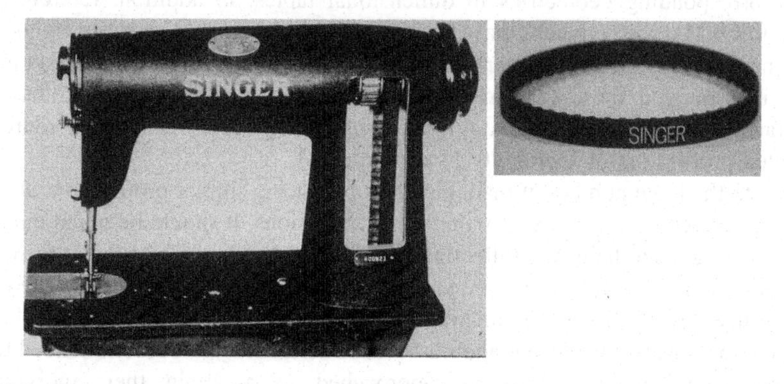

Fig. 2.3 First worldwide application of timing belts in Singer Sewing Machines

In Germany the development and launch of the T-profile metric pitch timing belts by the MULCO Group, Hanover began in about 1950. The belts were made of Contilan®, a cast polyurethane with a hardness of 90 Shore A and tension members of steel cords. The same production method is still used today for casting endless belts in closed moulds. T-profile drives are now standardized in DIN 7721 Part 1 for the belts and Part 2 for the pulleys [20]. In the 1960s the first use of timing belts for automotive overhead camshaft drives was successfully pioneered by *Hans Glas* Automotive, using MULCO belts. Through applications in the automotive industry and especially due to a wide acceptance in general engineering, the inventors decided to further develop other manufacturing processes for the production of timing belts. Following the introduction of cast and vulcanized endless belts, BRECO Antriebstechnik, Porta Westfalica (a member of the MULCO group) developed the thermoplastic polyurethane manufacturing processes for timing belts in 1968. The result was open length extruded belts in rolls, later injection moulded timing belts, and extruded endless belts. BRECO Company was also the pioneer belt manufacturer to produce endless joined timing belts from open lengths. As application demand increased, timing belt users called for higher power capacity, greater rigidity and improved accuracy. The resulting developments have led to new types of belts with optimized materials and significantly stronger tension members, or even new tooth profiles. The performance gains to date are impressive indeed and development is still ongoing.

Against the backdrop of consistent development of a successful market product, today's user can draw on a wide range of different types of belts, depending on the drive task to be solved.

> Different types of belts are not interchangeable. Each profile requires-apart from a few minor exceptions-a different pulley or toothform geometry.

The following pages consist of the current commercially available profiles, with their corresponding geometries in dimensional tables. In addition, reference is made to OEMs, who typically have registered their design and property rights for each new profile. After the eventual expiry of these rights, other manufacturers have incorporated these formerly patented profiles into their own production programs. This allows the user to rely on a sufficiently diverse number of sources. Patented profiles (as of March 2011) are indicated.

Due to the large number of manufacturers producing similar profiles, not all the detail measurements have exactly the same dimensions. It should be noted that, in particular, the back height h_r (affecting the belt height h_s) can be subject to some variation. Thus the deviations observed from the table values are averaged or standardized by rule-based calculation. Since it is to the benefit of the user to have reproducible measurements and their associated tolerances, it is recommended that the respective manufacturers are approached to ascertain their individual specifications.

The timing belt drive user can assume that, in each case, there is a consistency between the dimensionally interchangeable belt profiles that will run on the pulleys of each system. While the drive geometry is identical, the technical data can exhibit significant variations. The use of modified elastomers (for high or low temperature) and the application of different tension members (standard, enhanced, or particularly flexible cords) leads to a variation of property characteristics. In calculating the performance, one is therefore dependent on the technical data from the respective manufacturer.

2.3.1 Imperial Pitch, Standard Profiles

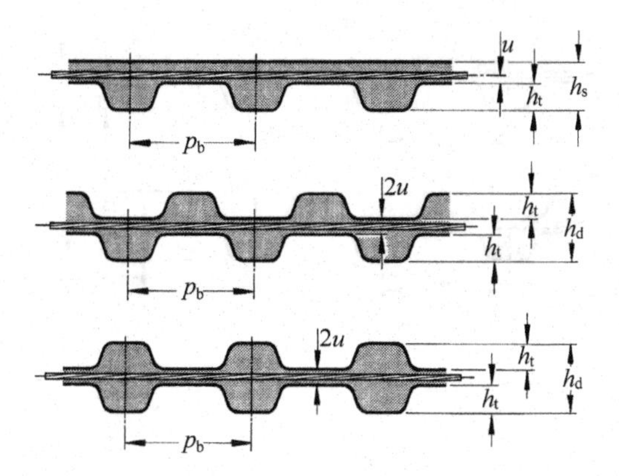

Pitch designation[a]	p_b Inch	p_b mm	h_s mm	h_t mm	h_d mm	u mm	Remarks[b]
MXL	0.08	2.023	1.140	0.510	1.530	0.225	DIN ISO 5296
XL	1/5	5.080	2.300	1.270	3.050	0.225	DIN ISO 5296
L	3/8	9.525	3.600	1.910	4.580	0.380	DIN ISO 5296
H	1/2	12.700	4.300	2.290	5.950	0.685	DIN ISO 5296
XH	7/8	22.225	11.200	6.350	15.490	1.395	DIN ISO 5296
XXH	11/4	31.750	15.700	9.530	22.100	1.520	DIN ISO 5296

[a] The pitch designation describes both the tooth geometry and the pitch
[b] DIN ISO 5296 [24]

These are inch pitch belts with a trapezoidal profile made from polychloroprene rubber and a glass fibre tension member with a polyamide tooth facing. They were developed around 1940 by the US Rubber Corporation, now the Gates Corporation, Denver, USA [37]. These belts are produced worldwide by almost all leading belt manufacturers. They are also available in polyurethane with steel or Aramid tension members in moulded endless belts, endless joined belts and as open metre lengths. The double-sided belts are available both as opposing-tooth and tooth-staggered designs.

2.3.2 Metric Pitch, Standard T Profiles

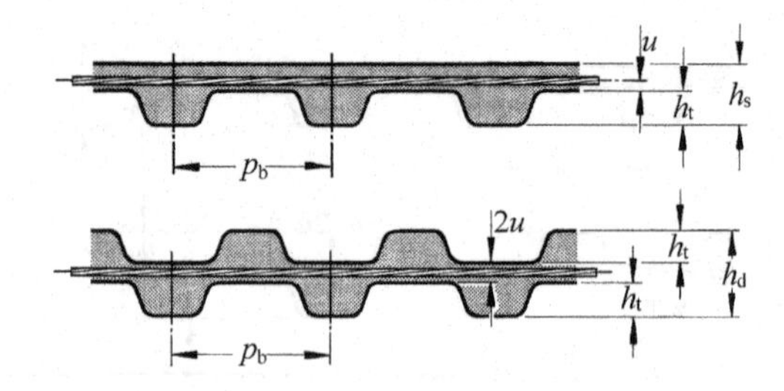

Pitch designation	p_b mm	h_s mm	h_t mm	h_d mm	u mm	Remarks[a]
T 2	2.0	1.1	0.5	–	0.3	–
T 2.5	2.5	1.3	0.7	2.0	0.3	DIN 7721
T 5	5.0	2.2	1.2	3.4	0.5	DIN 7721
T 10	10.0	4.5	2.5	7.0	1.0	DIN 7721
T 20	20.0	8.0	5.0	13.0	1.5	DIN 7721

[a] DIN 7721 [20]

With a trapezoidal tooth profile and metric pitch, the T section timing belts are made from polyurethane with steel or Aramid tension members. The symbol T stands for trapezoidal profile. This belt was developed around 1955 by Wilhelm Herm. Müller GmbH [83] in co-operation with Continental GmbH [15], both of Hanover, Germany. The MULCO group [84] distributed the belts under the brand name Synchroflex® in Germany, and later in Europe. In 1977, the specification was standardized in DIN 7721 [20]. These belts are found worldwide and available as moulded-endless belts, endless joined belts and as open metre lengths.

2.3.3 High Power Profile AT

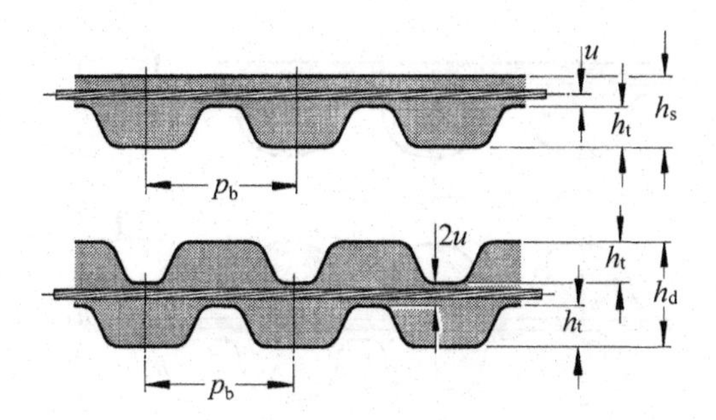

Pitch designation	p_b mm	h_s mm	h_t mm	h_d mm	u mm	Remarks
AT 3	3.0	1.9	1.1	–	0.18	–
AT 5	5.0	2.7	1.2	3.6	0.60	–
AT 10	10.0	5.0	2.5	6.7	0.85	–
AT 15	15.0	6.5	3.8	9.8	1.1	[a]
AT 20	20.0	9.0	5.0	12.4	1.20	–

[a] MULCO-Group designation AT 15. New profile introduced 2008

With a trapezoidal tooth profile and metric pitch, the AT section timing belt is a development of the metric T section timing belt. They are made from polyurethane with steel or Aramid tension members. They are characterized by a wider tooth section and significantly stronger tension members, compared to the standard metric T profile (see Chapter 2.3.2). A special characteristic of the AT profile is that the belt tooth rests against the base of the pulley tooth gap. The MULCO Group, Hanover, Germany [84] developed these types of belts and launched them under the brand name Synchroflex® AT around 1980. They are distributed worldwide and available as moulded endless belts, endless joined belts and as open metre lengths.

2.3.4 High Power Profiles H/HTD

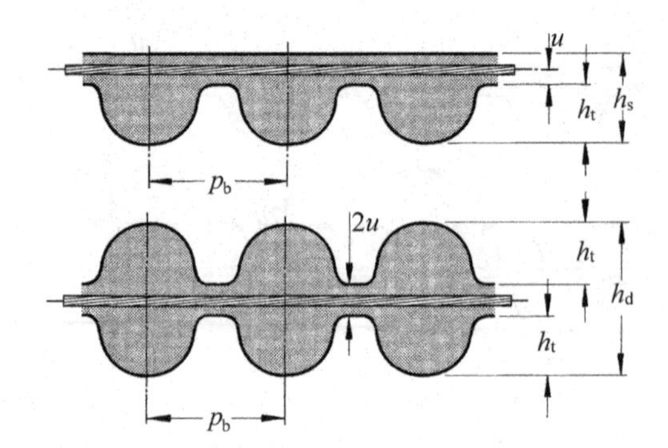

Pitch designation	p_b mm	h_s mm	h_t mm	h_d mm	u mm	Remarks
HTD2 M	2.0	1.5	0.70	–	0.250	–
HTD3 M	3.0	2.4	1.20	3.2	0.380	–
HTD5 M	5.0	3.6	2.10	5.4	0.570	–
HTD8 M	8.0	6.0	3.38	8.1	0.686	ISO 13050
HTD14 M	14.0	10.0	6.02	14.8	1.397	ISO 13050
HTD20 M	20.0	13.2	9.00	–	2.200	–

These belts are ISO 13050 [54] standardized profiles with the prefix H and are also known as HTD timing belts. HTD stands for High Torque Drive. The belts are made of polychloroprene rubber with glass fibre tension members and a polyamide fabric tooth-facing and were developed by Gates Corporation [37] (formerly Uniroyal) which in 1973 led the US market. The curved tooth with a circular geometry and the larger tooth height clearly increases the tooth load capacity and the resistance to tooth jump. Belts with this profile are found worldwide and many manufacturers are involved in their production. They are also made of polyurethane, optionally with steel cord or Aramid tension members. The user can choose from an extensive range of endless moulded belt lengths, open metre lengths and continuously extruded endless belts.

2.3.5 High Power Profiles R/RPP

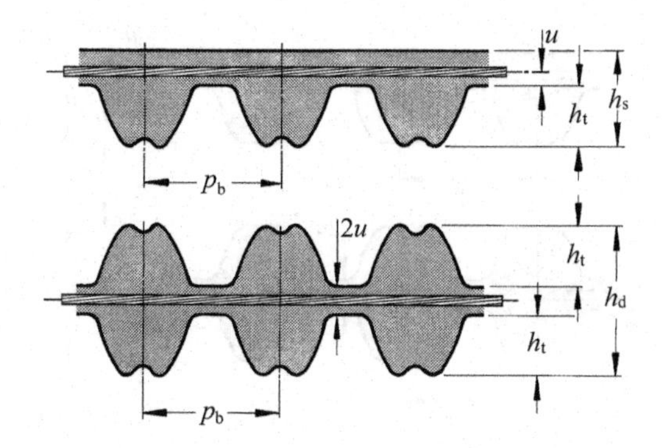

Pitch designation	p_b mm	h_s mm	h_t mm	h_d mm	u mm	Remarks
RPP2	2.0	1.5	0.73		0.270	
RPP3	3.0	2.4	1.15		0.380	
RPP5	5.0	3.8	2.00	5.14	0.570	
RPP8	8.0	5.4	3.20	7.80	0.686	ISO 13050
RPP14	14.0	9.7	6.00	14.50	1.397	ISO 13050

These belts are ISO 13050 [54] standardized profiles with the prefix R and are also known as RPP belts. RPP stands for rubber parabolic profile. This type of timing belt was developed with a double parabolic tooth profile in 1985 by Pirelli S.p.A., Pescara, Italy (modern-day successor Megadyne [78]) and they were made of polychloroprene rubber with glass fibre tension members and a polyamide fabric tooth-facing. Belts of this type are mainly distributed in southern Europe and many manufacturers are involved in their production. They are also made of polyurethane, optionally with steel cord or Aramid tension members. The user can choose from an extensive range of endless moulded belt lengths, open metre lengths and continuously extruded endless belts.

2.3.6 High Power Profiles S/STD

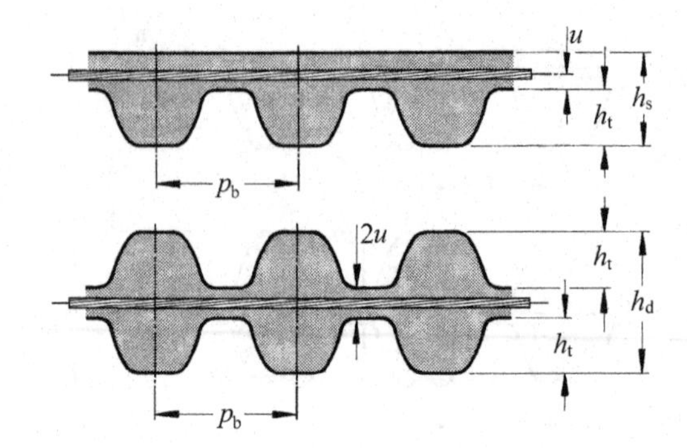

Pitch designation	p_b mm	h_s mm	h_t mm	h_d mm	u mm	Remarks
S2 M	2.0	1.4	0.76	–	0.254	–
S3 M	3.0	2.3	1.14	–	0.380	–
S4.5 M	4.5	2.7	1.71	4.18	0.380	–
S5 M	5.0	3.4	1.90	4.70	0.480	–
S8 M	8.0	5.3	3.05	7.50	0.686	ISO 13050
S14 M	14.0	10.2	5.30	13.40	1.397	ISO 13050

These ISO 13050 [54] standard profiles with the prefix S are also known as STD timing belts. STD stands for Super Torque Drive. This profile was developed in 1976 by Goodyear Corporation, Lincoln, Nebraska, USA [42]. This evolutionary belt with involute profile geometry is made of chloroprene rubber with glass fibre tension members and a polyamide fabric tooth-facing. Belts with this profile are found worldwide, with many manufacturers involved in their production. They are also made of polyurethane, optionally with steel cord or Aramid tension members. The user can choose from an extensive range of endless moulded belt lengths, open metre lengths and continuously extruded endless belts.

2.3.7 High Power Profile Omega

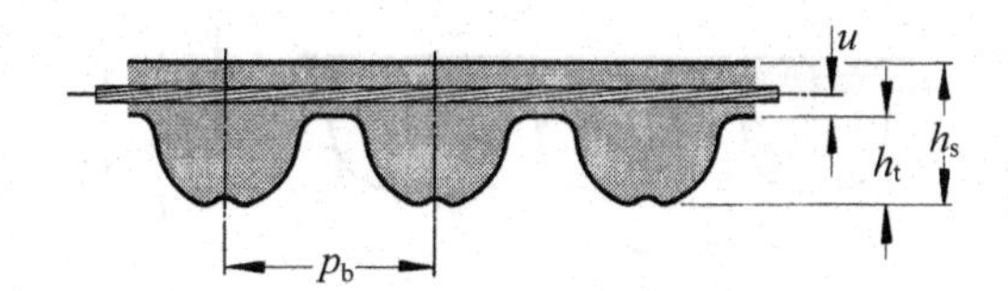

Pitch designation	p_b mm	h_s mm	h_t mm	h_d mm	u mm	Remarks
Omega 2 M	2.0	1.5	0.7	–	0.250	Similar to HTD profile
Omega 3 M	3.0	2.3	1.1	–	0.380	Similar to HTD profile
Omega 5 M	5.0	3.4	1.9	–	0.570	Similar to HTD profile
Omega 8 M	8.0	5.4	3.2	–	0.686	Similar to HTD profile
Omega 14 M	14.0	9.5	5.6	–	1.397	Similar to HTD profile

Timing belts of this type have similar geometric characteristics to the HTD pitches. They are interchangeable with them and run on the same pulleys. They were introduced by Optibelt GmbH, Höxter, Germany [94] in 1990 and trademarked as Omega® toothed belts. They are manufactured as endless belts made of chloroprene rubber with glass fibre tension members and a polyamide fabric tooth-facing. According to the manufacturer, this belt is also able to also run on RPP timing pulleys.

2.3.8 High Power Profile GT3

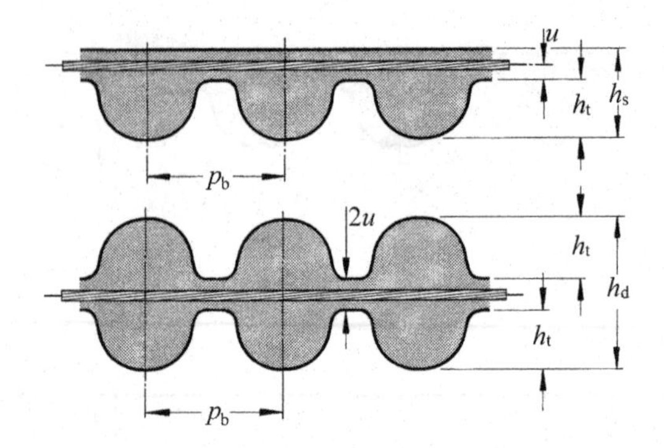

Pitch designation	p_b mm	h_s mm	h_t mm	h_d mm	u mm	Remarks
2GT3	2	1.52	0.71	–	0.255	
3GT3	3	2.41	1.12	–	0.380	
5GT3	5	3.81	1.92	–	0.570	
8GT3	8	5.60	3.40	8.17	0.685	As HTD profile
14GT3	12	9.91	5.82	14.43	1.395	As HTD profile

Belts with an arc-shaped tooth profile registered under the trademark PowerGrip® GT3. The manufacturer, Gates Corporation, Denver, Colorado, USA [37] introduced these belts to the market in 1991. They are made of chloroprene rubber with glass fibre tension members and a polyamide fabric tooth-facing. Belts in the pitches 2, 3 and 5 mm run only on the manufacturer's own "GT3" pulleys, while belts of 8 and 14 mm pitch will run on HTD pulleys. These belts are available as moulded lengths as well as endless belts by the metre (pitch designation LL-GT3) with glass fibre or steel tension members.

2.3.9 High Power Profile Polychain GT2

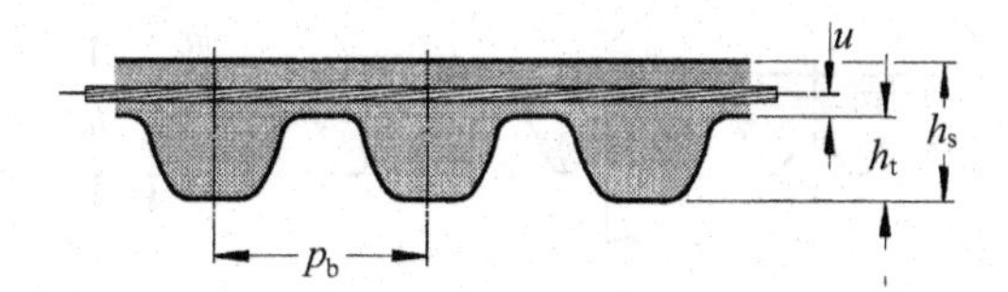

Pitch designation	p_b mm	h_s mm	h_t mm	h_d mm	u mm	Remarks
PC-8MGT2	8	5.9	3.40	–	0.8	–
PC-14MGT2	14	10.2	6.00	–	1.4	–

Belts registered under the trademark Polychain® GT2 [36]. They are made by Gates Corporation, Denver, Colorado, USA [37] of polyurethane with Aramid tension members and a tooth-facing of polyamide. The fabric tooth-facing is optimized by a special PTFE-based impregnated coating to give an extremely low coefficient of friction. The belt has a high power density with low internal losses and the toothform is particularly quiet. According to the manufacturer's guidelines the belt is not suitable for contraflexure with tension idlers on the back of the belt. Polychain® timing belts are available in a range of lengths either as endless belts or by the metre.

Since 2007 these belts are also available with carbon fibre as an alternative tension member. In this state, contraflexure and back idlers are allowed. They have the designation PCC GT2.

This profile is also made from polyurethane with steel tension members by Gates Mectrol [38] in open metre lengths with the designations HPL8 and HPL14.

Continental GmbH, Hanover, Germany [15] offers the Synchrochain® belt in 8 and 14 mm pitch with a new polyurethane compound and a low-friction fabric PE film tooth-facing. Contraflexure has always been allowed.

2.3.10 High Power Profile ATP

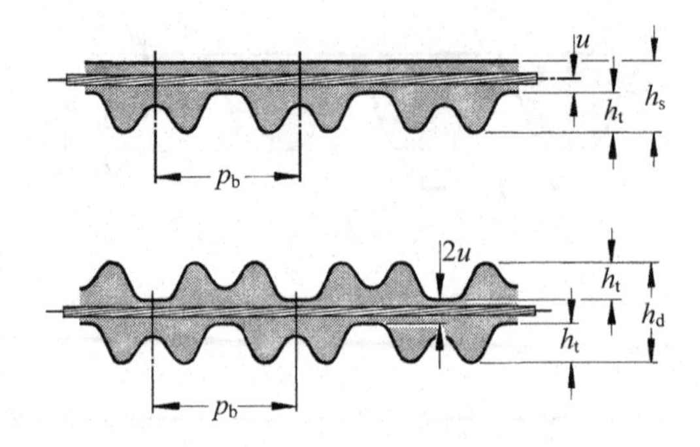

Pitch designation	p_b mm	h_s mm	h_t mm	h_d mm	u mm	Remarks
ATP 10	10	4.5	2.50	6.5	0.75	–
ATP 15	15	6.6	3.75	9.4	0.95	–

ATP timing belts are characterized by a double tooth section where the tooth load per pitch is split between two flanks. The ATP exhibits reduced polygonal effect and lower noise levels than other belts. This belt was first developed around 1992 by Wilhelm Herm. Müller, Hanover, Germany [83] and is distributed exclusively by the MULCO Group [84]. These belts are available in polyurethane with steel tension members in moulded closed lengths.

2.3.11 *Special Profiles Self-Tracking Belts*

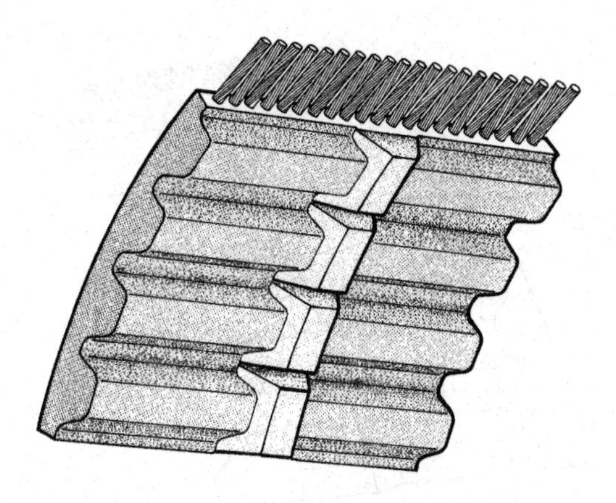

Self-tracking belts are distinguished by a central or offset vee-guide incorporated longitudinally along the length of the belt and usually located on the tooth side to provide lateral guidance. A corresponding vee groove has to be incorporated in the associated pulleys, idlers and guides. Self-tracking belts are particularly suitable for use in handling and transport applications.

The first self-tracking belts were developed around 1980 by BRECO Antriebstechnik GmbH, Porta Westfalica, Germany [9] and are distributed exclusively by the MULCO Group. Due to broad acceptance by users, all the major sizes of T and AT profiles are available as self-tracking timing belts. They are available in polyurethane with steel tension members as continuously extruded closed length and welded to finished length types.

The vee-guide usually has a relief at the tooth root of every pitch to maintain the belt flexibility.

2.3.12 Special Profile SFAT

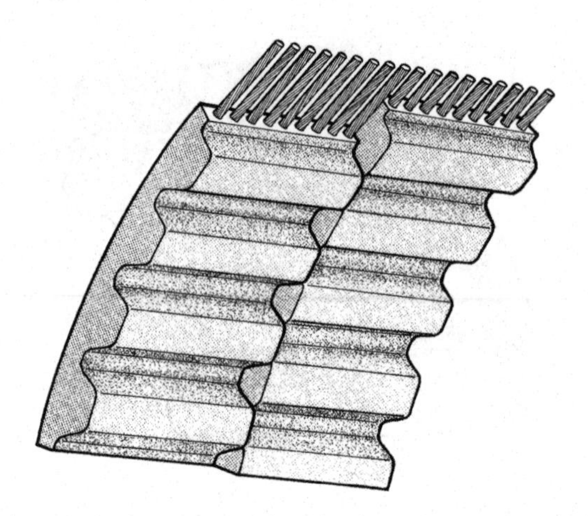

SFAT belts are characterized by having two rows of teeth offset by a half tooth in pitch. They are inherently self-tracking. The designation stands for self-tracking belt with AT tooth profile (see also Chapter 2.3.3). These belts were developed in 1985 by BRECO Antriebstechnik GmbH, Porta Westfalica, Germany [9] and are distributed exclusively by the MULCO Group [84]. They are available in 10 and 20 mm pitches and are made of polyurethane with steel tension members as continuously extruded closed length and welded to finished length types.

2.3.13 Special Profile BAT

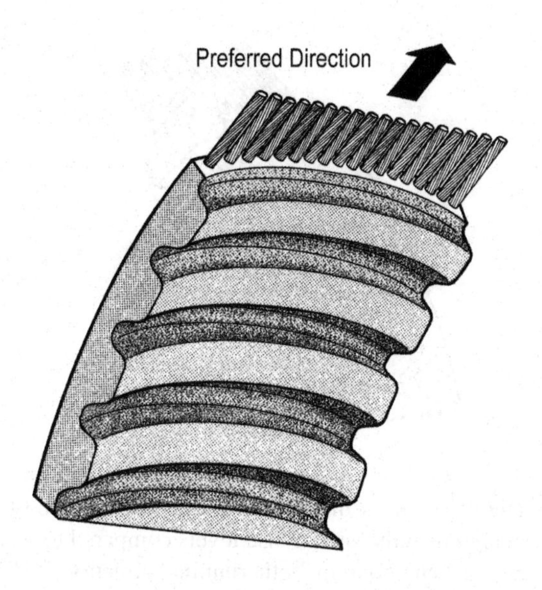

BAT belts are arc-toothed belts that run in matching pulleys and exhibit no polygonal effect. They are characterized by significantly lower running noise compared with other belts and show self-tracking behaviour in the preferred running direction. This belt was developed in 1990 by BRECO Antriebstechnik GmbH, Porta Westfalica, Germany [9] and is distributed exclusively by the MULCO Group [84]. They are available in the profiles BAT10 and BAT15 and are made of polyurethane with steel tension members as continuously extruded closed length and welded to finished length types. The designation stands for B (arc) tooth section with AT high power profile. The BATK, another variant of this belt type, incorporates a self-tracking guide allowing alternating running directions.

2.3.14 Special Profile Eagle

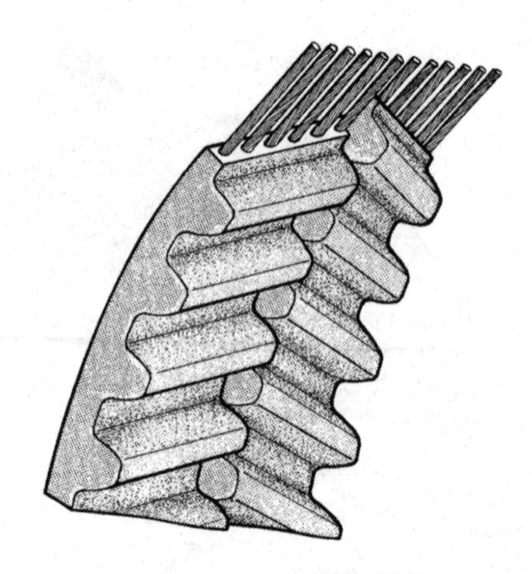

Double-helical Eagle belts run in matching pulleys without almost any polygonal effect. They exhibit significantly lower noise levels compared to other belt systems and show self-tracking behaviour in both running directions. These belts were patented in 1994 by Goodyear Tyre and Rubber Corporation, Lincoln, Nebraska, USA [42] and were launched in 1997 with the pitches of 8 and 14 mm. They consist of synthetic rubber with Aramid tension members and a tooth facing of polyamide fabric. The tooth geometry corresponds to the S and STD profiles (see Chapter 2.3.6). Each row of helical belt teeth is mutually offset by half a pitch.

These belts are also licensed by the patentee to Elatech [27] and Megadyne [78], both of Italy, and are made of polyurethane with steel or Aramid tension members and manufactured as open length and welded to finished length types. Patent protected.

2.3.15 Special Profile N10

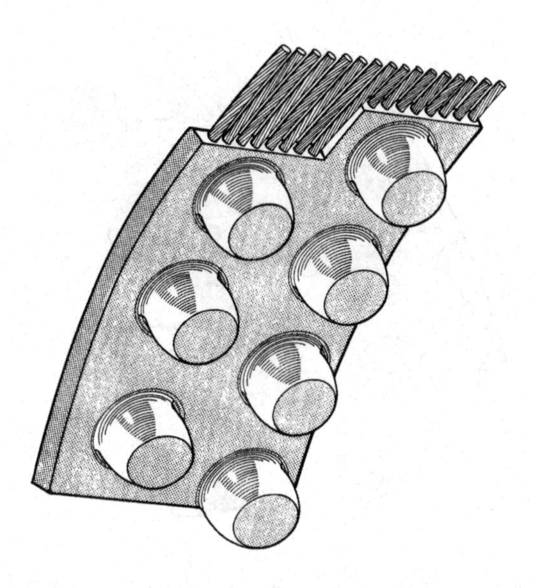

This studded belt combines its special tooth profile and matching pulleys to give a self-tracking solution. Continental Antriebstechnik GmbH, Hanover, Germany [15] developed this belt in 1998. It has the designation N10 (stud profiles at 10 mm pitch) and is made of polyurethane with steel tension members as open length and welded to finished length types. Patent protected.

Belts of this type are likely to be suitable for applications in handling and conveying equipment.

2.3.16 Special Profile ATN with Insert Attachments

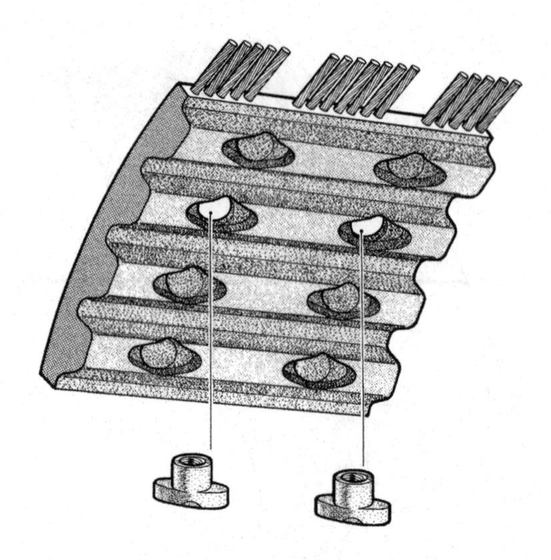

ATN is a family of timing belts with apertures for attachment inserts, developed in 2002 by BRECO Antriebstechnik GmbH, Porta Westfalica, Germany [9] and is distributed exclusively by the MULCO Group [84]. The profile designation indicates that the ATN belt has an AT tooth profile, while the N suffix stands for attachment insert belt. The threaded T-studs are inserted on the tooth side of the belt and the attachments are mounted on the back of the belt allowing easy removal and re-pitching. As delivered, the apertures are initially closed and, depending on the user requirements, punched through at the attachment pitch. Belts are available with the AT profile (see Chapter 2.3.3) in 10, 12.7 and 20 mm pitches made from polyurethane with steel tension members as open length and welded to finished length types. Another version of the ATN is supplied as a self-tracking belt (see Chapter 2.3.11). Patent protected.

2.4 Manufacturing Processes and Elastomers

The marketplace provides designers with timing belts of differing materials and manufacturing processes. What they all have in common is a composite structure that is composed mainly of elastomer-based materials with embedded, high-stiffness tensile cords called tension members. Each material has its own specific characteristics depending on the tasks it has to perform. The modulus of elasticity of the elastomer body and tension member materials are separated by about four orders of magnitude. Consequently, the tension members (Chapter 2.5) are coated with appropriate bonding agents to allow high elastomer adhesion. The materials and the manufacturing steps used, must be co-ordinated during the manufacturing process, so that the individual components are mutually compatible and complementary and their required characteristics are retained permanently. The selection of the body elastomer defines, in essence, all the possibilities of the design and the process steps. Below, after a brief description of the various manufacturing processes, the final product can be seen to depend on the individual manufacturing process which results in the possible options and quality features. The chosen elastomer and its processing, will also affect its design suitability and application areas. These areas are already defined by the rough division of production types between *open length* timing belts, manufactured by the metre for linear motion and conveying, and *endless timing belts,* more likely to be used in power transmission systems. Approximately 90% of all timing belts produced are used in the latter application. Where the tension member is spirally wound or spirally "embedded" in the body of the belt, the manufacturer will describe the semi-finished product as a "timing belt sleeve" or "endless sleeve".

2.4.1 Cast Timing Belts Manufactured From Thermoset Polyurethane

Polyurethanes are well-known for their ability to withstand large deformations and completely revert to the initial unloaded state of the elastomer. This branch of polyurethane timing belt manufacture depends on two-component castable polyurethanes which, in the processing condition, i.e. directly after mixing, have a low viscosity. Suitable mould designs use closed forms consisting of an inner core and an outer form as in Fig. 2.4. Generally, the inner core has the required tooth profile geometry and the inside diameter of the outer form forms the back contour. In the case of a double-sided belt, an equivalent outer form with tooth profile may be used with either in-line or opposing teeth as required.

The tension members are wound on the inner core and are supported on the so-called winding noses. For high performance belts, pairs of cords are used with alternating S and Z twists, which reduce the belt run-off forces due to the helical tension member and thus promote a smoother running belt drive. Through the

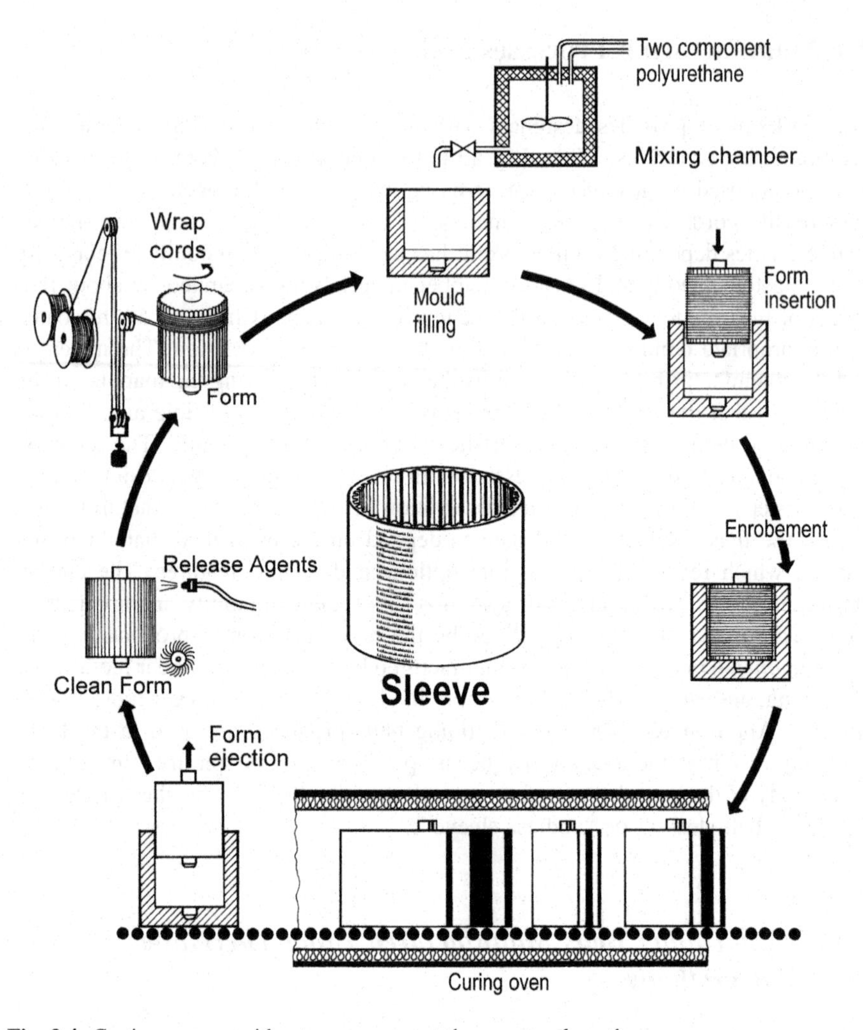

Fig. 2.4 Casting process with a two-component thermoset polyurethane

tension member winding load (preload force), the tolerance of the belt length can be influenced during the winding process. The outer form is then filled with a measured volume of casting polyurethane and the wound core introduced into the form. As the core is immersed, squeeze casting takes place, where the polyurethane flow direction favours the escape of trapped air from the bottom up. When subjected to heat the cross-linked polyurethane hardens in the tempering conveyor and after core removal and belt extraction, further processing takes place such as printing the pitch and belt length, cutting and trimming to individual belts, as well as specific quality assurance tests. A Shore hardness test will show the level of cross-linking, a tooth load test assesses the tension member adhesion and the process ends with quality control of the dimensional accuracy.

Belts made from two-component castable polyurethane are characterized, in particular, by pitch accuracy, casting accuracy and consistent mechanical performance throughout the entire length of the belt. The low viscosity of the polyurethane during the casting process allows the formation of fine detail and thus this manufacturing process is favoured for small pitch belts. Through capillary action, the interstices of the stranded cable in the tension members are almost entirely filled with polyurethane. The adhesion forces between the tension member and the elastomer are particularly good at mechanically anchoring the two components together. This also largely prevents mechanical friction between the filaments of the tension member during bending and flexing.

Cast timing belts made of polyurethane have narrow grooves with exposed tension members perpendicular to the tooth base between every tooth. This feature is formed by the previously described winding noses, which are used both to support and space the tension member on the mould core. Other than this the grooves have no further significance or function.

This style of polyurethane belt is generally available with steel or Aramid tension members with elastomer hardness in the range 85–95 Shore A. These belts are not available with an additional nylon fabric tooth facing. The use of glass-fibre tension members has not been developed further because de-moulding the belt against the winding noses makes the tension members brittle and the shear stress sensitive glass filaments are easily susceptible to damage. Cast polyurethane moulded belts are only available as endless belts. The manufacturing process offers a length range of approximately 50–5,000 mm.

The methods described above can be used in many different ways depending on the equipment and facilities available. Instead of squeeze casting, for example, “injection moulding” can also be used. This is where the wound core is already placed in the outer form and the low-viscosity two-component polyurethane is supplied through an injection line in the base of the mould. Similarly, in this case the actual casting process still takes place from the “bottom up”. The use of vacuum casting techniques in production systems makes further quality improvements possible, such as better casting accuracy for subtle contours and improved penetration of the tension members with polyurethane.

2.4.2 Synthetic Rubber Vulcanised Timing Belts

Chloroprene rubber (CR) is the most popular material for the manufacture of synthetic rubber endless timing belts and it is often found under the DU PONT trade name Neoprene®. CR has a good balance of properties including tensile strength, elongation at break, resistance to ageing, and maximum/minimum service temperature without achieving the maximum levels available from other types of synthetic rubber. In automotive applications there is increased use of HNBR synthetic rubber, particularly for camshaft drives, where high ambient temperatures often occur. HNBR stands for hydrogenated nitrile butadiene rubber. HNBR has improved heat resistance (+25°C in DIN 780078) against CR with a temperature limit of +130°C and with special mixtures, of up to +150°C.

The manufacturing process for producing a timing belt sleeve of synthetic rubber with embedded tension members is by layering the components on the mould core. First, the polyamide fabric tooth facing, already made in the shape of a tube of the required belt length and sleeve width, is pulled over the inner mould core. The tooth facing is elastic in its length direction which allows it to fill the tooth gaps when the rubber is forced through the tension members under pressure. Subsequently, the tension members, made of glass-fibre or Aramid cord, are wound on the mould core. The tension members are wound with a predetermined preload force across the tops of the mould core teeth and this holds the nylon fabric tooth facing in position. Finally skins (strip cut from raw rubber) of the non-cross-linked and plastically deformable synthetic rubber are applied in a uniform thickness around the mould core.

The prepared mould is then placed in an autoclave and exposed to both high temperature and pressure. Under these influences, the non-cross-linked rubber exhibits an increasingly viscous behaviour and the rubber flows between the tension members and deforms the tooth facing into the tooth gaps. As a result of the temperature and pressure for a predetermined period (approx. 30 min) the plastic synthetic rubber is cross-linked and vulcanized and enters an elastic and therefore dimensionally stable condition. For future reference, HNBR is a peroxide cross-linking type of synthetic rubber. Simultaneously, the elastomer bonds to both the tooth facing surface as well as the tension member via a bonding agent. The bond between the components is based primarily on chemical and physical forces.

The tooth side of the de-moulded belt is an exact replica of the tooth geometry of the mould core with high pitch accuracy. With this method, no outer mould is required and irregularities in the back thickness of the belt are equalized by grinding. The last operation consists of cutting the sleeve into individual belts.

Timing belts made of vulcanised synthetic rubber are available with glass-fibre or Aramid tension members (rarely with steel tension members) with the teeth generally reinforced with a polyamide fabric tooth facing. Belts produced by this process are available in lengths of approximately 100–5,000 mm.

2.4.3 Thermoplastic Endless Polyurethane Extruded Timing Belts

Manufactured using plastifiable polyurethanes, such as Desmopan® from Bayer, and available in a modulus of elasticity range of 10–650 MPa. For use in timing belts, the types of polyurethanes of interest have a modulus of 15–40 MPa which corresponds to a hardness of 85–95 Shore A.

Extruded polyurethane endless sleeves are formed between two oppositely arranged pulleys. Figure 2.5 shows an unfinished, partially extruded belt. The cords are wound under a pre-load between two pulleys, also called form-wheels, supported by the winding noses. The molten extruded polyurethane is supplied at the appropriate processing temperature from an extruder. The formation of the tooth geometry and back of the belt from the composite structure of the polyurethane and tension members, is carried out by the plastification process through a slot die whereby the melt is continuously expressed into the rotating production pulley.

The belt is progressively built tooth by tooth along with the back of the belt. The cooling section corresponds to an 180° rotation around the forming pulley. The incorporation of PAZ[1] is achieved by simultaneously feeding a polyamide fabric tape into the forming pulley during the extrusion process. The extrusion of a finished sleeve is then completed after a full revolution of the belt when the first formed belt tooth meets the last formed belt tooth. This production method means the back of the endless sleeve needs to be equalized by being completely ground. The last manufacturing step consists of cutting the sleeve into individual belts.

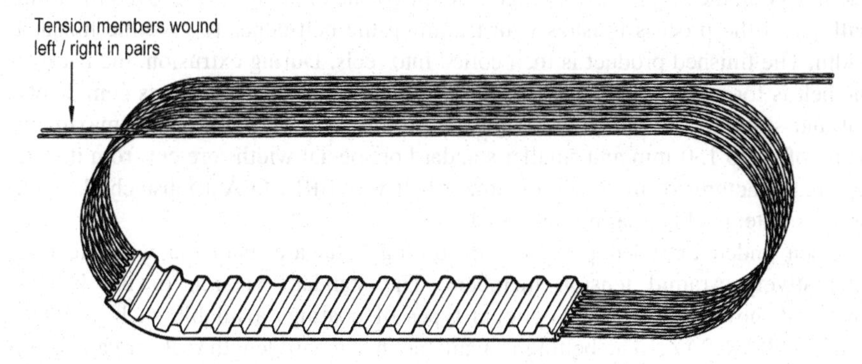

Fig. 2.5 Extruded timing belts from thermoplastic polyurethane

The special feature of this process is that by adjusting the centre distance on the production machinery, an infinite number of belt lengths can be achieved as long as they are divisible by the tooth pitch. The shortest possible length achievable is that length in which the form-wheels do not quite touch. In practice, endless timing belts of approximately 22 m long are possible.

This extrusion process was developed by BRECO Antriebstechnik GmbH [84] in about 1980 and they have refined the quality characteristics over the subsequent years with regards to both length tolerance and pitch accuracy. Today, pitch

[1] PAR Polyamide fabric facing on the belt back
PAZ Polyamide fabric facing on the teeth,
PAR-PAZ Polyamide fabric facing on both sides

differences cannot be differentiated between the first and last formed tooth compared to any other teeth or groups of teeth.

Extruded polyurethane endless sleeves can be produced with steel or alternatively, Aramid tension members. Standard versions of the belt are produced uncoated but are optionally available with a PAZ[2] tooth facing.

2.4.4 Extruded Polyurethane Open Length Timing Belts

The production of open length belting differs from that of extruded endless belts because the production machinery uses just a single form pulley. The term "wound" belt is no longer true because the tension members in this process are parallel to the edge of the belt, with the alternating adjacent tension members in S and Z twists. The plasticizing of the belt composite structure is accomplished in the same manner as the endless belts described above. After forming the belt back and the teeth, the belt passes through a cooling zone of about 180° around the form pulley and the process finishes with trimming the belt edges to give the finished width. The finished product is then coiled into reels. During extrusion, the back of the belt is formed by a moving flat metal belt and thus the belt back is completely flat and smooth. Belts of this design are produced in all pitches in a maximum width of up to 150 mm and smaller standard or special widths are cut from it. The first manufacturer of this style of timing belt was BRECO Antriebstechnik, with process patents [111] dating back to 1970.

Open ended extruded polyurethane timing belts are available with steel or alternatively Aramid tension members. The standard version of the belt is uncoated, however there are optional types of coatings available such as PAR, PAZ and PAR-PAZ.[3] The belting is available in standard length reels or can be cut to length as required. Theoretically unlimited lengths are achievable.

2.4.5 Other Manufacturing Processes

Other manufacturing processes are listed but because of their low market penetration, they are not discussed in detail.

- Tooth belts can be manufactured by injection moulding polyurethane into closed moulds. This process has relatively high tooling costs and is only suitable for volume production applications. The possible belt lengths are limited.
- Extruded, open length, synthetic rubber timing belts with steel, glass-fibre or Aramid tension members are also possible.

---

[2] See footnote one.

[3] See footnote one.

- Endless timing belt sleeves can be converted to open length belting by "spiral cutting". It should be noted that, with this production method, the belt teeth are not at right angles to the belt edges but at the helix angle of the cut. This angle changes depending on the sleeve length and the chosen belt width.
- All belts, no matter from which type of production process, can also be re-worked by further mechanical processing for special customer requirements. This re-work can include longitudinal and transverse machining on the back and/or tooth side of the belt, hole punching and grinding the belt edges to achieve high tolerance, special belt widths. Belt backings are used for handling and feeding where the friction properties are matched to the transportation task. Extruded belts and also some types of cast belts can also be fitted with welded profiles. Chapter 5 deals with specific examples of conveying technology solutions for indexing and separating.

2.4.6 Joining of Open Length Belts

This process uses a die-punch to cut the prepared belt ends for welding (Fig. 2.6). Two main finger cutting techniques are used, but all work on the principle of enlarging the welding area by increasing the weld face. The prepared ends of the belt are placed into the welding machine between the upper plate (smooth) and lower plate (toothed), put under pressure and subjected to a thermal cycle consisting of heating, welding and cooling.

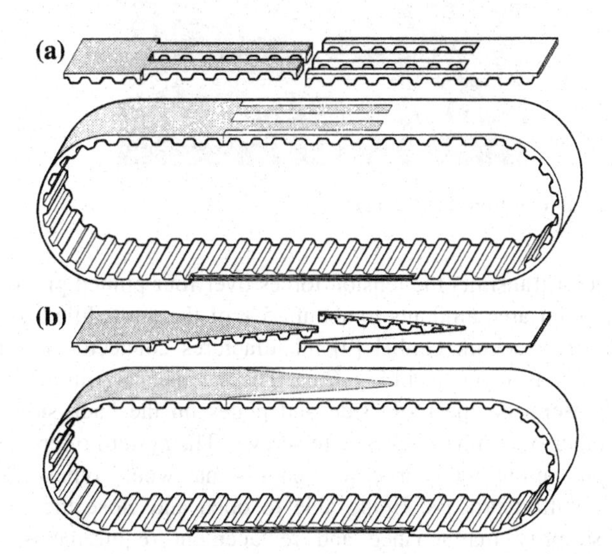

Fig. 2.6 Endless joining with **a** rectangular finger joint, **b** vee finger joint

The welding process does not give rise to any pitch abnormalities in the joint area in comparison with the rest of the belt. The finger length is calculated on a welding length of 9 pitches (10 teeth) and the tension members are cut through in this area. The resultant reduction in strength in the welding zone is about half the allowable tensile load F_{zul} compared with that of the standard belt. The first welded belts were offered by BRECO Antriebstechnik GmbH [9] in approximately 1970.

2.4.7 Timing Belt Mechanical Joint System

A timing belt mechanical joint (see Fig. 2.7) is a device which allows an open length belt to be threaded along its drive path through a machine and then be joined by mechanical means. The joint system is designed to be highly flexible and allow belt bend radii approximately equal to the recommended minimum pulley diameter for the relevant pitch.

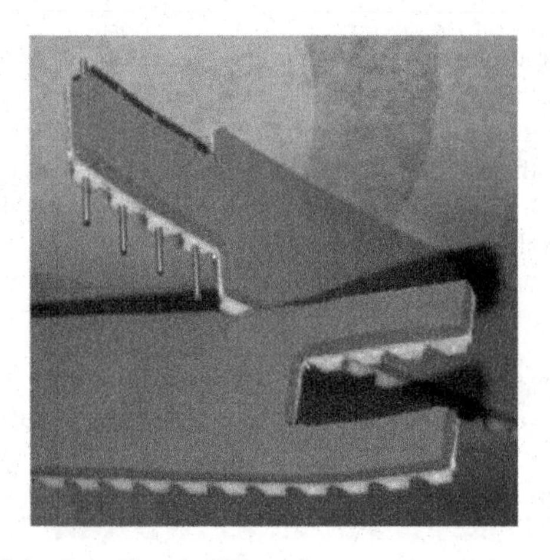

Fig. 2.7 Mechanical joint from Elatech [27]

The finger joint transmits the tension forces over four pinned pitches. The belt capacity in the joint area amounts to about 25% of the unmodified belt.

BRECO Antriebstechnik GmbH [9] manufactures connector systems of this kind with rectangular fingers as in Fig. 2.6a. The mechanical joint system consists of a series of inserts on the tooth side and plates on the back side, which are connected together over nine pitches with screws. The system offers considerable benefits for applications with large belt lengths that would otherwise require a significant installation process. Being able to open and close the belt can also considerably simplify maintenance and replacement requirements. Connector designs are available for T10, T20, AT10 and AT20 belt types.

2.4.8 *Applications and Usage*

It is axiomatic that with all the different profile geometries, materials and production processes, the varieties of timing belts will also have different physical properties. When drive problems are to be solved or where rotary or linear movements are to be implemented then, depending upon environmental and application conditions, the drive components have certain requirements. This handbook is designed to give impartial advice to enable an appropriate selection, and recommendations for manufacturer specific products are not given. In Chapters 3, 4 and 5 however there are suggestions for preferred belt types for each individual application, without limiting or excluding any other types. The examples are there to help ensure that the designer can base his work on proven designs.

There are always a certain number of operational areas and applications that cannot be clearly associated with a particular kind of belt. Here, each individual application should be assessed to ascertain which type of belt is most suitable.

Many manufacturers offer the same belt with different performance levels where the profile geometry matches and they are functionally exchangeable. This means belts with strengthened tension members and/or with modified elastomer mixtures (see Chapter 2.4.9). In choosing the most appropriate type of belt for new applications, this handbook always recommends the use of the profile with the highest power density. The consequence of this is that the belt can often be dimensioned with a narrower width and thus the total drive cost can be reduced due to the narrower belt and pulleys. Reducing the belt width usually also gives the benefit of lower noise levels (see Chapter 2.12).

2.4.9 *Power Increases and Further Development*

From 1975 to 1995, belt manufacturers expended considerable effort into increasing power outputs by the use of new profile geometries. Increased belt power density was achieved by the optimization of the tooth geometry and in particular, by enlarging the tooth volume.

These developments have led to a variety of different belts, each of which can only be run on specially made pulleys and with each newly introduced profile, interchange-ability decreases. From the users' and the machine builder's point of view it is to be hoped that the number of different types of belt now available does not increase further.

Any increase of tooth volume is inevitably accompanied by a reduction of the corresponding pulley tooth top face which supports the belt while it is in contact with the pulley (Fig. 2.8).

In practice, and in comparison with standard belts, this has led to a new failure mode, where the decreased tooth top face contact surface frequently leads to bridge wear in the belt (see Chapter 6). Based on this knowledge, this problem

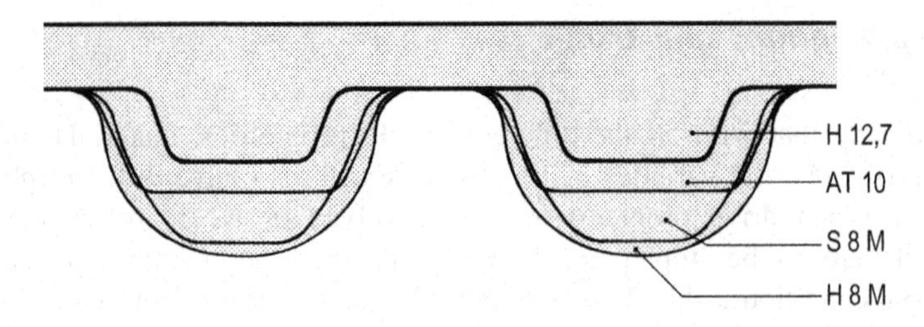

Fig. 2.8 Volumetric comparisons of selected tooth profile sections [89]

appears to limit profile design to a point, where a further enlargement of the tooth volume can no longer be expected.

Soon after the introduction of new profiles, further advances were made in the materials sector, specifically in engineering elastomers, tension members and their processing into the belt composite structure. Further developments are in hand and further performance improvements in belt drives are to be anticipated [89, 90].

High-performance timing belts can be further optimized by the use of rubber-based materials with specific amounts of short Aramid fibres added. These are known as granular reinforcements, which are embedded in the elastomer and thus raise the tooth load capacity to a higher level.

For applications in automotive technology, novel PTFE (Teflon) impregnated fabric designs are now used where the increased layers applied also enhance the profile and enable a low-friction tooth meshing. The additional elastomer fibre additives contribute very significantly to the timing belt's performance. For example, the Synchrochain® high power belt range from Contitech AG has a novel fabric friction compound in combination with PE film and further improvements are more than possible in this area.

In the high performance polyurethane belt market, tension member windings are now available in S-Z-coiling which are wound much closer together [87] and with increased hardness polyurethane mixes. Both measures result in significantly increased performance in conjunction with a decrease in tooth deformation.

All these implemented enhancements are reflected in elevated values for the transmissible power, improved stiffness and greater tooth load capacities. The basis for the calculation of belt drives is contained within the manufacturer's published technical data documentation. Developments in recent years are also reflected in the technical and promotional literature of those manufacturing companies that have optimized their belts for Speed, Torque and Force such as ISORAN® Silver and Gold®, Synchroforce Extreme®, Powergrip® GT3, Synchroflex® Generation III or HTD-Mustang. Providing current technical data for each type and make is not the purpose of this handbook. Direct contact with the belt manufacturer is recommended, if necessary, to obtain detailed advice on the company's specific product or web-design services [84].

The calculation of timing belt drives is not standardized and those standards that are available are solely based on standard sizes. Only the VDI-2758 guidelines [120], set out the basic equations and values, but it is applicable only to imperial (inch) pitch belts with trapezoidal profile and HTD-profile belts. This makes it clear that a comparison of timing belt drives without specific manufacturer's design data is not possible. Since different manufacturers use different methods for calculating the required belt drive, this makes quick product comparisons difficult. Such comparisons will not give completely identical solutions, but by using comparable assessment procedures, then uniform sizing criteria can be found [89, 90].

2.5 Tension Members

A timing belt is flexed at least twice per revolution at the diameter of each pulley in a two-pulley drive and then returns back to the straight span between the pulleys. Assuming the number of revolution at 3,000 min^{-1}, a belt will have already experienced about 1×10^6 flex cycles within 8 h running and will undergo further flexing with the use of any idler pulleys.

The tension member working load consists of the transmitted torque plus the bending stresses which give the total load. Thus, it can be seen that the belt is exposed to highly dynamic loads and it is necessary to have detailed knowledge of the internal stresses occurring in the belt and the associated wear mechanisms for optimization of timing drives. The loadings are very complex, since the individual components of the belt have very different material behaviours. Timing belt composite construction comprises of the base polyurethane or synthetic rubber, plus high-strength tension members made from steel wire, glass-fibre or Aramid fibres, which differ in their modulus of elasticity by about four orders of magnitude. The tension member is exposed to load cycles of varying tensions and superimposed bending stresses and due to the structure of multiple individual fibres (filaments), friction occurs between the filaments, especially with small pulley diameters. These relative movements and the pressures between the individual filaments are very large and lead to increased abrasive wear of the fibres.

The complex interactions of the tension member filaments have been little studied and projections for the tension member service life, or complete timing drives, are limited. Knowledge of failure mechanisms for tension members are of the utmost importance to belt manufacturers as this could lead to increased efficiency and reduce the levels of empirical testing. Timing belt end-users are also increasingly interested in additional information about lifespan in order to define maintenance intervals.

Steel tension members are made of cold drawn steel wires which are wound symmetrically in a helix on a common axis. In contrast to steel cables in other applications, where the wires are sometimes found in different cross-sectional shapes, steel tension members for timing belts consist exclusively of circular cords. The smallest wire diameters are as little as 0.04 mm and the nominal tensile strength of the wires $\sigma_{0,2}$-values are around 2,500 N/mm^2 [3, 19, 126].

The tension member can be considered to comprise of a simple stranded cable with a second stranding operation making the finished article. More levels of stranding are possible but they are not used in timing belt tension members. There are a variety of possible strand types, as in addition to the wire diameter and the number of wires per strand, both the number and arrangement of the strands can be variable. Figure 2.9 shows a common 7 × 7 tension member which means the central strand consists of 7 wires (one core wire + six wound wires) surrounded by 6 other strands, which in turn are also made up of 7 individual wires. The tension member outside diameter d_Z is understood to be the circular diameter of the cable.

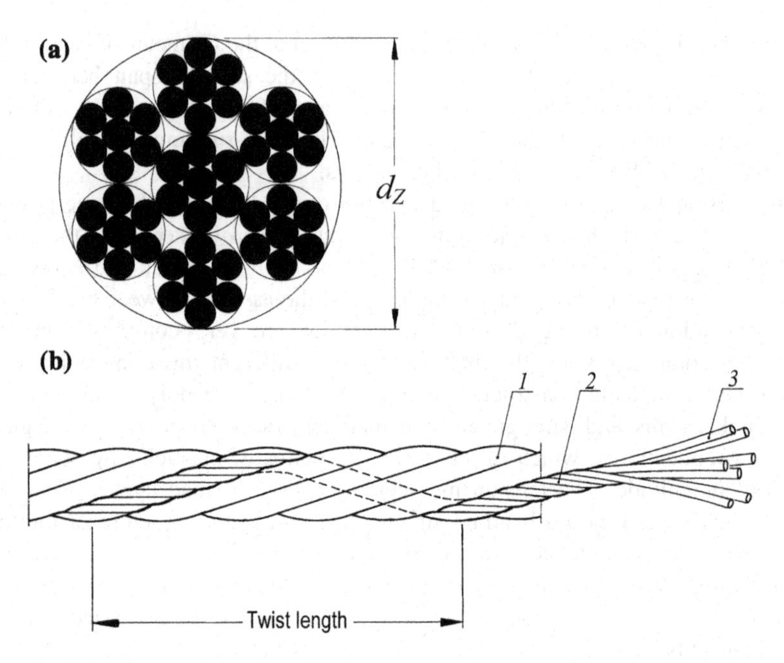

Fig. 2.9 Tension member construction 1 Cable/Tension member, 2 Cord, 3 Filament/Wire, d_Z circular diameter

In small precision timing belts, the tension member consists of a single strand of wire, while for higher performance belts; tension members of multiple stranded cables are used. In general, the smaller the tension member diameter and the thinner the individual wires, the more flexible the overall design. In addition to the number and arrangement of the wires, the lay length (pitch) and the lay direction is particularly important. The lay length is the distance or pitch a wire takes to make a complete wrap of a strand, or a strand the complete wrap of the cable. The lay length also refers to a multiple of the nominal diameter of the wire or rope. For steel tension members in timing belts, it is customary for the lay length to be between 6 and 12 times the strand or cable diameter. The lay length also influences the modulus of elasticity of the strands and cable. While the E-module of a single straight wire is around 210,000 MPa, this decreases in the tension member to about 190,000–140,000 MPa. This reduction is caused by the movement of the wire in the combined cable. The location of each individual wire in a strand or in a cable follows a helix, and increasing tension compels every individual wire to take the shortest possible route. Approximations can thus be calculated of the associated flattening of the individual wires at the contact points and as a result, the additional flexibility.

The twist direction is differentiated as left-handed or S-twist, and right-handed or Z-twist. Since this designation may be used for both strands, and for the entire cable, the combinations are shown in Fig. 2.10. It is customary to designate the twist direction of the strand with lowercase letters and that of the cable with uppercase letters.

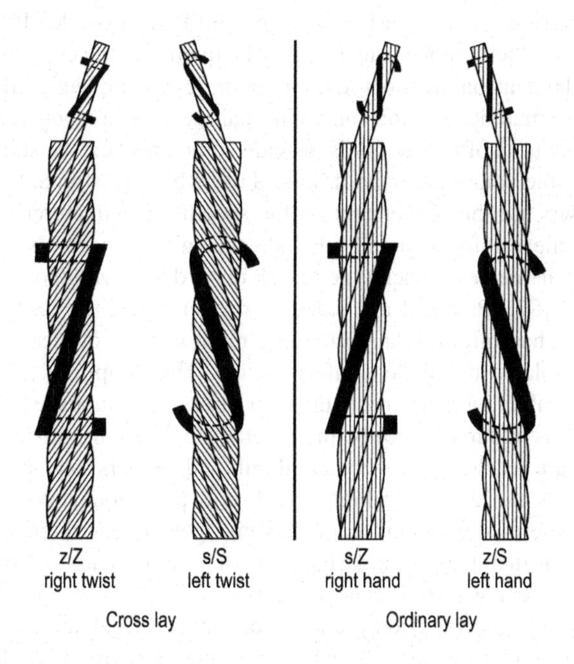

Fig. 2.10 Lay types

Also noteworthy, are parallel twisted strands. In this situation, the wire's positions, one above the other, have the same length of lay and same twist direction. This has the effect that, contrary to the cross lay types, the outside wire positions fit into the gaps between those wires lying under them. Inevitably, this gives a more favourable packing factor (cross section factor) and additional support effects arise amongst the wires as a result of line contacts. The parallel twist and the type of the contact of the individual wires, however, leads to instabilities under load and bending, so this construction is seldom used for timing belt tension members.

Current tension member research focuses on endurance tests in cable bending machines [3, 19], in which the tension member, under a given load, is repeatedly driven over one or more pulleys. Here, the number of bends achieved is a measure of lifespan and, next to the axial force, the ratio of tension member diameter to pulley diameter is crucial for the number of bends achieved. Other parameters such as bending length, temperature, pulley material and toothform also affect results.

Inside the belt body, the tension member is completely embedded in the comparatively soft base material and has no direct contact with the timing pulley. Furthermore, the base material, assisted by a bonding agent, is in direct contact with the tension member, i.e. the relative movements of the filaments are largely avoided. The balanced formulation of the bonding agent is thereby a key consideration for overall function between tension member and elastomer. The choice of the manufacturing process (cast two-component PUR, extruded PUR or pressed and vulcanized synthetic rubber) also affects the quality of the composite structure. The gaps in the tension member are more or less completely filled with the elastomeric base material and thus an additional mechanical anchorage is created. Although fatigue tests of tensile cords provide some basis for the suitability of the bending ability, the values are only of limited use when transferred to timing belts. Timing belt users are thus dependent on the specific manufacturer's information. In most cases the catalogue details include the minimum number of teeth and minimum bend diameter for each type of belt drive design with both "mono-" and "contraflexure" [87]. Practical application experience has shown that the manufacturer's researched information on the minimum number of teeth and minimum diameter is reliable with sufficient safety factors. This simplifies the drive design and a separate calculation on the equivalent stress (a composite of tension and bending stress) is redundant. Only in extreme applications, in which the geometrical limits are reached (e.g. very small pulley diameters) and at the same time the timing belt is run at a very high speed, is a post-elongation of the tension members encountered due to high frequency changes of bend direction. Otherwise steel cord tension members do not change their original length. Drive load fluctuations always occur within the elastic range.

The most widely used tension members are made of galvanized steel cord. If used in corrosive environments, special tension members are available in stainless steel but their stiffness and bending values are significantly lower than the standard galvanized versions.

The technology of tension member manufacture is subject to continuous development. The steel cord manufacturer understands, as a system supplier, the special requirements for the use in different timing belt profiles and elastomer applications. The cord finishing with adhesion agents and the quality of optimization is now also their responsibility [118]. Brass coating offers the best adhesive surface for chloroprene rubber belts and zinc coatings for polyurethane belts. Additionally, it is now possible to supply the cords in an elastomeric embedding matrix. The intermediate spaces of the pre-treated filaments of such cords are 100% filled. This prevents the relative movement of individual filaments against each other and thus prevents tribological wear due to single wires touching. The tendency to fine abrasion, accompanied by fretting, is entirely eliminated by this process. In this way, optimized cords give the composite structure of the timing belt a significant increase in adhesion due to good mechanical bonding of the material components with each other. Such optimized tension member systems improve the length stability as well as reduce the tendency to corrode.

Glass-fibre tension members differ from steel in the fineness of the filaments, the twist, the weight per metre and the composition of the components used. Today's conventional single filament diameter is between 5 and 9 μm and the trend is to ever finer filaments (2 μm) due to better material utilization [17]. Chemically pure glass (quartz glass) is used, consisting only of SiO_2. Then, with the addition of red lead, borax, kaolin or feldspar, a glass for almost any desired purpose can be produced [53, 113]. The best known standard yarn type for glass-fibre is made of E-glass which is known for its dielectric qualities, that is, its low ionic conductivity. This glass has been produced since 1930 and is still the most widely used for timing belt reinforcement. The addition of aluminium oxide and borax gives a relatively high filament strength, and magnesium oxide which reduces crystallization and allows processing at higher temperatures. The addition of titanium oxide, zinc oxide or zirconium oxide improves, but not completely eliminates, hydrolysis.

The second component of the glass cords is a coating of great importance (Fig. 2.11). Each individual cord is plastic-coated as protection against hydrolysis which also serves as a bonding agent in the elastomer composite.

The third component of the glass-fibre tension member is the embedding matrix of the dip coating. This is usually resorcinol–formaldehyde latex and other

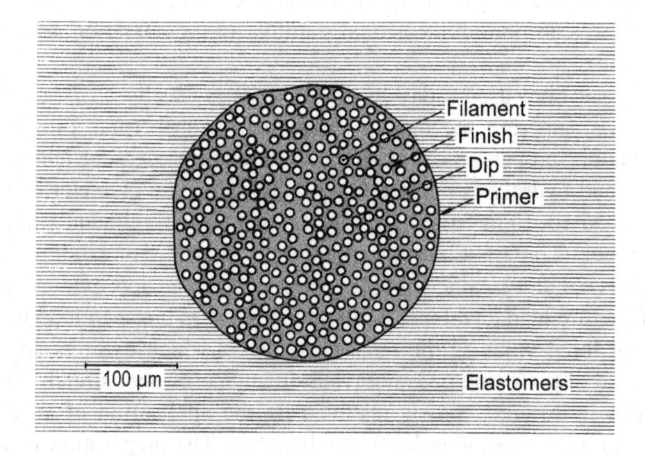

Fig. 2.11 Glass-fibre cord construction

adhesion promoters. The dip coating and its matching formulation takes care of the chemical-physical protection of the glass strands and it embeds the brittle and extremely sensitive individual filaments in an elastic layer and protects them from mechanical contact with each other. The dip coating covers both the individual filament and the entire cord and then the cord receives a final coating layer of adhesion promoters.

Due to the embedding matrix process, glass-fibre tension members are significantly different from other types of tension member. Glass-fibre is dipped as a single untwisted yarn to ensure a complete covering of the filaments. This is followed by a multi-stage twisting process to achieve the particular desired structure.

Example: A cord in a construction dtex 1,400 × 1 × 3 × 12 with a twist of S/S/Z140/80/40 means that the yarn has a nett weight of 1,400 g/10,000 m with a left-handed twist of 140 turns/m with a second stage of three left-handed twists of 80 turns per m and a third stage of twelve times right-handed twist with 40 turns/m. Starting with 1,400 dtex, this yarn is twisted a total of 1 × 3 × 12 = 36 times, giving the cord a total value of 50,400 dtex. The nett weight, excluding the weight of the untreated glass-fibre tension member, can now be understood. Titre is a value used in the textile industry for yarn and is defined as cord mass per unit length. Such a cord construction is also available in a Z/Z/S winding and the usual cord deployment in timing belts is in side by side pairs that have the tension members lying in mirrored lay directions.

Tension members of man-made fibres are most often used in conjunction with elastomers in a composite structure such as those found in tyres, hoses and belts. In automotive fan belts, for example, twisted cords of Polyester have become generally accepted, set to a specific pre-stretch by a temperature-induced high shrinkage force while fixed at a predetermined belt length. The result is the well-known maintenance-free Vee-belt, where re-tensioning is no longer necessary. While such stretch and shrink characteristics are desirable in some drive elements, a timing belt needs absolute dimensional stability during operation, as a necessary requirement.

The man-made fibre used today in timing belts is usually an Aramid, for this drawn yarn has superior mechanical, chemical and thermal properties. The favourable thermal properties of Aramid aids both in processing and application (positive heat and hydrolysis resistance up to 250°C). The diameter of the single filament is around 10–15 microns. The tensile strength is significantly higher when compared to steel cord, but adversely exhibits greater elongation values. The fracture strain is about 4%. Since a toothed belt requires an approximately constant pitch over the entire load envelope, a maximum permissible span load limit F_{zul} is defined as that force value in the tension members which exhibits an extension value under load of 0.4% and does not result in any permanent stretch. The lack of lateral stiffness in Aramid cord is resolved by the application of an elastomer-bonding system with a resin/hardener combination. The preparation is so configured that the textile flexibility is unaffected. This is followed by further refinements in which the Aramid fibre is processed in single or multiple stages to achieve the desired structure of twisted cord.

Example: A cord construction of dtex 1,610 × 1 × 3 with a double twist Z/S150/100 means that the yarn has a net tension member weight of approximately 1,610 g/10,000 m. The first step is a simple right-handed twist of 150 turns/m and the second twisting stage is a triple left-handed twist with 100 turns/m. The cord thus has a total titre of 4,830 dtex.

Carbon-fibre tension members are well-known in the manufacture of composite materials. Their major applications are in highly stressed structural components for air and space technology. Depending upon the type of fibre, the elastic module reaches figures of 220–700 Gpa (in comparison steel is 210 Gpa). The filament diameters range from 5–11 μm. This basic fibre cannot simply be used for the manufacture of tension members because of its poor transverse rigidity. Due to the development of special coating materials (surface treatments) a fibre product is now available with good bending ability and textile properties. The raw material for belt use is made of carbon filaments in a multi-stage twisted and matrix embedded cable. The adhesion agent provides superior mechanical characteristics for tension, stiffness and bending ability. The first timing belt to be equipped with tension members of this kind was from Gates [37] in the spring of 2007, in the Polychain PCC-8MGT timing belt. It is particularly worth mentioning that the length/temperature coefficient of this tension member is almost zero, or even a slightly negative value.

Summary

All the described tension members (steel, glass-fibre, Aramid, carbon fibre) exhibit considerable service life and outstanding length constancy under continuous dynamic stress in *the direction of load.* With force types transverse to the direction of load, as occurring during shearing or buckling, the belt composite structure can provide very little resistance, however this sort of force direction is virtually impossible under normal operating conditions. It is conceivable that from the point of dispatch to installation that handling defects could happen. A kinked belt can be considered significantly damaged and such belts should not be accepted for use.

2.6 Forces in Timing Belt Drives

A timing belt exhibits the same behaviour in its interaction with a machine regardless of type or profile geometry. Figure 2.12a shows a timing belt drive at standstill or idling under ideal conditions and it can be seen that the pre-tension F_V is equally distributed throughout the belt, as there is no torque transfer. Figure 2.12b shows the same belt transmitting power with the right-hand pulley being the driving pulley and the left-hand the driven pulley. The driven pulley is shown

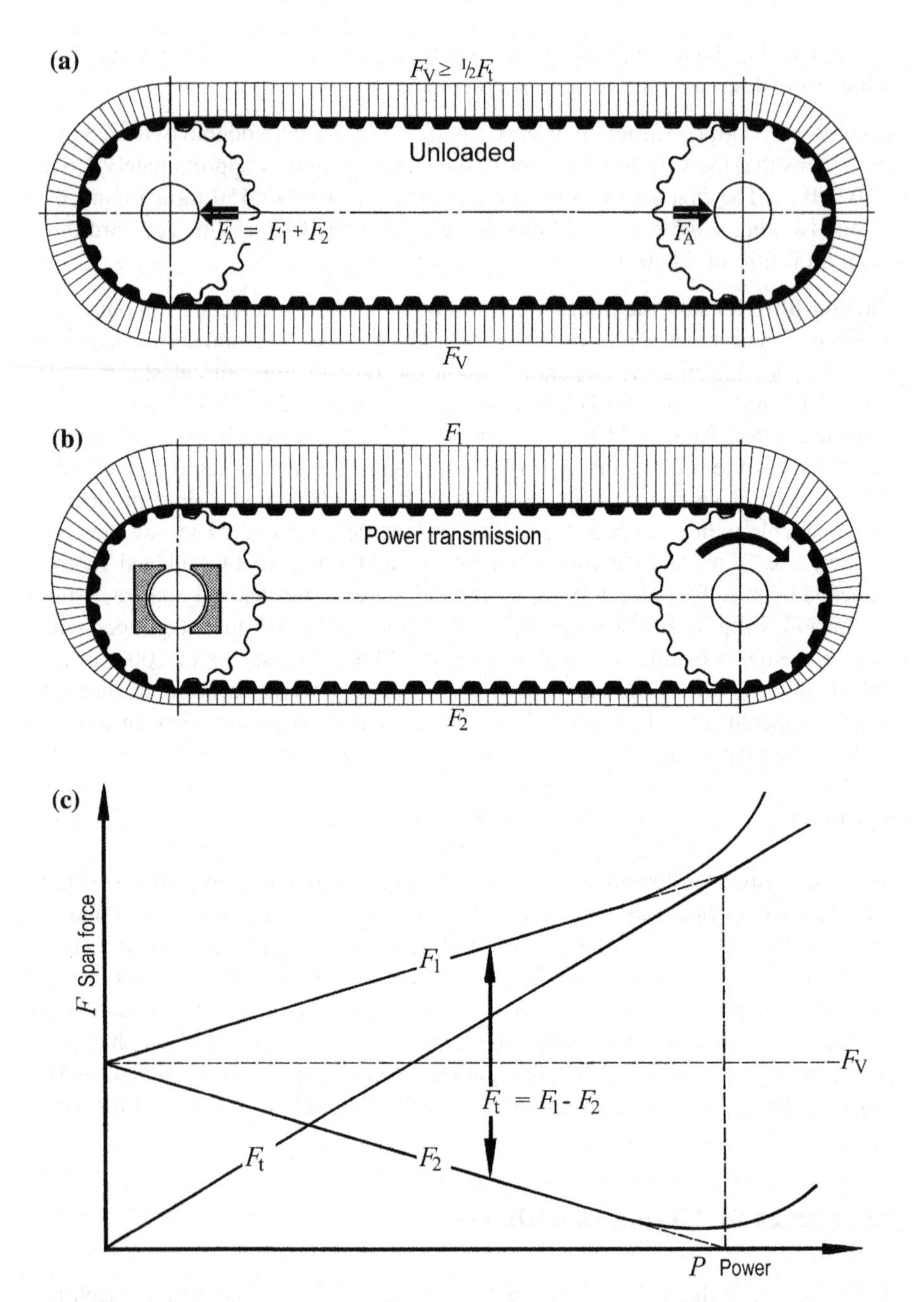

Fig. 2.12 Forces in timing belt drives: **a** unloaded condition, **b** transmitting power, **c** graph: Span force curve as a function of power. *F* Force, F_A Axial force, F_t Tangential force, F_V Pre-tension, F_1 Force in loaded span, F_2 Force in unloaded span, *P* Power

with a brake symbol to indicate power use. When the drive is working and the driven pulley is transmitting power, then force increases in the loaded span and the resulting belt elongation is seen on the idle side with a corresponding force reduction. The graph in Fig. 2.12c shows the force curve in the belt under both loaded and idle conditions. Where power = 0, i.e. under no-load conditions, the strand forces F_1 and F_2 meet at their starting point of F_V. The more power is required, then the greater the difference in tension between that driven and idle strand. This differential force is the peripheral force, also known as the tangential force F_t:

$$F_t = F_1 - F_2. \tag{2.11}$$

Torque and power is transmitted from the right to the left-hand pulley only by this peripheral force. The graph, as well as the observation model for idle and power transfer, is of fundamental importance for further examination and, in particular, the initial pre-tension to be applied is examined more closely.

2.7 Force-Effect Mechanisms

We can see the importance of the applied pre-tension load F_V in Fig. 2.12a–c. Reliable power transfer is only possible as long as a residual tension remains in the idle side of the belt. Thus, the timing belt should be pre-tensioned to a value of

$$F_V \geq \frac{1}{2} F_t \tag{2.12}$$

Here the symbol "≥" should be understood to be "equal or slightly larger", so as not to give an unnecessarily high preload in the belt, shafts and bearings. When the power increases beyond a critical point, the slack side will respond by sagging and as a result there is acute danger of belt tooth jump. Compact drives with relatively short belts will better overcome minor overloads but drives with long belts and less preload tend to see more tooth jumping. The tooth jumping danger zone is always at the point of entry of the belt into the driven pulley where tensile forces in the belt are at their lowest.

In this context, the work of Lothar Köster [67] is important, because it first described the meshing behaviour and the wedge-effect in the convergence of the belt and the driven pulley. The formation of this wedge-effect (Fig. 2.13) occurs when the radial component of the existing pre-tension load, at the point of entry, is not sufficient for the belt tooth flank to fully mesh in the tooth gap against the frictional resistance of the meshing teeth. Only with an increase of the tensile force in the arc of contact can this resistance be gradually overcome until the belt teeth finally fully mesh into the corresponding tooth gaps. Stroboscopic studies have shown that, under rotation, there is only a gradual meshing of the belt taking place with the tooth gaps in the convergence zone of the driven pulley. Thereby a wedge-shaped disengagement of the belt forms on the pulley and the belt teeth

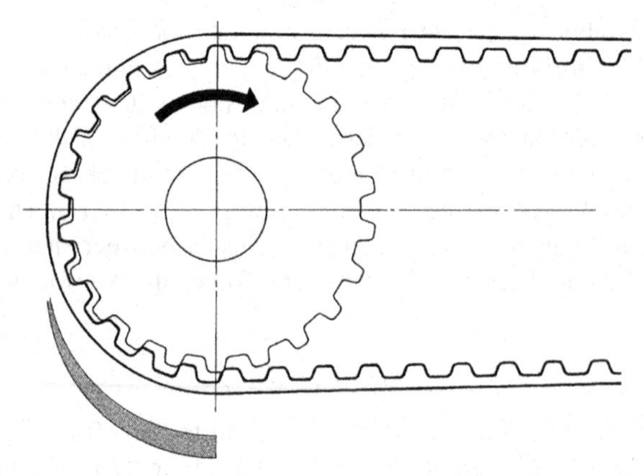

Fig. 2.13 Wedge-effect in the convergence zone of the belt and the driven pulley

located in this area will only carry partial loads. The wedge-effect is all the more pronounced the closer the drive is operated to its power limit, the greater the F_t/F_2 power ratio is, the lower the preload F_V and the greater the opposing friction resistance of the meshing process.

In comparison, compact drives with relatively short belt lengths have high values of stiffness, making the belt elongation low, and therefore the wedge-effect is reduced. According to manufacturer's recommendations [87], drives of this kind can be run with significantly lower pre-tension (see the power drive in Fig. 3.37 in Chapter 3.9).

Additionally, the relationship between the belt and pulley pitch impacts on both the smooth running of the belt and the formation of the wedge-effect at the driven pulley. The belt can also undergo significant load-elongation, which leads to small changes in pitch, depending on the load conditions. The belt section in the convergence zone of the driven pulley will experience a decrease in traction with the increasing size of the torque to be transferred; the belt pitch in the convergence zone is reduced. Therefore, the incoming tooth of the belt can climb the working flank of the pulley, with the belt experiencing tooth flank compression, in order to transfer the tangential force from the belt to the pulley. The belt acceleration also gives a pitch increase at the teeth on the driven pulley [80], so that the belt/pulley meshing errors continue to increase until equilibrium between the frictional forces and the span tension is established.

Extreme pitch changes may, in conjunction with the formation of the wedge-effect, also cause abnormal wear and tear to the belt. Frictional forces increase at meshing due to non-optimum tolerances and low pre-tension hence the tendency to wear.

Metzner and Krause specifically expand upon the relationship of the pitch consistency between belt and pulley in a chapter of their book concerning force-effect mechanisms [77]. A critical assessment of smoothness, efficiency and belt wear is derived therein from the load distribution on the teeth in the angle of wrap.

The optimum solution would be if each single tooth could enter its respective tooth gap without load or friction, or at least with reduced friction, while the tooth surface pressure increases on the working flanks up to the belt tooth leaving the pulley tooth gap. The tooth meshing on the driver pulley takes place under elevated tension while conversely on the driven pulley it takes place under reduced belt tension. These locally acting forces result in pitch changes in the belt and could be offset by increasing the driving pulley outside diameter and reducing that of the driven pulley. Endurance tests using a corrected outside diameter (of about 0.1–0.15 mm) have shown significant increases in performance [77]. Such corrections of the outside diameter mean being unable to use standard pulleys, which limits interchange ability and should only be applied in very exceptional circumstances.

The formation of the wedge-effect is accompanied by an increase in the axial loads as the belt does not run on its ideal line due to the neutral line of the belt being moved outwards. The belt is stretched so strongly and builds such high tensile forces that the elongation load also acts on the shafts. This effect can be seen in the diagram in Fig. 2.12c at the operating point of the load border and, in particular, the slack side force F_2 increase into the overload range is clearly visible.

This knowledge of tooth-jump conditions can be used by drive fitters to test run the drive by closely observing the slack side and the meshing of the driven pulley. If the belt meshes with the driven pulley smoothly, i.e. does not climb, and if the belt slack side shows no inclination to flutter (transverse vibrations, see Chapter 2.11) then the belt is properly pre-tensioned. Therefore, the pre-tension must relate to the maximum effective drive torque. Generally, it is recommended to create the initial conditions for the largest transfer of torque and use a test run to simulate this operation.

Belt assembly and the setting and measuring of pre-tension are treated separately in Chapter 2.17.

Model analysis shows that the tension member is only resilient to a certain degree. If this value is exceeded, at any point during the drive working, then the belt must yield. As a consequence, it briefly jumps over a large length of the arc of contact as the ideal line length yields. Thus, the belt is either damaged (stretched or teeth torn off), or in short and compact drives it cannot elongate any further and will rip immediately it jumps the pulley teeth. During drive design, it is imperative to ensure that the maximum tensile force F_{zul} in the belt is not exceeded.

The maximum tensile forces are formed in the loaded belt span, and they are calculated from the relationship:

$$F_{max} \approx F_V + \frac{1}{2} F_t. \tag{2.13}$$

The span force affects the axes and with drive ratios of 1:1 the shaft loads amount to:

$$F_A = 2 \cdot F_V \approx F_1 + F_2, \tag{2.14}$$

where the approximation character, "$\approx$" refers to each point in the operating conditions. In the lower and medium load range, and with a properly adjusted preload force, "=" can be used. If the timing belt drive is operated at the upper performance limit, or if this limit is exceeded, or the pre-tensioning force is set at too low a level, then the drive smoothness is compromised and F_{max} and F_A may well take on higher values.

The maximum tensile forces in the belt must not exceed:

$$F_{max} \leq F_{zul}. \tag{2.15}$$

This knowledge of the wedge effect, pitch variances and friction conditions during meshing can also be used in assessing damage to the belt (see Chapter 6.2 about damage). It can, therefore, be used to judge a toothed belt transmission which is always operated at its maximum. For example, if tooth-wear shows on the load-bearing tooth side of the belt and on the working flanks of the driven pulley then it is clearly facing a drive overload. Incidentally, the wear pattern is very similar if the belt has too low a pre-tension. To properly assess a drive using the above criteria, the belt's running direction should be clearly marked before removal.

A possible solution for meshing difficulties is to lubricate the belt. Köster [67] proposes such measures and he indicates that he also found a significant increase in performance. However one should ascertain beforehand whether the particular belt is oil-resistant.

In the above considerations, the correct pre-tension choice should always take into account any "braking" mode. In particular, clarification is needed of how the motor is controlled, and how the inertial masses in the drive train are braked before and after the timing belt drive. Also of interest, is that some braking situations, i.e. emergency stops, are normally significantly larger than the torque at start-up and whether the braking torque acts fully on the drive belt. During braking, the drive load moves to the belt slack side and the driving pulley becomes driven and vice versa. However, the basic considerations remain unchanged, i.e. the maximum transmittable torque is to be used as the basis upon which to calculate the necessary pre-tension.

While calculating the maximum forces in the tension member and the associated shaft loads in Eqs. 2.13–2.15, it is also practicable to consider the installation process. In a real drive, an exactly required pretension F_V is not available. Rather, the fitter works in a boundary region between F_{Vmin} and F_{Vmax}. Thus, it is necessary during drive design, where appropriate, to build-in a safety margin. The subject of belt installation and the adjustment of the pre-tension are treated in detail in Chapter 2.17.

2.8 Pre-tension in Multiple Shaft Drives

A multiple shaft drive is defined as when a belt is coupled with three or more pulleys. With multiple shaft drives, it is not important whether any of the pulleys are involved in the transmission of power or if individual pulleys "only act" as deflection or idler pulleys. By running the belt over at least three pulleys, the geometric relationships are different from the two-shaft drives previously considered. The pre-tension is especially influenced by the length difference between loaded and unloaded parts of the belt length (Fig. 2.14) [97].

If the loaded part of the belt length is denoted by l_1 and the unloaded part of the belt length l_2, then the belt length ratio is calculated from

$$\lambda = l_2/l_1. \tag{2.16}$$

The pre-tension for multiple shaft drives is

$$F_V = \frac{1}{1+\lambda} \cdot F_t. \tag{2.17}$$

If the loaded part of the belt length is "very long" and the unloaded length "very short" (Fig. 2.14b), then the belt elongation in the loaded belt length must be reflected in the "very short" belt length by a tension reduction during power transmission. In this case the tension in the unloaded part of the belt length would decrease according to the required torque. The short length, however, has the capacity to absorb only a finite and extremely limited elongation and thus this belt section would tend to sag. Such drive designs, when compared to two-shaft drives, require a higher pre-tension (see the conditions to Eq. 2.12), namely

$$F_V \geq 1 \cdot F_t. \tag{2.18}$$

The tangential force F_t is therefore not driving the driven pulley by an increase in pulling power but almost exclusively by the power loss in the unloaded belt section.

The transmittable tangential force F_t is formed from the difference in forces F_1–F_2, as described in Eq. 2.11, only with the difference that the force increase in the loaded belt section and the force decrease in the unloaded belt section is divided asymmetrically, which is expressed in the different slope angles of the load diagrams (Fig. 2.14b).

With a direction of rotation reversal (Fig. 2.14c), the situation described above changes in such a manner that one "very short" loaded section faces a "very long" unloaded section. Theoretically, this drive arrangement requires no or only a very small initial pre-tension. For such applications it should be considered that the load conditions are reversed during braking. Thus, this drive should be pre-tensioned to $F_V \geq 1 \cdot F_t$ on safety grounds. However, if only one load direction whilst running is possible, then the pre-tension $F_V \geq 0 \cdot F_t$ can be ignored. The sign "≥" here stands for "the same as or a little larger than zero".

Multiple shaft drives, as shown in Fig. 2.14d, are usually driven from a motor pulley and the driven pulleys absorb parts of the power transmitted. It is rare that the power transmitted is shared evenly between all the driven pulleys. If one of the driven pulleys is excessively braked or jammed then the entire torque must be able to be taken up by that pulley. If no knowledge exists of the individual pulley loadings then a pre-tension of $F_V \geq 1 \cdot F_t$ is recommended, however should the

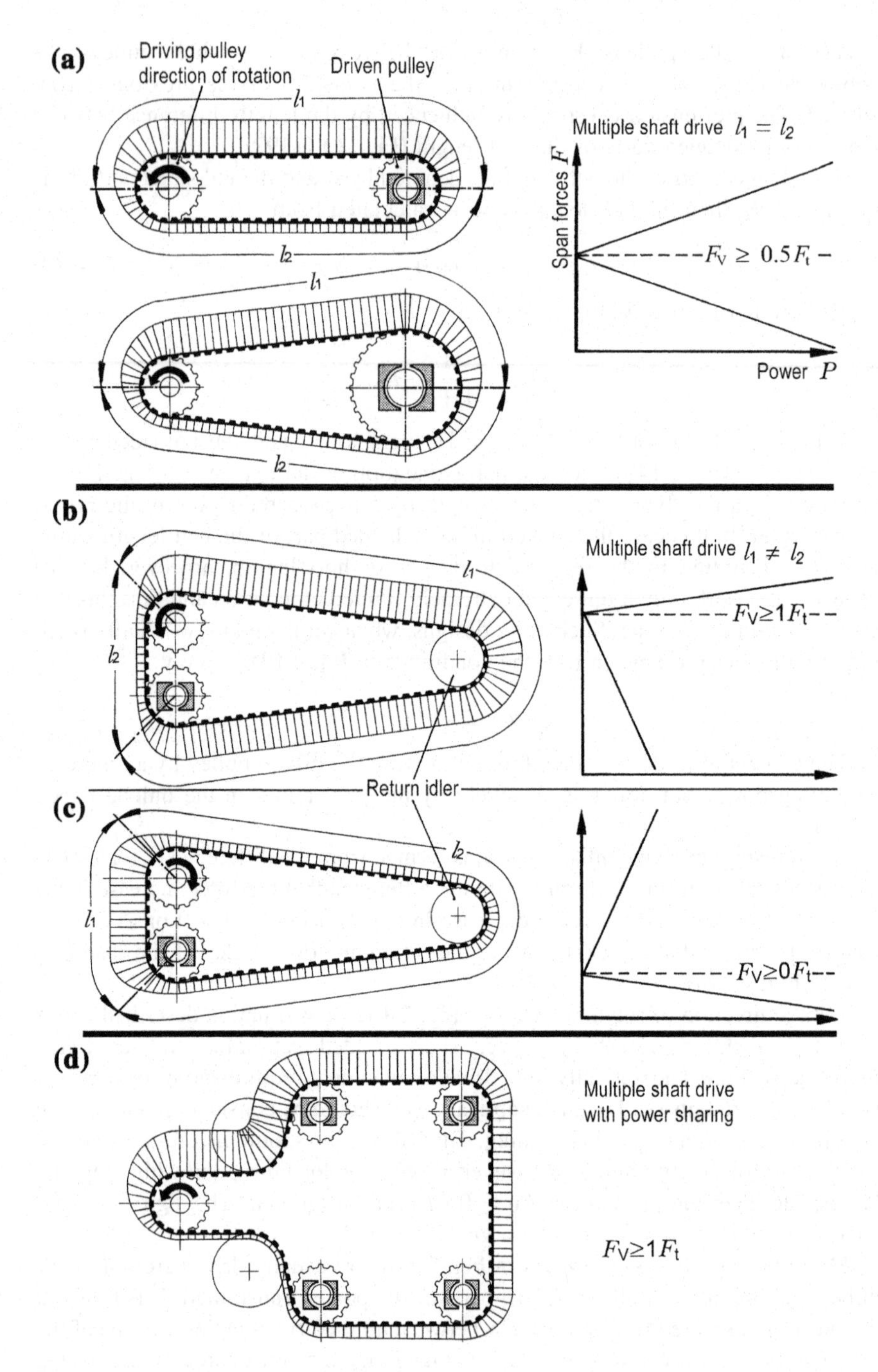

Fig. 2.14 Influence of pre-tension due to the difference between loaded and unloaded part of the belt length: l_1 loaded part of the belt length, l_2 unloaded part of the belt length, F_V pre-tension, F_t tangential force

power distribution on the individual driven pulleys be evenly shared, then the option exists of reducing the pre-tension to $F_V \geq \frac{1}{2} \cdot F_t$.

For the multiple shaft drive in Fig. 2.14d, the load case should be considered in each different part of the belt length related to the motor pulley or any driven pulley. Multiple shaft drives can result in a variety of values for l_1 and l_2. The appropriate pre-tension is calculated in the same manner as described above.

2.9 Tooth Load Capacity

The arc of contact of the belt teeth in mesh transmits power through flank contact with the pulley teeth. The driving pulley transmits power to the belt and from the belt to the driven pulley.

The belt tooth load capacity is understood to be that load which the timing belt is able to tolerate without damage over the long run [87]. In most cases, manufacturers' catalogues indicate such a maximum load capacity value as the specific peripheral force per tooth and per cm belt width F_{tspez}. Multiplication by the number of teeth in mesh z_e and with the belt width b gives the transmittable peripheral force

$$F_t = F_{tspez} \cdot z_e \cdot b. \tag{2.19}$$

In the calculation model, the maximum value of the multiplier z_e is limited, i.e. while there may be a lot more teeth in mesh on the pulley, the number of teeth in mesh used in the drive design process has a maximum value (for example, twelve teeth). This value has a theoretical origin as belt teeth in the arc of contact on the pulleys pass through different load zones (Fig. 2.15). The belt teeth entering driving pulley generally have a low loading but as they move along the arc of contact the load increases on the working flanks. This is the cause of the pitch changes in the belt as described in Chapter 2.6. On the other hand, when a belt tooth meshes in the driven pulley, it turns in the working direction of the belt teeth and the first incoming tooth is usually the most heavily loaded.

The diagram in Fig. 2.15 assumes a symmetrical load distribution in the driver and driven pulleys. The actual load distribution depends on the tolerance of all the components involved from the pre-tensioning force in the belt to the current power requirement from the total drive.

The manufacturer's intended purpose is that the tooth load curve, over the arc of contact, acts on all belt teeth as evenly as possible. A tension member without any elongation would be a desirable but is not achievable in real life. Belt manufacturers, being aware of this problem, use stronger tension members than the calculated peripheral forces actually require.

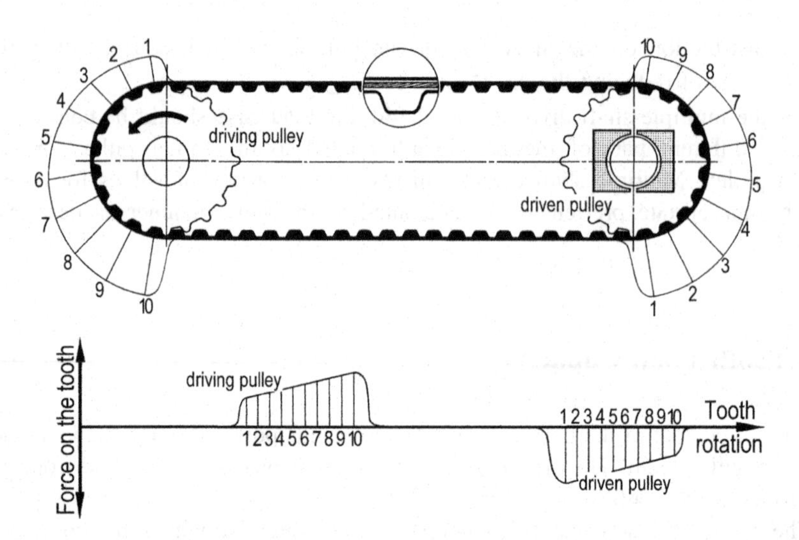

Fig. 2.15 Tooth load curve: Load values for a single tooth over one revolution. The tooth numbering is in the running direction

Better tension members also benefit the tensile stiffness, by aiding the load transfer from the belt into the pulleys and vice versa, where the load is distributed more evenly over the arc of contact. Additional desirable effects are significantly increased power, with greater safety factors for both the tooth load capacity the tension members.

In currently available literature, the belt tooth load capacity is derived from the wear strength of the elastomer used (also occasionally referred to as the surface pressure). Clearly, the better term here is tooth load capacity because it best describes the complex function of the task to be performed by the belt tooth. The term includes the suitability of the materials used (Shore hardness, tear strength, etc.), the tooth geometry, the tooth root radii, the bonding of the elastomer to the tension member and, if present, the fabric facing on the tooth side of the belt. Of ultimate significant influence is the associated tooth gap geometry in the pulley, the sizes of which vary significantly depending on the timing belt profile, the corresponding manufacturing processes, and elastomer types with different quality characteristics (see Chapter 2.4). In calculating the capacity of the belt, the user is therefore dependent on data from the product manufacturer. In some literature, the specific tooth bearing load is indicated as either F_{tspez}, M_{spez} or P_{spez} [87] where the values progressively rise as speed increases.

In general, belt manufacturer's values for tooth load capacity are reliable since they have been evaluated as safe on test-beds, by endurance testing and other empirical testing. The manufacturer is ultimately concerned about product liability and the supplied drive components must also be able to integrate with less than favourable tolerances from other components in the system. Typically, catalogue values for the tooth load capacity are given with an "abundant" safety factor, and this means little or no risk of overloading leading to the failure of the belt drive. However, a significant excess of the allowable tooth load capacity, will allow different damage modes to occur, depending on the type of overload. One must be able to differentiate between long term damage to failure and failure due to a momentary single power overload. Chapter 6.2 deals with failure modes and analysis.

Load Sharing Between Belt and Pulleys

The load transfer characteristics in the pulley arc of contact have been the subject of much research which, in particular, looks closer at the flank contact load bearing area of the individual teeth. The works of Köster [67], Metzner [80] and Koyama [69 and 70] deal experimentally with the relative displacements due to single tooth geometric pitch deviations between belt and pulleys. Also evaluated is the elongation behaviour due to the pre-tension force, as well as length variations due to the active dynamic forces in the running belt. These studies give a sufficiently accurate model of the effective contact behaviour of the paired working flanks and show the considerable irregularities in the load distribution on the flanks of the individual teeth.

Further results are shown in the work of Librentz [76], which includes the parts of the frictional peripheral force derived from the *Eytelwein* cable friction equation. In two-pulley experiments, Librentz considered the contact in the arc of contact and also the radial pressure on the pulley outside diameter, where the corresponding tolerances between belt and pulley allow the rotating belt to run at a high speed. In the arc of contact, the belt meshing can be on the "wrong" flank as the belt moves within the backlash of the tooth gap. The model here is to consider the tolerance of neutral line compared with the belt support outside diameter, between the tooth pairs, under load conditions. Incorrect tooth gap positions occur, particularly at part load on both the driving and the driven pulley. In addition, the belt can occupy tooth gap positions in which there is no tangential contact between the belt and pulley teeth. Torque is transmitted purely by friction in this case. The "correct" flank will engage instantly with increasing load. These belt tooth movements in the pulley tooth gap amount to a few $^{1}/_{10}$ mm and give the danger of "tribological bridge wear" due to the belt support force (see damage illustrations in Chapter 6.2). The movement of the belt tooth in its tooth gap will increase with changing dynamics. One possible measure against this type of wear is to narrow the tooth gap or to use a Zero backlash-free tooth gap c_{m1} to reduce or prevent abrasion.

The different belt tooth positions in the pulley tooth gaps are therefore dependent on tolerances, load, friction and pre-tension. Additionally, belt teeth in the arc of contact go through different load zones (see Fig. 2.15) and the resulting sliding motions of the teeth determine the wear behaviour. The position of adjacent flanks is only detectable with considerable measurement effort and for practical belt use a prognosis of the actual tooth position is hardly possible.

The type of strain on the belt tooth, and the analysis of the operating behaviour taking place in the arc of contact of a rotating two-pulley drive, exhibits the highest levels of process complexity. However, detail judgements of mechanisms taking place are possible through knowledge of the refinement, wear characteristics and the positioning behaviour. Real-life applications experience is also available to the designer and often the realistic view of the higher loaded segments is to describe the mechanisms of action in a simplified way. Here, the loaded flanks in driving and driven pulley are clearly defined in their tangential direction.

Chapter 2.13 provides calculation options, for the estimation of the torsional angle deviations φ in the driven pulley, based on the position of the driving pulley. The results of these calculations usually refer to an assigned tooth position. Thus, it should be noted that the actual deviation can vary substantially should the position of the teeth be unknown but may also turn out to be much lower.

The common factor between all belt profiles is that the tooth load capacity is the dominant determinant of size and the result for the calculated minimum belt width is that value most affected.

2.10 Belt Guiding and Flanges

A self-guiding function is completely missing from timing belts as they generally seek to run-off the pulleys. The typical result of this behaviour means that they must be guided, either by special belt variants such as the "self-tracking belt" (see Chapter 2.3.11) or be contained by pulley guide elements.

Compare, for example, the directional stability of the vee and ribbed belt, which has a clearly defined profile in the belt running direction. A flat belt's directional stability is achieved through a *crowned* surface on the highest diameter of the pulley, and thus, it guides itself independently on these pulleys.

Should the outside diameter of timing belt pulleys be crowned, then a timing belt could possibly show a tendency to straight-line running, but only under no-load conditions. Even at low power the timing belt transmits load through the pulley tooth flanks, which are nearly perpendicular to the outside diameter, so should also consequently be crowned. Besides the fact that this kind of tooth geometry is very difficult to produce such a measure would restrict the performance of a normal timing belt. Experiments with crowned timing belt pulleys have proved less than successful.

To achieve straight-running timing belts a drive must use some form of guidance. In general, toothed pulleys are provided with flanges.

This run-off characteristic is explained by the fact that the tension members, embedded in the belt consist of a wound filament bundle and each individual filament in the belt describes a helix around the tension member axis. Under load each single filament strives to take the shortest possible route but the surrounding elastomer prevents the tension member untwisting. Instead, the effect is a rotational torque which is implemented as a run-off force on the belt. These forces are higher as the tension member diameter increases, the number of tension members in the belt increases, the belt width increases and the more torque is transmitted. The impetus to run-off can be compensated by having adjacent tension members arranged in pairs, which are alternately of right/left twist, which reduces the impetus to run-off and leaves just small residual forces. Not all timing belt manufacturers use this tension member construction.

Another type of run-off behaviour, generated by the belt itself, comes from the spirally wound steel cord in joined timing belts. The cords are wound in the belt running direction to a corresponding pitch angle. The run-off forces triggered are much lower in the lay direction than that of the continuous tension member belt generated forces.

As the transmitted power changes (e.g. from no-load operation to rated output), it can be observed that the toothed belt can also change sides on the pulley. While this side-to-side behaviour is affected firstly by the cylindrical accuracy of the pulley outside diameter, the pulley teeth parallelism dominates under performance conditions. Thus, tolerances in both the belt and the pulleys contribute to the running-off behaviour. Further causes come from shaft misalignment and poor alignment and offset of the pulleys. Assembly errors can produce different edge tensions and the belt is inclined to run towards the largest tension. The allowable variations for pulley alignment, angularity and offset are dealt with at the end of this chapter, see Fig. 2.22 and Table 2.4.

The use of flanges, their geometry, and the method of attachment to the pulleys, is such that one part of the design process is ensuring sufficient side load resistance and that cost-effectiveness is achieved (Fig. 2.16).

A timing belt is securely restrained when it is run between flanges. The standard for timing pulleys [22] recommends coined flanges with an in-feed chamfer within the specified angle range.

Tables 2.2 and 2.3 contain a shortened and simplified version of this standard where the minimum height h for flanges and the minimum width b_f, (the clear width between the flanges) are indicated.

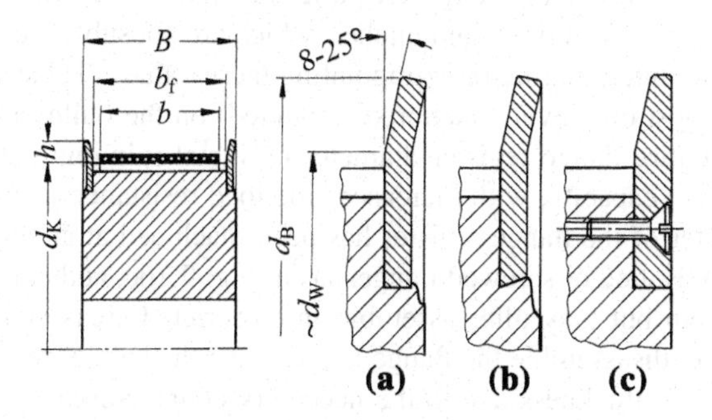

Fig. 2.16 Examples of flanges and mounting methods: **a** crimped, **b** undercut crimped, **c** screwed

Table 2.2 Flange heights

Pitch code	Flange minimum height h (mm)
T 2.5	0.8
T 5	1.2
T 10	2.2
T 20	3.2

Table 2.3 Inside width between flanges

Belt width b (mm)	Minimum width b_f between the flanges (mm)
4	5.5
6	7.5
10	11.5
16	17.5
25	26.5
32	34
50	52
75	77
100	102

The minimum values listed here are defined so that they allow the timing belt width b enough room for tolerances and flexing inside the clear width b_f between the flanges. Manufacturers of belts and pulleys understand the nominal width B of the timing pulley as the distance between the faces adjacent to the pulley teeth. After deduction of the flange thickness and width addition needed for clearance, the recommended minimum width b_f is found in Table 2.3. The clear width between the flanges is not dimensioned as a rule. It should be noted that due to the belt in-feed angle that the flanges are not completely flush with the pulley faces. The above values relate to metric pitch timing belt pulleys to DIN 7721 Part 2 [20]. Flange dimensions for further belt types and pitches, which are not subject to this standard, can be derived by interpolating or extrapolating these values (Table 2.3).

Flanges are most effectively anchored (mounted) on the pulley by rolling over (crimping) the entire flange register diameter by metal spinning (see Fig. 2.16a). Higher attachment capacity, with increased side load resistance, is available when the inner diameter of the flange register has been machined at an angle of 15–30° (see Fig. 2.16b). Bolted or screwed flanges, as in Fig. 2.16c, add complexity to the attachment system but they offer assembly in constricted areas and further facilitate the ability to disassemble the flange and remove the belt. Screwed flanges are usually of a greater thickness due to the necessary countersinking and the in-feed ramp is machined instead of having a coined angle.

In a timing belt drive, each participating pulley and idler could be fitted with guide elements in the form of flanges. It is recommended, however, to use them only as often as necessary to effectively guide the belt. The functional arrangement of flanges takes into account some special features such as those shown in Figs. 2.17, 2.18, 2.19, 2.20, and 2.21.

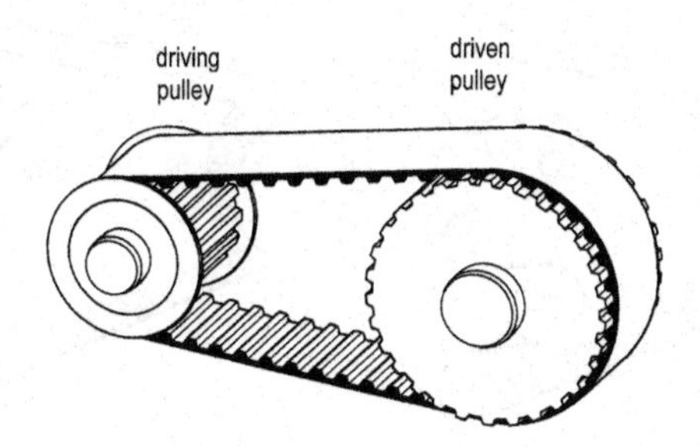

Fig. 2.17 Two-shaft drive

For two-shaft drives it is always recommended to have the flanges on the driving pulley as the guided belt will hold position on the driven pulley. With centre distances greater than 15 times the belt width, additional lateral support at the driven pulley can be used, and flanges on both pulleys are useful from centre distances greater than 30 times the belt width.

Guiding with flanges should be avoided on those axes of a drive where small tension member forces at the belt in-feed side are to be expected. This situation applies regularly to driven pulleys. Chapter 2.7 describes the in-feed wedge effect, as well as the deviation of the belt from the ideal line in Fig. 2.13. A belt section under small span loads entering a flanged pulley would be lifted by the one-sided frictional force effect as the tension member load increases in the direction of the raised belt side. Through this coupled mechanism of a single-sided belt lift, paired with increasing impetus of the belt to run-off to the side of higher stress, may lead to a self-reinforcing abrasion effect and the eventual destruction of the belt.

The preferred method of flanged pulley belt guidance in multi-shaft drives, as in Fig. 2.18, takes place at those pulleys where the in-feeding belt has as large a free span length as possible. The direction of belt travel should also be considered.

Belt drives in the form of the Greek letter Ω are particularly used in connection with drive units where the double belt deflection gives the desired large angle of belt wrap for efficient power transmission. In drives such as Fig. 2.19, however, due to the close successive deflections with very short belt spans, the flanges on the driving pulley would consequently develop strong lateral belt forces. Therefore they are moved onto the idler pulleys. This gives significantly lower frictional forces on the flanges and the belt edges.

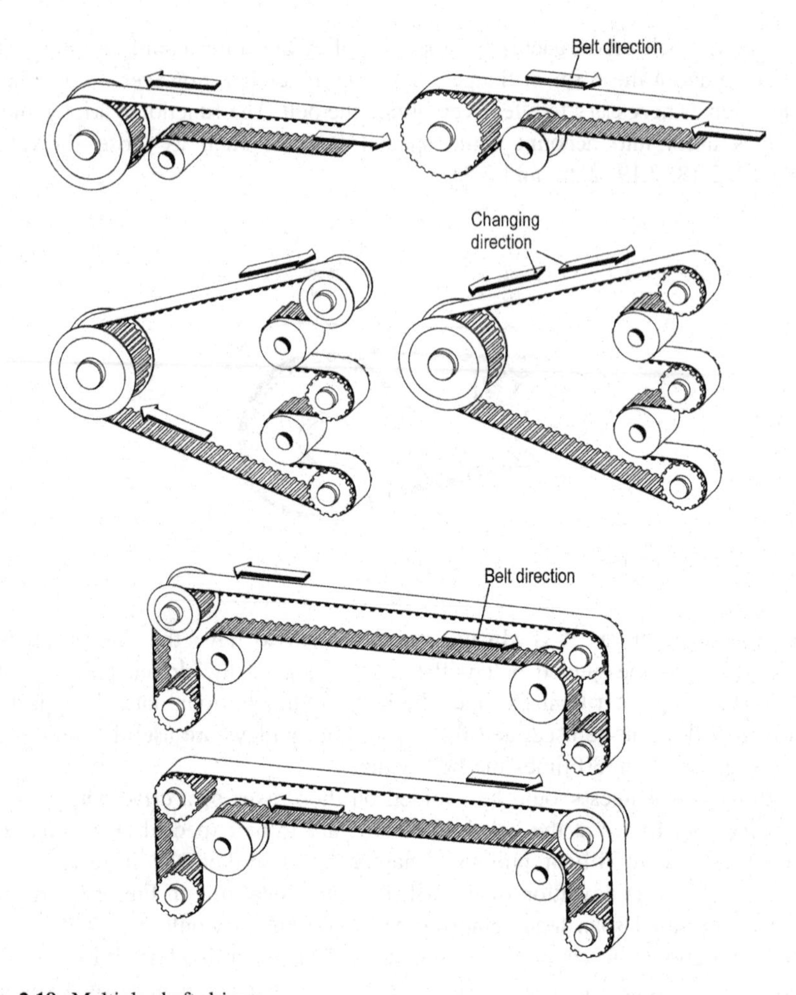

Fig. 2.18 Multiple shaft drives

Fig. 2.19 Omega drive layout

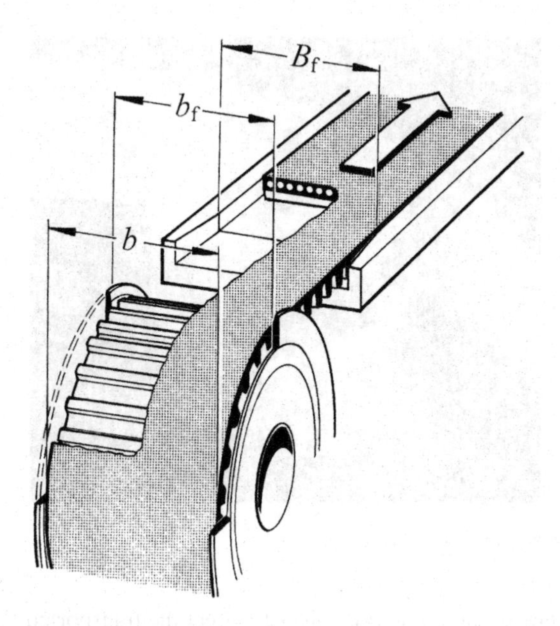

Fig. 2.20 Timing belt guide for conveying

The in-feed situation with edge-guided support rails, as in Fig. 2.20, is of particular importance. It is important to prevent the belt being positively driven to vertical edges. Therefore, it must be ensured that the clearance width b_f between the flanges is smaller than the clearance width B_f between the edge-guides of the support rail. A belt already within the guide rail cannot generate significant run-off force and, without excessive friction, will not jump the edge-guided support rail. Starting from the belt width b, with respect for the possible tolerances of all the assemblies, the rule applies:

$$b < b_f < B_F. \tag{2.20}$$

The application of Eq. 2.20 is subject to conditional use of standard pulleys but with restrictions, as the inside measurement between the flanges b_f is dependent on the application drive conditions. It thus should not be too large; however, it is a measurement that is not usually documented by the manufacturer. For the desired guide rail in-feed "the clearance width between the flanges should be narrow" and therefore, dimensioned separately or toleranced, if necessary. Thus, toothed pulleys are only suitable for this type of application as special designs, or as an alternative, the flanges of standard range designs reworked and attached in corrected positions.

Another alternative guide element suitable for both retrofit and in limited installation space is the easy-drive® [63]. This system was introduced into the market and patented in 2006. It can be used to solve problems in the conveying

Fig. 2.21 Split guide ring [63]

field because there are no projecting edges to affect the transported goods. The belt is guided by a toothed side machined groove supplement (Fig. 2.21).

Summary

The measures described in this belt guidance chapter should be given serious consideration over the choice of the arrangement of flanged pulleys for each type of drive design. However, it may lie in the nature of the application, that specific requirements and other constraints lead to compromise solutions rather than the recommended flange arrangement. To what extent toothed belt guidance has been treated in the past, consists only of favourable "discretionary suggestions", which are differently solvable in addition to the following:

The belt guide arrangement shown in Fig. 2.17 is achieved by flanges on the driving pulley. As this example always has stable belt in-feed conditions, the guide method here is selected with regard to the overall function. It is conceivable, however, that a limited installation area would impose a different solution on the designer. The flange layout selection is not always chosen from the perspective of cost-effectiveness.

Example: Motor and drive shafts are increasingly produced as a smooth shaft and, thus, mechanical clamping elements or heat-shrinking press fits are more frequently used. Flanges attached by crimping or rolling-on are problematic in this case since heat-shrinking the pulley onto the shaft can damage the pulley flange attachment. In addition, space restrictions are often crucial as flanges also need construction space. Belt guidance in this situation is then better solved by a guide rail.

Nearly every design comes about only through a compromise solution. However, if the responsible engineer knows the solution needed, and through compromise finds it, and keeps it under particular scrutiny, then negative surprises will not be encountered.

Rules for Choosing the Flange Layout:

- Flanges are preferably fitted to the driving pulley.
- Directional stability is needed where the belt enters a pulley after a long span length.
- In multi-shaft transmissions belt side guidance is used where the span forces under operating conditions are as large as possible.

Pulley Alignment

The conditions needed for friction and wear resistant directional stability are the careful adjustment of axes and pulleys. Deviations from the ideal alignment cause one-sided edge tensions in the belt and as a consequence the belt runs off towards the side with the largest tension. With assembly discrepancies, the frictional pressure can increase excessively against the flanges and lead to belt edge abrasion and running noise. There are three kinds of problem as in Fig. 2.22.

- Parallel offset of the pulleys
- Angular deviation of the axes
- Twisting of the axes

Any assembly discrepancies will affect the smooth running of the belt, either individually or in combination, and could increase the run-off impetus or even equalize it, see Table 2.4. If the edge stresses caused by angular deviation are not corrected and act unilaterally, there is a risk of premature fatigue fractures in the tension members. Twist is an axial inclination perpendicular to the angle of deviation. Offsets that affect the belt in- and out-feeds are like a parallel misalignment of the pulleys, whereby the sum of both deviations must remain the same or smaller than each single deviation. The wider the belt and the more rigid the belt construction, the more remarkably a toothed belt reacts to assembly errors of this type. Too high run-off pressure is recognizable if the belt back bows to ascend the side of the pulley flange. Further characteristics of this type of disturbance are visible edge abrasion or a bowl-shaped deformation of the belt back. A slight contour of the belt against the pulley flange is normal.

To prove the angular alignment of the axes and pulleys, measuring equipment such as a straight edge can be used. For widely spaced shafts and pulleys, an Easy-Laser® device is sold, which displays a laser line and target to indicate the direction for correction.

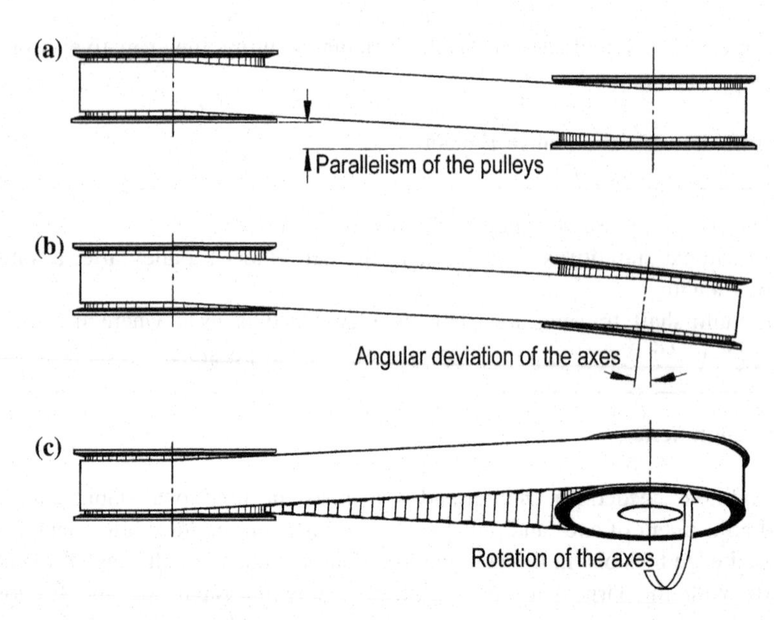

Fig. 2.22 Installation problems through **a** parallel offset, **b** angular deviation, **c** twisting

Table 2.4 The installation instructions for [33, 36 and 94] recommend the following maximum deviations

Parallel offset of pulleys	Maximum ± 5 mm per metre of centre distance[a]	
Angular deviation of the axes	For belt widths of ≤ 25 mm >25 ≤ 50 mm >50 ≤ 100 mm > 100 mm	Maximum deviation ±1,00° ±0,50° ±0,25° ±0,15°
Axial offset	An axial offset twists the belt around its middle. Such a rotational shift is allowed within predetermined rules (see angle drives in Chapter 3.10). As in Fig. 2.22c axial offset causes deviations with the same effect as a parallel misalignment of the pulleys. Maximum permissible deviation is 5 mm per metre of centre distance [a].	

[a] The allowable sum of the deviations of parallel offset + twist is ≤ ± 5 mm

2.11 Irregularities, Vibration and Dynamics

Under favourable conditions, timing belts have the ability to reduce vibration [120], because their elasticity combines with their self-damping properties. Mostly it is the effective power or average torque values that are transferred in drives over which periodic variations are superimposed. These variations come from either the outside of the drive in the form of torque or angular fluctuations or they are created from the drive components themselves. A timing belt typically operates in a system with upstream and downstream components and the entire drive train, as well as parts thereof (e.g. unloaded belt spans), can form a vibration system. The following chapter describes how to detect vibration and which measures to apply to reduce it.

Torsional Vibration

Torsional vibrations can be compared to span vibrations (transverse oscillations) but they are difficult to see by observation because the oscillating motion is about the axes and the rotational motion of the drive is superimposed. In belt drives, these are known as longitudinal vibrations because their dynamic effects follow the tension member direction. They are excited by either periodic, shock or random drive torques [25]. A belt drive consists of at least two rotary axes which tend to stimulate vibration, especially in opposite directions, when periodic disturbances, either in the same or approximate frequency, contribute to the natural frequency. To assess the torsional response, you must first transform the real system into a useful simulated system. As a model, you are looking to set-up two rotating bodies, with the belt connecting them, as an elastic member (Fig. 2.23b). This results in a spring-mass system coupled via a flexible belt, wherein the deflection behaviour is used to define the parameter k (stiffness) (see Chapter 2.13). With known values for the rotating bodies, the mass moments of inertia Θ_1 and Θ_2 can be determined to give the natural frequency of the system. The timing pulleys, as the rotating bodies, are assumed to be rigid and the timing belt as an elastic massless element. The pulleys are rigidly coupled to the shafts and shaft components with their corresponding moments of inertia. The following equations relate to one- and two-mass systems. For three or greater mass systems then computer-aided simulation programs as in [25, 58] should be used.

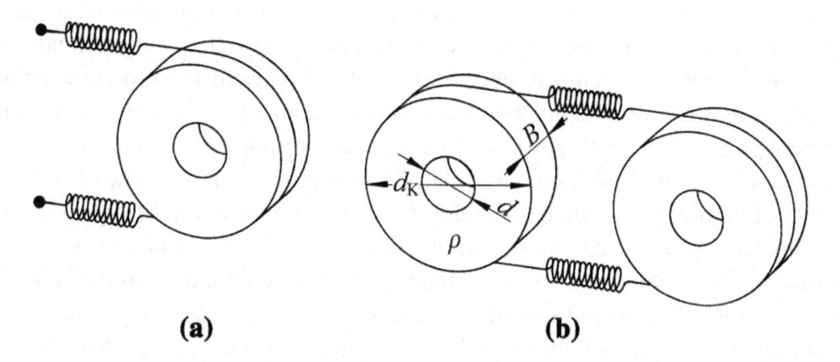

Fig. 2.23 Vibrating system **a** with one rotational mass, **b** with two rotational masses

A rotating mass as shown in Fig. 2.23a vibrates with the natural angular frequency of

$$\omega_e = \sqrt{k/\Theta} \tag{2.21}$$

with the natural frequency of

$$f_e = \omega_e/2\pi. \tag{2.22}$$

The two belt-coupled rotational masses in Fig. 2.23b have a natural angular frequency of

$$\omega_e = \sqrt{k\frac{\Theta_1 + \Theta_2}{\Theta_1 \cdot \Theta_2}} \tag{2.23}$$

with the mass moments of inertia Θ together of each side of the shaft and the pulley. They are calculated by

Shaft $$\Theta = \frac{1}{32} \cdot \rho \cdot \pi \cdot l \cdot d^4 \tag{2.24a}$$

Pulley $$\Theta = \frac{1}{32} \cdot \rho \cdot \pi \cdot B \cdot \left(d_k^4 - d^4\right), \tag{2.24b}$$

with density ρ in kg/m^3, pulley width B in m, outside diameter d_K in m, bore d in m and shaft length l in m.

An "internal excitation" occurs when periodic irregularities arise from the belt itself. The more the irregularities increase, the more the frequency approaches to the natural frequency of the system. If they match, the two frequencies are in resonance. A reduction in rotational mass will reduce the vibration effect and an enlargement of rotational mass will give an increase. If the rotational mass changes are unequal between the pulleys then a reduction in vibration of one pulley will be at the expense of the other. "Outside excitations", especially near resonance, can be reduced by increasing by the relevant mass moments of inertia.

Vibrations of this type can significantly disturb upstream and downstream drive components and additionally stress them further by repeated longitudinal load changes in the tension member. Furthermore, the drive will tend to show bridge wear (see Table 6.1 of Chapter 6). The force fluctuations caused by the relative movements act on the teeth pairs with tribological sliding wear motions.

Resonant torsional vibrations are relatively simple to combat, since the mass-moments of inertia Θ and the belt rigidity k, relevant to the natural frequency, also remain unchanged with changing operating conditions. They can be remedied with increased rigidity values, e.g. by strengthening the tension member in the belt. The torsional stiffness is also influenced by the size of the toothed pulleys. Increasing or reducing the rotational masses allows further possibilities of changing the natural frequency in the desired direction.

Span Vibration

The up and down vibration of the belt span is triggered by excitation forces in accordance with their frequency and with the natural frequency of the belt span or a harmonic of it. Transverse vibrations, as such oscillations are called, are where the deflection of the belt span is in a plane perpendicular to the belt surface. Increasing amplitude manifests itself as longitudinal tension fluctuations in the tension member which are converted to torsional vibrations in the pulleys. Both modes of vibration are therefore separate or combined for effect. Transverse vibrations are formed from the output of the excitation forces and are usually at the frequency f_e of the fundamental frequency of the belt section of interest, and this is calculated from:

$$f_e = \sqrt{\frac{F}{4 \cdot m_{spez} \cdot b \cdot l_T^2}}, \tag{2.25}$$

with span load F in N, specific belt mass m_{spez} in kg/mm belt width and per m belt length, belt width b in mm and span length l_T in m.

Transverse vibrations are often generated by the meshing frequency or by periodic disturbances in the transmission components which are due to quality variations in both the belt and pulleys. There are also possible harmonics of the appropriate number of vibration nodes to consider. Possible belt irregularities can be caused by manufacturing variations in the belt and/or the tooth stiffness, altered tension member location and varying belt thicknesses in conjunction with belt back idler pulleys. Non-circular or eccentrically mounted pulleys can also significantly contribute to vibration excitation.

Irregularities in the belt drive can be mitigated through careful selection of accurately manufactured drive components. Pulley concentricity and accurate mounting of the pulleys are also desirable. Often irregularities are external, such as rotary field oscillations of the motor or periodic fluctuations in the torque delivery. The possible causes of vibration excitation are diverse and while they can be reduced, they can never be completely eliminated.

Slight disturbances are sufficient in many cases to encourage any unloaded spans to vibrate. For example, vibration sensitivity increases with the increased length of span, and unloaded belt spans (slack side) are particularly vulnerable. Since with every variation in load the associated span load changes, thus, a belt and every observed free belt section experiences large ranges of different natural frequencies. When vibration phenomena are detected, then a change of pre-tension can be considered an appropriate measure. With renewed observation under operating conditions, it can be expected that the susceptibility to vibration will have shifted to another range to allow the machine to operate in a non-critical operating condition. If this does not achieve the desired result, one can combat the natural oscillation of the specific belt span by using a tensioner. A suitable method of preventing the oscillation of the problematical belt span is also to provide a device located close to the centre of the belt section which nearly touches the belt back. As a result each vibration peak will be inhibited by direct contact.

Such a component should be shaped to the back of the belt and covered with a low-friction plastic or felt surface.

Stick-Slip Effect

A non-uniform motion, that is encountered at very low speeds or while creeping, in which motionless and motion phases periodically change direction. Viewed from the drive motor, it arises in those transmission parts which are subordinate the toothed belt. The descriptive name of stick-slip effect has arisen from the character of the irregular movements that occur as stick-slip motions. The causes of this are found in the flexibility behaviour of the belt and in the interaction of the variable coefficients of friction in the driven part of the drive. Coefficients of adhesion friction at standstill and friction in motion have usually significant differences. When a drive first starts with a very slow movement, the tension forces in the belt will be established and then, due to the friction reduction, the movement continues in the driven part until the stored tension in the belt is reduced. A specific stoppage time is needed in slow drives to allow the tensions in the belt to settle before the next movement cycle.

Such periodic irregularities are not uncommon in low-speed drives. These effects occur in particular in connection with long belts coupled with high coefficients of friction in the driven components. To achieve a smooth motion for all drive members, at least either one of the following measures should be adopted or in combination thereof:

- Increase the belt stiffness
- Reduce friction
- Raise speed or number of revolution

Dynamics

The increasing demands for shorter cycle times in production environments results in the necessity to optimize the dynamic properties of drives. Dynamics are to be understood as time-dependent changes, i.e. acceleration and deceleration of the motion of bodies under the forces acting upon them. As timing belt drives have a low mass, they are clearly better able, compared with other transmission systems, to withstand the special load conditions encountered during starting, braking and torque reversal. The arc of contact of the belt/pulley gives a load balancing effect across many teeth and the resilience of the elastomeric belt teeth gives shock-free and damped transitions. Thus, one solves one of the special requirements of drive applications where the rotary inertia of the masses or the spring-mass behaviour is of concern.

If a rotating body having a mass moment of inertia of Θ and having an angular acceleration of $\dot{\omega}$, then an acceleration torque M_B applies

$$M_{\mathrm{B}} = \Theta \cdot \dot{\omega}. \tag{2.26}$$

It is only in rare cases that the value of the angular acceleration is available. Thus, from the speed range of Δn from 0 to $n_{\max}$ and with the corresponding acceleration time of t_{B} we assume the acceleration value for the entire speed range as a constant. The acceleration torque is thus calculated from[4]

$$M_{\mathrm{B}} = \frac{\Theta \cdot \boldsymbol{\Delta n}}{9.55 \cdot t_{\mathrm{B}}} \tag{2.27a}$$

From an equation transposition and with known drive data one can determine the acceleration time from[4]

$$t_{\mathrm{B}} = \frac{\Theta \cdot \boldsymbol{\Delta n}}{9.55_{\mathrm{B}}} \tag{2.27b}$$

The mass moment of inertia of a shaft is calculated by

$$\Theta = \frac{1}{32} \cdot \rho \cdot \pi \cdot l \cdot d^4. \tag{2.28}$$

The mass moment of inertia of a timing belt pulley, considered as a hollow cylinder, is determined from

$$\Theta = \frac{1}{32} \cdot \rho \cdot \pi \cdot B \cdot \left(d_{\mathrm{K}}^4 - d^4\right). \tag{2.29}$$

The moment of inertia of several bodies, or more parts of a body in relation to the same, or more shafts that find themselves in the context of the same speed, is equal to the sum of all moments of inertia of each body

$$\Theta_{\mathrm{ges}} = \Theta_1 + \Theta_2 + \Theta_3 + \Theta_n. \tag{2.30}$$

The total mass moment of inertia Θ_{ges1} is based on the *driving* shaft 1, which is linked to a driven shaft 2 by a ratio of i and is calculated by

$$\Theta_{\mathrm{ges1}} = \Theta_1 + \frac{\Theta_2}{i^2}. \tag{2.31}$$

The total mass moment of inertia Θ_{ges2} is relative to the *driven* shaft 2, which is connected to driving shaft 1 at a ratio of i and is calculated by

$$\Theta_{\mathrm{ges2}} = \Theta_2 + \Theta_1 \cdot i^2. \tag{2.32}$$

During investigation of the inertial masses and their effect on the ability to accelerate the toothed pulleys, they are accepted as rigid, and the belts as massless

[4] The Eqs. (2.27a and 2.27b) are numerical value equations. They are used with the specified dimensions: acceleration torque M_{B} in N·m, speed range Δn in min^{-1}, acceleration time t_{B} in s, moment of inertia Θ in $\mathrm{kg \cdot m^2}$.

elements. The shafts and shaft parts are rigidly coupled to the toothed pulleys and should be added to the respective rotational masses.

Example: A 960-HTD-30 belt drives the reel of a wire winding machine with a three-phase squirrel cage motor. The technical information for the motor is: power $P = 6$ kW at a speed $n = 1{,}430$ rpm, rated torque $M_{nenn} = 40$ N·m, starting torque $M_{max} = 80$ N·m, moment of inertia of the motor $\Theta = 25 \cdot 10^{-3}$ kg·m², steel drive pulley $z_1 = 36$ with $d_{K1} = 90.30$ mm, and pulley width $B = 36$ mm and bore $d = 24$ mm H7. The specifications on the machine side are: steel tooth pulley on the reel $z_2 = 72$ with $d_{K2} = 181.97$ mm, and pulley width $B = 38$ mm and bore $d = 40$ mm H7, moment of inertia of the driven machine $\Theta = 400 \cdot 10^{-3}$ kg·m². It is necessary to calculate if the final speed of the machine (reel) within $t = 1$ s is possible.

The mass moments of inertia of small and large pulleys with numbers of teeth $z_1 = 36$ and $z_2 = 72$ are calculated from Eq. 2.29:

$$\Theta_{kl} = \frac{7.8 \cdot 10^3 \cdot \pi \cdot 0.038}{32} \cdot \left(0.0903^4 - 0.0024^4\right) \mathrm{kg \cdot m^2} = 2.17 \cdot 10^{-3} \mathrm{kg \cdot m^2},$$

$$\Theta_{gr} = \frac{7.8 \cdot 10^3 \cdot \pi \cdot 0.038}{32} \cdot \left(0.18197^4 - 0.004^4\right) \mathrm{kg \cdot m^2} = 31.83 \cdot 10^{-3} \mathrm{kg \cdot m^2}.$$

The mass moments of inertia of shaft 1 comprise of the inertia of small toothed pulley and drive motor and shaft 2 of the large pulley and reel of wire together. To calculate this one uses Eq. 2.30:

$$\Theta_1 = \left(2.17 \cdot 10^{-3} + 25 \cdot 10^{-3}\right) \mathrm{kg \cdot m^2} = 27.17 \cdot 10^{-3} \mathrm{kg \cdot m^2},$$

$$\Theta_2 = \left(31.83 \cdot 10^{-3} + 400 \cdot 10^{-3}\right) \mathrm{kg \cdot m^2} = 431.83 \cdot 10^{-3} \mathrm{kg \cdot m^2}.$$

The total mass moment of inertia related to the motor shaft (shaft 1) is determined from Eq. 2.31:

$$\Theta_{ges1} = \left(27.12 \cdot 10^{-3} + \frac{431.83 \cdot 10^{-3}}{2^2}\right) \mathrm{kg \cdot m^2} = 135.1 \cdot 10^{-3} \mathrm{kg \cdot m^2}$$

The acceleration time is calculated from Eq. 2.27b where M_B is the average of the starting and rated torque.

$$t_B = \left(\frac{135.1 \cdot 10^{-3} \cdot 1430}{9.55 \cdot 60}\right) \mathrm{s} = 0.337 \mathrm{\ s}$$

Summary: The calculated acceleration time is less than a second.

Centrifugal Force

When a timing belt revolves around a pulley centrifugal forces build up which applies the tension force F_Z to the tension members. This is calculated from the weight per metre m_m in kg/m and the peripheral speed v in m/s:

$$F_Z = m_m \cdot v^2. \tag{2.33}$$

The resulting centrifugal force loads are to be considered as a preload which reduces the maximum allowable tension member load to:

$$F'_{zul} = F_{zul} - F_Z. \tag{2.34}$$

Centrifugal forces which act as part of the load on the tension member are only increasingly noticeable from rotational speeds >50 m/s.

2.12 Noise Behaviour

The running noise of a toothed belt drive can be perceived from about belt speeds of 1 m/s and are annoying from 3 m/s. From 10 m/s, the resulting sound level brings their use into question. Particularly striking is the tonal character of the radiated noise.

Noise Generation

A timing belt's tooth meshing region is recognized as the source of noise generation in the drive, where the meshing frequency and its harmonics are clearly pronounced. On the other hand, the frequencies due to the rpm of the pulleys and the belt's rotational speed have no influence. The number of impulses for each time period results in the meshing frequency and thus the noise pitch.

The increasing noise level is a complex procedure due to the overlay and development of several factors. In the cooperation with pulley manufacturers, Böttger [6] could prove the following causes of noise generation:

Air Displacement

During belt meshing, it is mainly air displacement between the belt and pulley tooth gaps that is involved in the generation of noise. The volume of air being displaced during the tooth meshing consists of two partial volumes and these pulsating escaping air masses cause the noise-producing air flow. The noise level produced is formed from the radial speed component, the meshing speed of the belt tooth into its pulley tooth gap and/or the belt impact speed on the tip diameter of the toothed pulley. Thus the rpm is the parameter which most strongly affects the level of radiated noise.

Polygon Effect

The meshing of the belt and the impact on the tooth head is a polygonal (intermittent) process which excites the belt spans with transverse vibrations at the meshing frequency. The smaller the number of teeth in the pulley, then the stronger the polygonal effect's impact where these induced oscillations can increase the noise generated.

Friction

Pitch differences between belt and pulley can cause flank friction generated noise during meshing where the noise level can rise by over 10 dB. On the other hand, the tooth meshing speed can be reduced by this frictional action. This leads to a dampening of the polygonal span vibrations and a reduction of the noise output is possible by up to 5 dB. Furthermore, radial belt support forces cause increased levels of frictional noise in connection with the belt motion around the arc of contact.

Resonance

The tooth mesh frequency can trigger transverse vibrations at the natural frequency of vibration of the belt span. An acoustically critical case is when the frequency of the air trapped in the wedge-shaped meshing area, between belt and pulley tooth gap, approaches the resonance of the oscillating span. The increase in noise level in these cases has values above 5 dB.

Belt Width

A dominant value affecting the noise level is the chosen belt width, as it proportionally influences the amount of air displaced. As a result of the investigations in [6] it was found, among other things, that the radiated levels of sound remain almost constant with load changes. For the generation of displacement air flow it is unimportant whether the radial force results from the tooth meshing, from the pre-tension or an equivalent tangential force. However, the associated span forces change depending upon load conditions so that the freely vibrating belt sections go through large ranges of different frequencies of vibration. In resonance with the meshing frequency, noise peaks can occur close to and within limited speed and load ranges.

Noise Reduction Measures

Knowledge of the main causes of noise generation allows the application of targeted reduction measures. Conversely, there are many applications that do not apply to these considerations on the grounds of low speed. Noise reduction measures should be applied as below

for pitch profiles <10 mm with a number of revolutions of approx. 750 min^{-1}
for pitch profiles ≥10 mm with a number of revolutions of approx. 400 min^{-1}.

For noise reduction to be particularly effective it needs the following primary measures. These are the methods that prevent or reduce noise at source. If these methods are exhausted then the result is only a question of reducing sound propagation through secondary measures (e.g. by containment or damping).

Narrow Belts

To reduce air displacement one should primarily aim at the narrowest belt width by the use of high power belt toothforms or by the use of types with extra-strong tension members. As the drive safety factor is proportional to the belt width, this must be taken into account with the use of narrower belts. By choosing the largest possible pulley diameter the necessary belt width is further reduced. This measure will also lower the bending loads on the tension members and is simultaneously associated with lower shaft loads. When halving the belt width the expected noise reduction is approximately 12 dB.

A high-quality belt with a large power rating is generally more expensive to buy. However a narrower belt width will more easily reach the target for noise reduction and narrower pulleys result in less installation space, which is often cheaper overall.

Toothform Sizes

Timing belts with large toothforms produce higher noise levels than belts with small toothforms. However, in a comparison of drives with the same power, the acoustic disadvantages of large toothforms are compensated by the lower belt widths possible. In addition, the lower meshing frequencies are of benefit for both the resonance behaviour as well as the subjective noise levels.

Belts with a Low Coefficient of Friction

Polyurethane belts with polyamide fabric tooth facing were initially developed to reduce friction during belt and pulley tooth meshing. Subsequently, it was discovered that these belts also give a substantial reduction in noise levels, which can be up to 9 dB. This also has an effect in the exit area of the belt on the pulley in that the structured fabric surface favours the separation of the tooth flank at the point of contact and the inrush of air. Timing belts made of synthetic rubber have, in general, always used this type of tooth coating. Other manufacturer provided enhancements are belts with tooth coatings of PTFE coated polyamide fabrics or self-lubricating polyethylene foil, for optimized low coefficients of friction (see Chapter 2.3.9).

Drives with Two Narrow Belts

A reduction of the noise level by 4–9 dB is attainable by splitting the total belt width into two or more single belts. By using two narrower belts, this results in the belt width being permeable, so to speak, and the pulsating air compression noise decreases considerably. The use of flanges between the individual belts will not cause any further noise reduction.

Increasing the Belt Mass

A critical resonance phenomenon occurs when the meshing frequency coincides with the natural frequency of the enclosed air in the tooth section and this additionally coincides with the natural frequency of the belt span. The natural frequency of the belt span is that frequency that excites the belt span to resonate with transverse vibrations. By increasing the belt mass, the natural frequency of the belt span can be moved significantly from the critical resonance range and give noise level reductions of about 6 dB. Such measures can be achieved by simply coating the belt back. In addition, some manufacturers offer a thick belt back option in their product range which roughly doubles the belt mass.

Punched Belts

In order to reduce the compression of the air located in the tooth gaps, the belt body can be perforated. In [6], different perforations, openings and grooves along with combinations of different measures and arrangements are recommended. With these methods noise level reductions can be achieved of up to 10 dB.

Arc-drive Belts

Through an arc-shaped tooth design, the belt tooth meshing is delayed and the polygon effect suppressed as far as possible. This special toothform (see Chapter 2.3.13), when compared to standard toothform belts, shows noise level reductions of approximately 15 dB.

Double Helical (Herringbone) Belts

When using a double-helical belt, in accordance with Chapter 2.3.14, then a noise level reduction of approximately 18 dB can be expected.

Prediction of Noise Levels

For the first time, the studies of Jansen [59] have given a computational approach for acoustic noise emission estimates based on a variety of regularly examined drive configurations. Even if a causal dominant noise source, such as the impact impulses of the belt on the pulley outside diameter have been identified then the derived dependencies (as those in [6]) are very similar (the biggest factor being belt impact speed due to rpm, power and belt width). From Jansen, centre distance and drive ratio are excluded because of their negligible influence. The pulleys are assumed to be "normal" and they are made of metallic materials of a normal surface roughness. The number of teeth is not less than the manufacturer's specified minimum number of teeth and no greater than threefold that number. The expected noise level L_{WA} is therefore clearly a function of power P and the speed of the small pulley n_1

$$L_{WA} = 60 + 0.0018\frac{n_1}{n_0} + \left(21.52 + 0.0014\frac{n_1}{n_0}\right)\log\frac{P}{P_0} + \Delta L. \quad (2.35)$$

L_{WA} Expected noise level in dB (A)
ΔL Belt type-specific noise increase or reduction in dB (A)
P Nominal power in kW, $P_0 = 1$ kW
n_1 Number of revolutions, $n_0 = 1\ \mathrm{min}^{-1}$

The table giving the belt-specific specific noise increase or reduction ΔL demonstrates the fundamental differences in the noise levels between the different belt types and pitches. For the determination of the related figures in [59], a large number of measured and calculated noise levels were compared. The results of this work with appropriate additions are reflected in Table 2.5. The values determined from Eq. 2.35 are understood to be forecasts with corresponding uncertainties. For the toothed belt user, the situation depends on the fact of whether, during the design phase, noise reducing measures have already been incorporated.

To ensure that all belt types can be compared for acoustic performance on an equal basis, it is appropriate to consider the nominal power. This is calculated as the design power and it is the timing belt drive power without the inclusion of a safety factor. Such a timing belt would be pre-tensioned, with the nominal pretension, according to the requirements for transmitting the rated power regardless of whether the maximum capacity would be used.

Evaluation of the Noise Calculation

The respective results of a noise level calculation can be related to a case study carried out with considerable scatter. All previously described effects of noise sources and their mathematical treatment are based on model considerations that do not apply equally to every individual application. Also in Eq. 2.35, the additive effects of the belt-specific noise increase or reduction ΔL is based on simplistic assumptions.

The value of the total noise level emitted is also influenced by resonant components of the surrounding structure. Instigated by the primary sound source, the structure or housings can transmit structure-borne noise and in turn contribute to the radiation of additional secondary noise.

Table 2.5 Belt type-specific noise increase or reduction in dB (A)

XL	L	H	XH	XXH
+10	+9	0	0	0

T 5	T 10	T 20
+4	-1	-1

AT 5	AT 10	AT 20
-2	-7	-8

HTD 5	HTD 8	HTD 14	HTD 20
-2	-8	-9	-10

STD 5	STD 8	STD 14
-3	-9	-10

RPP 5	RPP 8	RPP 14
-3	-9	-10

Arc-drive belt BAT 10
-22

Double helical belt	
S 8	S 14
-27	-28

For the control of structure-borne sound transmission it is necessary to reduce secondary noise by the use of insulating materials (e.g. rubber) as a vibration absorber. For airborne sound attenuation, machine enclosures are suitable, which are also lined with sound-absorbing foam or fibrous materials.

2.13 Transmission Accuracy, Rotational Stiffness

The stiffness, also known as the mechanical deformation resistance, is often the focus of interest of a component or assembly to an externally applied force. The stiffness of a body depends on both its material and geometry. In a belt drive, there is a distinction between the translational and rotational stiffness. The size of the force per direction relates to the neutral line of the belt and the torque per radian on the overall drive. The inverse of stiffness is compliance.

The transmission accuracy is analyzed to see how this affects the belt compliance behaviour between the driving and driven pulley as to the angular deviation, and how the elastic behaviour affects the dynamics of the drive. Users will want to know the extent of the expected compliances as early as possible in the design phase, in order to carry out drive-specific optimization. These calculations give a reliable assessment of the angle of twist and the torsional stiffness (also called the torsional rigidity). Furthermore, information must also be provided for the dynamic behaviour of the spring-mass system. This chapter addresses this issue with a two-shaft drive in a quasi-stable state for the rotary/rotational motion.

Positional deviations in rotational-linear movements in linear motion are the subject of consideration in Chapter 4.3.

The use of toothed belt drives for accurate angular applications and the transmission behaviour has already been thoroughly investigated in several papers [29, 50. 89]. The influence of fluctuations and running radius deviations occurring under stress and strain of the loaded span, as well as the relative motion between belt and pulley, have been particularly well documented.

Tension Member Behaviour and Cable Stiffness

The tension member has a nearly linear, slightly progressive, stress-strain behaviour within the allowable load range. Despite the damaging effect of constant load reversals, it is assumed that no changes occur in the tension member, provided the manufacturer specified values of F_{zul} are not exceeded. Thus, sufficiently repeatable behaviour predictions are possible. Two-shaft drives are characterized by equal length belt spans and the loaded span will experience a force increase during an increase in power. The belt elongation will increase due to this force on the slack side of the belt and will respond with an equally large power loss. In summary, the resulting stress-strain behaviour for the entire belt drive approximates to a linear relationship. Thus the size of the force increase of the linear spring stiffness can be used to find c_B. For timing belt users, it is often difficult to obtain reliable elongation/stiffness values as manufacturers rarely publish them. Comparing individual makes with each other, in spite of different pitches and different types of tension members can result in significant matches. A good starting point is to use the maximum peripheral force F_{zul}, a value that is documented in most catalogues. The corresponding elongations are 2–4 mm/m or 2–4‰ and if they are not available then the relevant manufacturer should be consulted. From these values, the belt-related stiffness over a 1 m length can be determined and is referred to as the length-specific or tension member stiffness

$$c_{Bspez} = \frac{F_{zul}}{\Delta l_{(1m)}}. \tag{2.36}$$

With the value c_{Bspez} it is now possible to calculate the elastic behaviour (accepted as linear for the entire load range) of the tension members for arbitrary timing belt lengths, see Eq. 2.39.

Size of the Timing Pulley

If a rotationally rigid toothed belt drive is to be achieved, then it is always recommended to use the maximum available installation space, by employing the largest possible pulleys, since their diameters affect the rotational rigidity squared.

Related to the accuracy between driving and driven pulley, this means a reduction of the angular deviation reciprocally to the square of the diameters. The designer can make an important contribution to the reduction of angular deviations by selecting a favourable drive geometry. The increase in pulley diameters is accompanied with smoother running characteristics and at the same time smaller bearing loads and a decrease in the polygon effect.

Belt Width

Increasing the belt width will give similar improvements in torsional and angular accuracy. In order to optimize the elastic behaviour of the belt spans and the belt teeth, belt width increases are a valuable option for the designer.

Pre-tension Load

The force effect mechanism described in Chapter 2.5 is equally valid for accurate angular transmissions. With too small applied pre-tension and loss of the remainder stress in the unloaded span under torque load, the rigidity in the entire toothed belt transmission would suddenly halve itself. The recommended and correctly applied initial pre-tension is thus of essential importance for accurate angular transmission. Beyond that, investigations [123] have shown that each further increase of the initial pre-tension is accompanied with an increase in the belt tooth stiffness.

Belt Tooth Stiffness

The compliance of the belt tooth, explained by the investigations in [62], is due to the material properties of the elastomer itself, under the assumption that the transfer torques act as deforming shear forces on the tooth body. Elastomers are incompressible and therefore do not deform when constrained on all sides. It is necessary to allow them to deform in certain directions. The restriction of the deformation potential for elastomers means that their E-modulus is not a true material parameter and is influenced in general by a form factor. In this sense, the tension members prevent the spread of belt teeth radially outward. It can be assumed, that the deformation properties of the belt tooth are dependent on the actual working forces of the drive. Thus, an increase in pre-tension, particularly in belts with the AT-profile, which are supported in the pulley tooth gap, causes an initial load and as a result, leads to a corresponding broadening of the tooth profile. From this correlation and supported by test measurements, the specific tooth stiffness c_{Pspez} is derived in N/m belt width per engaging tooth. The load dependent compliance is assumed to be linear.

Additionally, a selective increase in the pre-tension will cause positional changes to the respective belt tooth positions in the tooth gaps closest to the in-feed or out-feed locations (Fig. 2.24). The tooth engagement becomes essentially backlash-free through the change in pitch and the fact that driving and opposite pulley tooth flanks actively push away from each other, within the same arc of contact, on the belt tooth flanks. Due to this situation, the forces at the point of pulley contact direct themselves against each other, giving what can be called a pre-tension situation. As a result, the belt to pulley meshing can be assessed as backlash-free. Deformation forces on the belt teeth lead to corresponding elastic displacements. To achieve this quasi anti-backlash function, it is recommend to use the AT belt range and the associated toothed pulleys with the "Zero" toothform up to 24 teeth, the "SE" toothform from 25 to 48 teeth and the Normal toothform from 49 teeth and up.

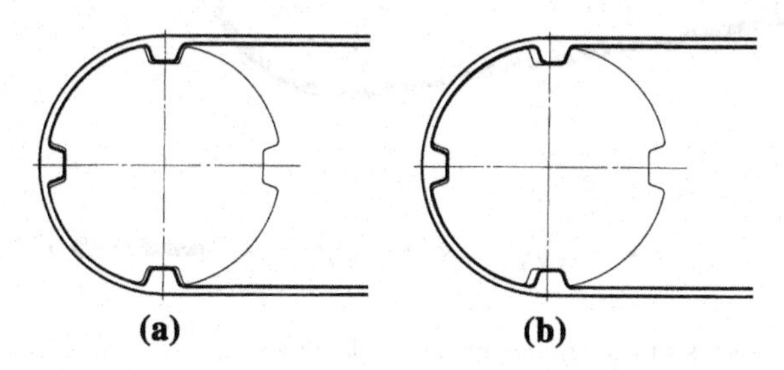

Fig. 2.24 Flank position in the arc of contact **a** normal pre-tension, **b** increased pre-tension

For the optimisation of a rotationally rigid and angularly accurate transmission a "moderate" increase of the pre-tension is recommended

$$F_V \approx 1.5 \cdot F_t \quad \text{to} \quad F_V \approx 2.5 \cdot F_t. \tag{2.37}$$

Narrow Tooth Gaps

Some belt manufacturers offer high accuracy drives with specially machined "narrow" tooth gaps or so-called "Zero" tooth gaps on the associated pulleys. The possibilities for their use are subject to restrictions on speed and the number of teeth of the pulley, so it is recommended that the specific manufacturer's advice should be sought. Further comments on the use of narrow tooth gaps are treated in Chapter 4.3 and Table 4.3 in the context of linear positioning. The definition of the tooth gap backlash is included in Fig. 2.28 in Chapter 2.15.

Step-by-step Calculation

The calculation of torsional stiffness and torsional angular accuracy can be illustrated by observation of the model of a two-shaft drive in a quasi-stationary state, illustrated in Fig. 2.25, where one imagines the driving pulley clamped and a torque applied to the driven pulley. Under these conditions a twist angle is formed due to the flexible behaviour in the belt span, as well as tooth deformation in the arc of contact in the small and large toothed pulleys. The value φ_{max} is that angle which is formed when the drive is loaded with its maximum allowable torque. When any other part load is applied, a corresponding proportional angle of rotation φ is formed. The calculation consists of several steps.

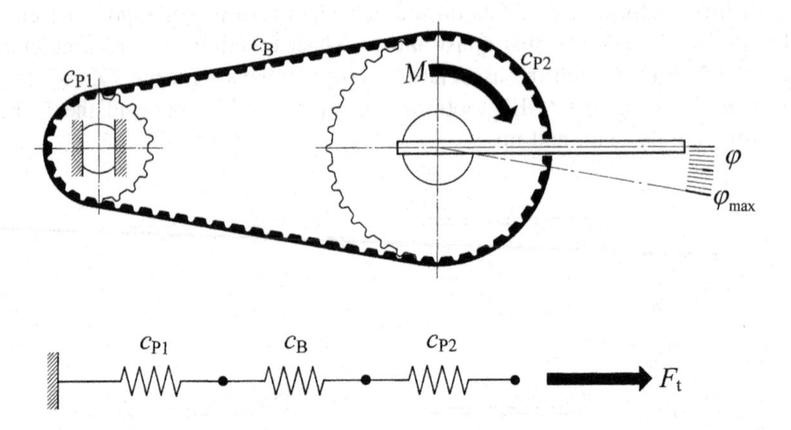

Fig. 2.25 Model analysis to determine the torsional stiffness

Firstly the corresponding total translational stiffness at the neutral line is determined

$$\frac{1}{c_{ges}} = \frac{1}{c_B} + \frac{1}{c_{P1}} + \frac{1}{c_{P2}}, \tag{2.38}$$

whereby the relationship of the tension member stiffness is found here

$$c_B = \frac{4 \cdot c_{Bspez}}{l_B}, \tag{2.39}$$

and the relationship of the belt tooth stiffness is

$$c_P = c_{Pspez} \cdot z_e \tag{2.40}$$

entered here. The number of teeth in mesh in Eq. 2.39 takes into account a maximum of $z_e = 24$ teeth. The only data for the specific tooth stiffness are to found in the available literature of the thesis by J.Vollbarth [73] for the belt profile AT5. The complete program of AT stiffness values are listed in Table 7.2, Chapter 7.3. It should be noted that the length of the belt used for Eqs. 2.38 and 2.39 is the full length l_B. The tension member elongation not only takes place in the belt spans but also in the arcs of contact but with only half the effect. This difference is not taken into account to simplify the calculation, since the belt tooth meshing has within its mesh length about a 10-fold greater compliance compared to the tension member. The proportional error is therefore negligible. With the following equation, the deformation takes place with the rotary value, rotational stiffness k.

$$k = \frac{d_{W2}^2 \cdot c_{ges}}{4}. \tag{2.41}$$

The unit of measurement for k is N m/rad and one rad (Radian) $= 180°/\pi$. To understand torsional stiffness we rotate pulley z_2 by 1 rad compared to pulley z_1. The stiffness of the belt is generated by the counter-torque to pulley z_2 and the value for stiffness is k.

By the conversion of the angle of twist, the transferrable torque can be shown as a linear result. If the drive is loaded with the maximum permissible torque (maximum catalogue value) then the result is a twist angle of

$$\varphi_{max} = \frac{180}{\pi} \cdot \frac{M_{max}}{k}. \tag{2.42}$$

As a rule, for small compliances, optimized belt drives should be properly dimensioned with the appropriate safety factors which give much smaller part loads in the drive. The angle of twist can be calculated from the torque to be transmitted as

$$\varphi = \frac{180}{\pi} \cdot \frac{M}{k}. \tag{2.43}$$

Conclusion

The torsional stiffness k is an important parameter in power transmission for the assessment of drives. It is both a prerequisite for the calculation of twist angles as well as for determining residual vibrations in the drive-train (see Chapter 2.11). In applying the equations provided here, together with the technical data, it is to be noted that all results can be fraught with uncertainties. The data for the tension member and tooth stiffness are, as a rule, subject to tolerances. Thus, analysis has revealed that of some manufacturer's maximum allowable tensile force F_{zul} specifications that the actual measured values are 20% or more higher than the stated catalogue values. Presumably these high values are due to possible risk management and to ensure safety. Furthermore, tolerances in the belt's Shore hardness affect the tooth stiffness. Similarly, the current pre-tension and the belt length tolerance also affect the result. As a conservative estimate, it is therefore proposed to use a minimum ±30% uncertainty in calculations.

In most cases during the design phase, users can estimate enough data to forecast the expected magnitude of the angular deviation. The diagram in Fig. 2.26 allows for the timing belt profiles AT3, AT5, AT10, AT15 and AT20 with the ratio $i = 1$, giving a direct reading of the estimated size of the angle difference between driving and driven pulley. The chart value is at speed $n = 0$ and at maximum torque from catalogue calculation. In accordance with the pro-rata operating torque the relationship is:

$$\varphi = \frac{\varphi_{max} \cdot M}{M_{max}}. \tag{2.44}$$

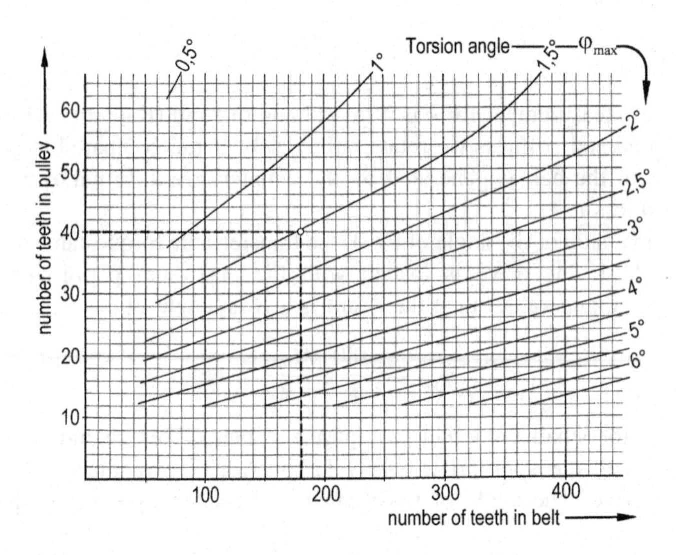

Fig. 2.26 Estimation of the twist angle for the AT belt profile with a ratio i = 1 Key: - - - - - - O example

Example: A drive has a belt and two pulleys 25 AT10/1800 with $z_1 = z_2 = 40$. The calculated torque from the catalogue without a safety factor is $M_{max} = 140$ N·m with an operating torque of 35 N·m to be transmitted. From the diagram in Fig. 2.26 a twist angle for the drive of $\varphi_{max} = 1.5°$ is shown. Due to the significantly lower operating torque an actual twist angle is given by Eq. 2.43 as:

$$\varphi = \frac{\varphi_{max} \cdot M}{M_{max}} = \left(\frac{1.5 \cdot 35}{140}\right)^{\circ} = 0.375°.$$

2.14 Drive-Train Mechatronics

A toothed belt drive is usually part of several transmission components coupled together into a drive-train. The actuator design shown in Fig. 2.27 is to be assessed for dynamics, vibration behaviour and compliance where the moment of inertia, torsional stiffness and angle of rotation of each individual component must be found. However, the angle of rotation only has a meaning as a value derived from the torsional stiffness. The following relationships describe the theoretical context for the determination of total torsional stiffness k_{ges}, total angle of rotation φ_{ges} and total moment of inertia Θ_{ges} as related to the transmission of power through a drive structure consisting of five elements [96]:

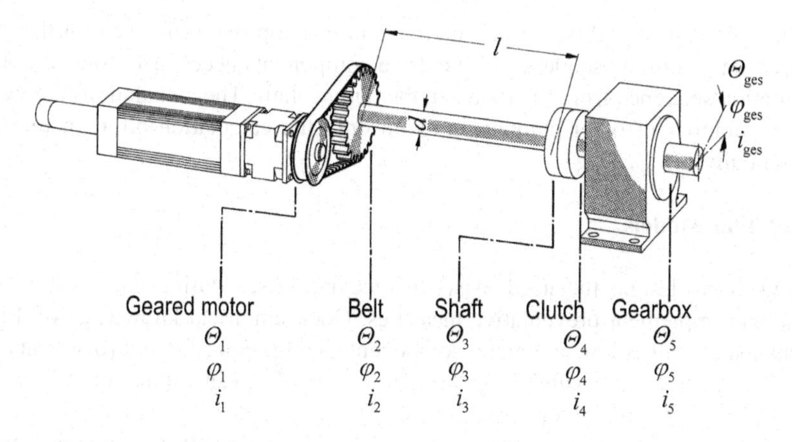

Fig. 2.27 Five component powertrain

$$\frac{1}{k_{\text{ges}}} = \frac{1}{k_1(i_2 \cdot i_3 \cdot i_4 \cdot i_5)^2} + \frac{1}{k_2(i_3 \cdot i_4 \cdot i_5)^2} + \frac{1}{k_3(i_4 \cdot i_5)^2} + \frac{1}{k_4 i_5^2} + \frac{1}{k_5}, \tag{2.45}$$

$$\varphi_{\text{ges}} = \frac{\varphi_1}{i_2 \cdot i_3 \cdot i_4 \cdot i_5} + \frac{\varphi_2}{i_3 \cdot i_4 \cdot i_5} + \frac{\varphi_3}{i_4 \cdot i_5} + \frac{\varphi_4}{i_5} + \varphi_5, \tag{2.46}$$

$$\begin{aligned} \Theta_{\text{ges}} = {} & \Theta_{\text{Motor}}(i_1 \cdot i_2 \cdot i_3 \cdot i_4 \cdot i_5)^2 + \Theta_1(i_2 \cdot i_3 \cdot i_4 \cdot i_5)^2 + \Theta_2(i_3 \cdot i_4 \cdot i_5)^2 \\ & + \Theta_3(i_4 \cdot i_5)^2 + \Theta_4 i_5^2 + \Theta_5. \end{aligned} \tag{2.47}$$

It should be noted that, how essential the influence of the ratio i is, to the order in which the respective individual components are arranged. However, by calculating the case study above, the result of exercise is to find the impact of each link in the drive chain. This finding is of particular interest, if the individual parts are to be optimized, as the designer will then be able to carry out cost-effective changes as big improvements can be found with relatively little effort.

The following relationships are used to calculate the torsional stiffness k in Eq. 2.41 and for the angle of rotation φ in Eqs. 2.42 and 2.43 and the moment of inertia Θ in Eq. 2.24a, 2.24b. The corresponding equations for the drive shaft are:

$$k_{\text{Welle}} = \frac{d^4 \cdot \pi \cdot G}{32 \cdot l}, \tag{2.48}$$

$$\varphi_{\text{Welle}} = \frac{180}{\pi} \cdot \frac{M}{k_{\text{Welle}}}, \tag{2.49}$$

$$\Theta_{\text{Welle}} = \frac{1}{32} \cdot \rho \cdot \pi \cdot l \cdot d^4, \tag{2.50}$$

where shaft diameter d in m, modulus of rigidity of steel at room temperature $G = 7.9 \cdot 10^{10}$ in N/m^2, shaft length l in metres and density ρ in kg/m^3.

The calculation of this five-part drive chain is composed of the sum of the five components, with the sequence of the drive components necessarily following the installation sequence from the motor to the output shaft. The calculations are valid for any number of other components with only the calculation effort increasing substantially.

Simulation Models

During drive design, the need exists to recognise risks during the construction phase and implement preventative measures. Good empirical knowledge of drive mechanisms will allow practical component sizing. Specific improvements in complex systems are feasible only through the use of model simulations using the appropriate software. Often drive-trains consist of a dozen or more individual components and the performance ramifications of multi-shaft drive requirements require additional consideration. Furthermore, different load cases need to be judged, such as acceleration and braking behaviour, as well as the operational behaviour over the entire speed range. If necessary, software programs can be used (e.g. [57]), with graphically-interactive drive layouts and editable parameters. The big advantage to the designer is that the effect of each change to the drive components can be seen on the screen, and it is possible to consider the technical and economic feasibility of completely new drive concepts in a very short time.

2.15 Timing Pulleys, Tooth-Form Geometry

The production of timing belt pulleys requires a variety of machining and non-machining processes and equipment, either alone or in combination. As timing pulleys do not need hardened teeth, the production process is considerably simpler than that of spur gears. The most important production methods are described in Table 2.6 along with the relevant dimensions.

A timing belt pulley has a central bore and meshes with the timing belt by a toothed outer surface which runs to the pulley edges (see Fig. 2.28a). Depending on the role it has to fulfil, a pulley can additionally be fitted with belt guiding elements, called flanges, the use and assembly of which is dealt with in Chapter 2.10. The flange retaining diameters are produced by turning, during manufacture. The central bore is preferably produced to a minimum tolerance of H7.

The generation of teeth on the outer surface of the pulley is achieved by shaping or hobbing as the pulley blank is rigidly held by a centred shaft on the gear cutting machine. Pulleys are made from semi-finished metallic and non-metallic materials, of sufficient strength, which are easily machined and commercially workable. The materials used include aluminium alloys, steel, grey cast iron and plastics such as PA, PE and POM. Volume pulley manufacturers usually only offer products in the first three materials.

Mass production processes, such as die-casting in zinc and aluminium alloys, as well as plastic injection moulding, should also be considered by the designer where the quantities are sufficient. Chapter 3.8 deals with the implementation of such manufacturing processes which eliminate the necessity to machine the pulley teeth.

The isostatic pressing and sintering of pulleys using powder metallurgy is very attractive for use in mass production because of its high productivity, especially in automotive camshaft drives. However, sintering involves very high tooling costs.

Additionally, pulley blanks can be made by common casting methods such as sand and gravity die-casting. The central bore is then machined and the tooth profile formed by hobbing or shaping.

In producing the pulley teeth, a distinction must be made between the differing machining processes by the use of topping or non-topping cutters [77, 26]. If, during the gearcutting process, the tooth outside diameter is not machined simultaneously with the teeth then there are likely to be problems with the functionally important tooth tip radius r_t. Additionally, the pulley outer surface must have been previously machined to the exact outside diameter and thus it can be seen that the use of separate processes for the machining of the pulley teeth and the outside diameter can lead to a loss of quality. Pulleys produced by this method are primarily for prototype production and test versions, as the proper gearcutting tools (shaper cutters or hobbing cutters) are easy to produce and allow geometric changes with little effort.

Gearcutting

The most common gearcutting process for producing pulleys is hobbing with a topping cutter. This process generates the teeth and the outside diameter in one setting and guarantees low run-out (eccentricity), a constant transition radius and a constant tooth tip shape where the (positive) belt tooth will ultimately form the corresponding pairing in the (negative) pulley tooth gap. The preliminary turning of the pulley blank is to a relatively coarse tolerance. The correspondence between the pulley, the belt pitch and particularly the drive running characteristics depends greatly on the accuracy achieved on the pulley outside diameter where the belt engages in the arc of contact.

In principle, the pulley gearcutting process is similar to that of the manufacture of spur gears. The tooth gap geometry is not created by a form but is generated by the tooth profile of the hobs. The result is inevitably an involute toothform. When the pulley is imperial (inch) pitch, according with the DIN ISO 5294 standard [22], the pulley tooth gap geometry is established by the cutting geometry of the hob. The length of the pulley tooth flank is significantly shorter than that of a spur gear as the flank virtually starts at the tooth base radius r_b and ends at the tooth tip radius r_t. Thus, with a decreasing number of pulley teeth there is a trend to an increasingly curved pulley tooth flank. This situation is accepted however, as it means that only one hob is needed per pitch used.

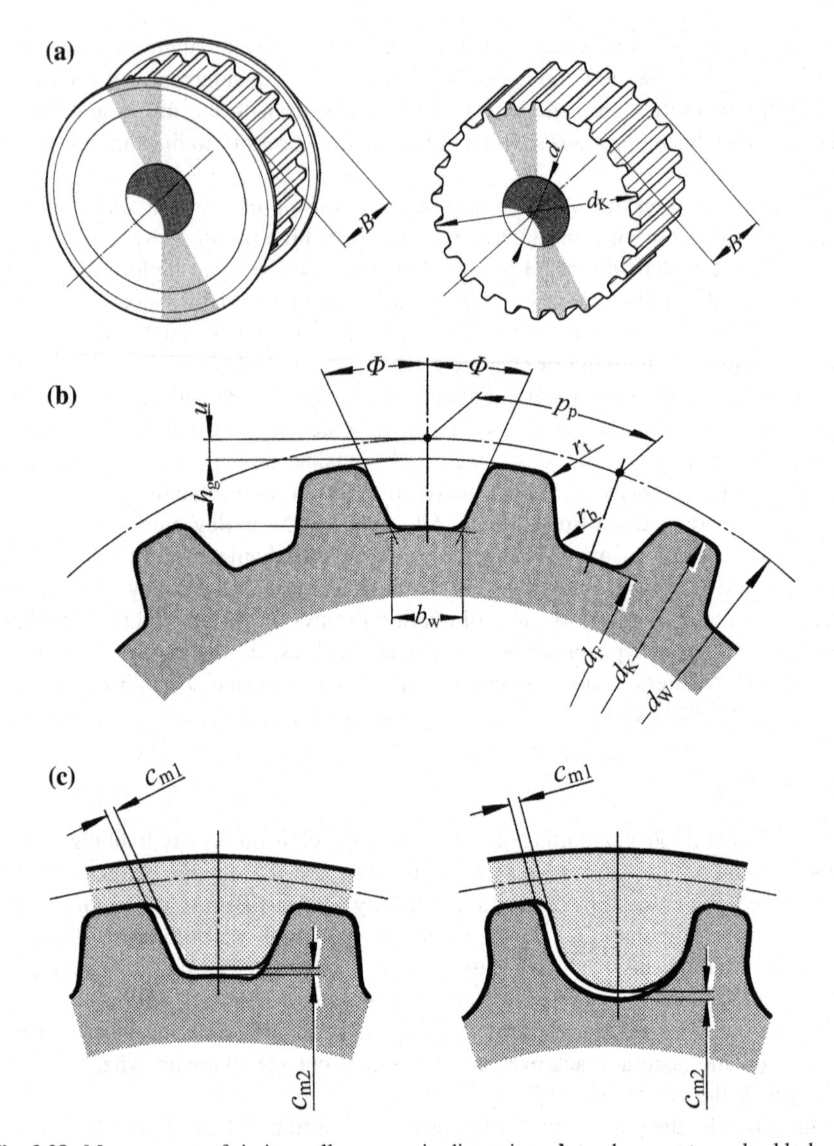

Fig. 2.28 Measurement of timing pulleys: **a** main dimensions, **b** tooth geometry, **c** backlash

It should be noted that in the DIN ISO 5294 standard, that imperial pitch belts and their accompanying pulley tooth geometry are specified with straight flanks.

For metric pitch timing belts with the T profile to DIN 7721 standard [20] and the high-power AT profile the associated pulley toothforms also have straight flanks. To produce these tooth gap geometries, multiple hobs per pitch are used. The range of hobs is used to manufacture fixed ranges of teeth to create the required straight tooth flank. All larger or smaller numbers of teeth machined with the same hob therefore will have flanks with a slightly concave or convex face.

Table 2.6 Main dimensions of timing belt pulleys

Character	Designation (Units)	Remarks
B	Pulley width (mm)	The pulley width relates to the distance between the two adjacent faces of the teeth. If the pulley has flanges then it relates to the distance over the flanges instead.
d	Bore diameter (mm)	The bore in the pulley is concentric to the outside diameter of the teeth and it usually serves to accommodate the drive shaft.
d_W	Pitch circle diameter (mm)	The pitch circle diameter lies in the middle of the tension member and through that line of arc formed around the timing pulley centre is where the belt pitch p_b and the pulley pitch p_p are the same.
d_K	Outside diameter (mm)	The outside diameter is the measurement of the outer surface of the timing pulley around which the timing belt wraps.
d_F	Root diameter (mm)	The root diameter is the diameter of the pulley which connects the base of the teeth.
b_w	Tooth gap base width (mm)	The tooth gap base width corresponds to the linear distance between the intersections of the flank extensions of the tooth gap on the tooth root diameter.
h_g	Tooth gap depth (mm)	The radial distance between the tip diameter and root diameter is the tooth gap depth.
Φ	Flank angle (°)	The tooth gap angle 2Φ is formed from the total angle between the two flanks. The half-angle is the flank angle.
r_t	Tooth tip radius (mm)	The tooth tip radius connects tooth flank and the outside diameter of the pulley.
r_b	Tooth base radius (mm)	The tooth base radius connects tooth flank and the tooth base of the pulley.
p_p	Pulley pitch (mm)	The pulley pitch corresponds to the distance between two adjacent tooth gaps on the pitch line.
u	Pitch line distance (mm)	The radial distance between pitch circle and the tip circle is called the pitch line distance.
c_{m1}	Tangential clearance (mm)	The distance between the unloaded pulley and belt tooth flanks resulting in tangential play if the loaded belt tooth flank rests against the working pulley tooth flank.
c_{m2}	Radial clearance (mm)	The shortest radial distance between the belt tooth head and pulley tooth base is the radial clearance.

Designations and remarks partly taken from ISO 5288 [55]

The required numbers of hobs are designed so that the deviations resulting from the hobbing process are very small and remain well below the allowable tolerance limits. This does not lead any performance degradation in the timing belt drive.

Due to these deviations from the exact toothform geometry, both the tooth flank and the tooth base develop a slightly convex or concave form during the tooth machining. The influence of these deviations is irrelevant in belts where the belt tooth gap rests on the pulley outside diameter, such as the T profile. The unusual feature of the AT profile (see Chapter 2.3.3) is that the belt tooth rests against the pulley tooth gap base. The ideal AT pulley toothform has target geometry of a flat support surface for the belt tooth. Again this needs correcting with a range of cutters, meaning the belt tooth support surface deviations remain insignificant.

The pulley toothforms for the curvilinear timing belt profiles HTD, STD and RPP are described in ISO 13050 standards [54] where the geometry of each toothform cutter is specified. Non-involute toothforms change slightly from one tooth number to the next and because of this, the toothform hobs can only cover a certain range of teeth before needing another different hob to achieve the optimum profile geometry. Each toothform is generated from a range of hobs covering specific numbers of teeth. Thus, for example, for the HTD8 M toothform profile there are three hobs and for the production of the HTD14 M toothed pulley range, six hobs are required.

Quality Assurance and Control

Because of the large variety of belt pitches, each with its own toothform geometry, which varies depending on the number of teeth in the pulley and that of the gear cutting tool, it can be considered that an accurate dimensional inspection of a pulley tooth profile is usually only possible by the pulley manufacturer. This involves special imaging equipment for comparison against the required target profile from the belt manufacturer. In such cases, a measurement report from the pulley manufacturer is recommended.

It is generally only necessary to check the outside diameter, tooth tip radius and pulley concentricity. Depending on the type of pulley/shaft attachment it can be important to rigidly secure the pulley to the shaft in case, due to installation conditions, axial loads act on the pulley body.

Quality control of the toothform geometry is needed due to the difficulty in measuring the spherical curves generated by the hob. The root diameter d_F of the toothed pulley, for example, is functionally critical for the proper running of all belt types with the AT profile. This diameter can be measured with the pulley flanges attached but only with considerable effort. Further difficulties exist with odd numbers of pulley teeth, where the tooth gaps are not diametrically opposed to each other. Therefore the outside diameter is preferred as the basis for measurement and the hob geometry is left to generate the exact depth of the tooth gap in the pulley.

Interchange Ability of Belt Systems

Each type of timing belt requires an optimum tooth shape matched to the specific toothform geometry of the pulley. The H and S and R profiles form a belt group with similar geometric characteristics. In particular their pitch line distances, u, for 8 and 14 mm pitches (ISO 13050) are exactly the same size. The elastomer belt teeth, which can be compared with the teeth of spur gears, can accept much larger deviations to the pulley flank geometry. Table 2.7 gives information about the operating characteristics and interchange ability between the belt and pulleys of different curvilinear 8 and 14 mm pitches.

Extensions to the ISO standard for 3 and 5 mm pitches are planned for belt types H, S and R. These pitches (in the absence of a standard) can be assumed to have

Table 2.7 Interchange ability table for 8 and 14 mm pitches

Timing belt / Timing pulley	H HTD	S STD	R RPP	Omega	GT3	PC-MGT2
H HTD	●●	○	●	●●	●●	–
S STD	○	●●	●	○	○	–
R RPP	●	○	●●	●	○	–
GT3	●	○	○	○	●●	–
PC-MGT2	○	–	–	–	–	●●

Key:
●● excellent running characteristics
● good running characteristics
○ conditionally able to run (emergency operation only)
– not able to run

similar operating characteristic relationships and interchange ability as per the above chart. One of the basic requirements of a belt and pulley running together is that the u-values of the differing belt/pulley types are matched (see the related dimensional tables in Chapter 2.3). For an authoritative statement of interchange ability we would recommend approaching the respective belt manufacturer.

2.16 Tangential Drives

A special property of timing belts is the ability to mesh tangentially, where the pulley and the belt work together as a rack and pinion. This type of drive geometry will work only under the special conditions discussed below. It is obvious, with this particular type of tooth engagement, that the drive performance is reduced to a much lower level than normal, since load-sharing does not happen over multiple teeth in the arc of contact. However, this application is of particular interest when the modified belt-pulley design leads to new drive geometries (see the applications described in Chapter 3.1, Figs. 3.15, 3.16, 3.17).

One of the measures needed to prevent deflection of the belt away from the pulley is that a support opposite to the point of engagement is necessary. This is achieved by a low friction support shoe or, better still, by a rail supporting the back of the belt. If the belt is undertaking the linear motion then it must be prevented from wandering off the support rail (Fig. 2.29). Furthermore, it is assumed that due to the fact of tangential engagement, that only a *single* belt/pulley tooth pair is involved in the transmission. However, according to new research, the resilient elastomer tooth can distribute the tangential force over *several* teeth. The way that the force is transmitted over such a group of teeth is dealt with at the end of this chapter.

A further requirement for tangential engagement involves a tooth profile alteration v_t due to the different pulley/belt meshing compared to the normal

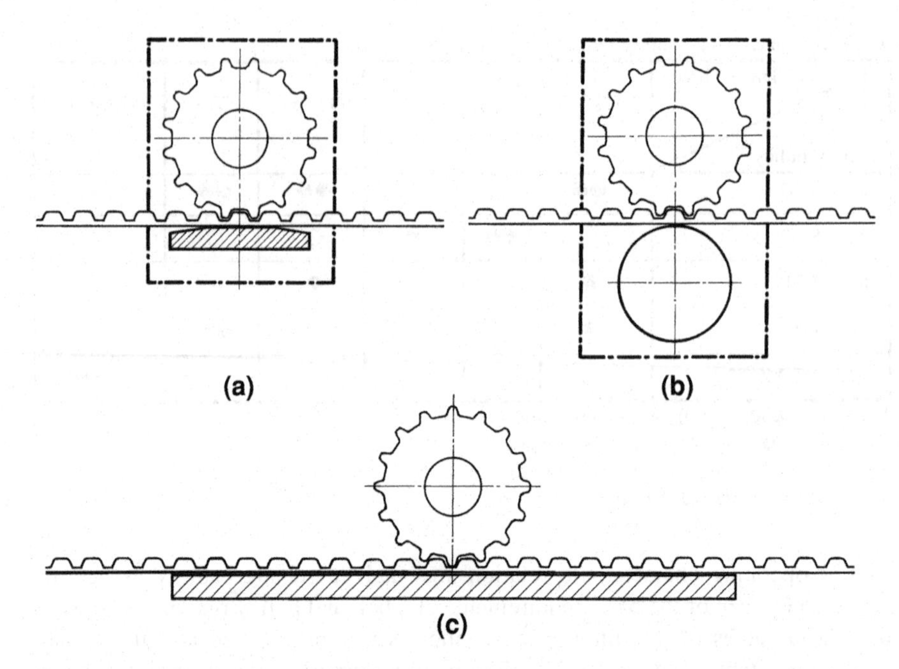

Fig. 2.29 Tangential engagement with **a** support shoe, **b** support roller, **c** support rail

meshing function. The belt, as well as the pulley gap profile, remains unchanged and it is only the outside diameter d_K that changes to the new outside diameterd'_k. A different pulley tooth-cutting hob/cutter is needed.

The necessary profile alteration v_t can be derived as follows: The belt section is considered to be in the pre-tensioned state. All points along the tension member level are considered to have the same value for the tooth pitch p_b. The belt pitch line is parallel to the toothed area in such a way that the belt tooth height h_t divides the head height h_a and the foot height h_f into a ratio of 1:1.25.[5]

The new line can also be described as the pitch line (Fig. 2.30), at which the pitch circle diameter of the pulley engages tangentially. This diameter changes with the tooth gap profile such that the pitch line for both the belt and the pulley are equal. This is achieved by the profile alteration of the outside diameter from d_K to d'_k and is calculated from the relationship

$$d'_K = \frac{z \cdot p_b}{\pi} + 2 \cdot h_a. \tag{2.51}$$

The values given are similar to the geometry of rack and pinion drives. Equations 2.52–2.57 are used to calculate the other values that need to be compliant with Fig. 2.30. It is especially worth noting that, in the changed pulley, the head height h_a is derived from the belt on the pulley. This has no further reference to the tooth height or tooth base height of the pulley.

[5] Notes regarding the head/foot height ratios can be found chapter below.

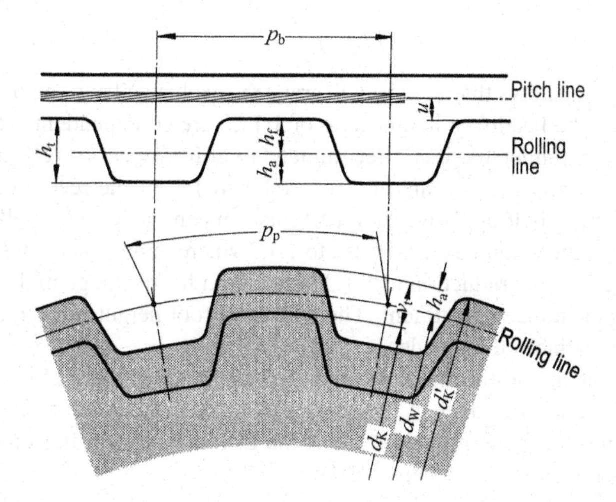

Fig. 2.30 Profile alteration for tangential meshing. The neutral line changes to the pitch line, the outside diameter d_K is increased to d_k', p_b belt pitch, p_p pulley pitch, h_t belt tooth height, h_a belt tooth head height, h_f belt tooth foot height, u U-value, v_t profile alteration for tangential meshing, d_W rolling line

In order to achieve the optimal conditions for a tangential drive assembly, special attention should be given to the head clearance c_t in conjunction with the other tolerances. For the force transmission to be effective as possible from the pulley tooth to the belt flank (or vice versa), it must happen exactly in the predetermined pitch line. As a consequence of the tooth profile, angular radial forces result under operating conditions and the belt tooth strives to leave its pulley tooth gap. Any support roller at the belt back should be arranged opposite to the pitch point as it both takes up the radial forces pushing the belt away from the pulley and is used to adjust the head clearance c_t. Thus, the belt back height h_r and its tolerance should be particularly considered as the belt back height uniformity can affect the running behaviour. The belt thickness accuracy should be also examined to see whether reworking, by grinding of the total thickness of the belt, is necessary. The adjusted head clearance (Fig. 2.31), is calculated from

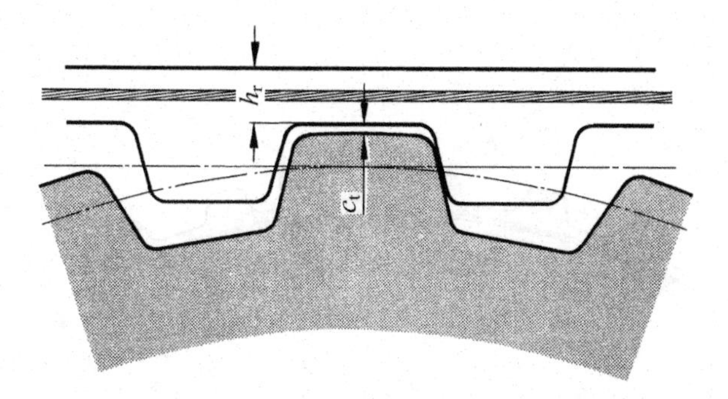

Fig. 2.31 Head clearance set for tangential engagement, c_t head clearance, h_r back height

$$c_t = h_f - h_a. \tag{2.52}$$

Equation 2.52 shows that the head clearance is exclusive from the profile height of the belt and the head/foot height ratio of 1:1.25 are co-dependent. The value of the ratio is to be understood as a recommendation in respect to [26] for involute spur gears working in ratios ranging from 1:1.1 to 1:1.3. The recommended ratio of 1.25 for timing belt applications represents a mean value, for small pitches of less than 5 mm there can be an increase to 1:1.3 where appropriate, and for pitches larger than 12.7 mm a reduction to 1:1.2 is favoured for the tangential engagement and the head clearance adjustment. Thus the head/foot height ratio h_a/h_f becomes an additional variable to consider.

To determine the profile alteration and head clearance in Eqs. 2.51 and 2.52, the required individual variables h_a and h_f need to be calculated beforehand. The output variables for the tooth height h_t and the U-value u are taken from dimension tables for each profile (see Chapter 2.3):

$$v_t = u + h_a, \tag{2.53}$$

$$h_t = h_a + h_f, \tag{2.54}$$

$$h_a : h_f = 1 : 1.25, \tag{2.55}$$

$$h_a = \frac{h_t}{2.25}, \tag{2.56}$$

$$h_f = \frac{h_t}{1.8}. \tag{2.57}$$

Additional Profile Alteration

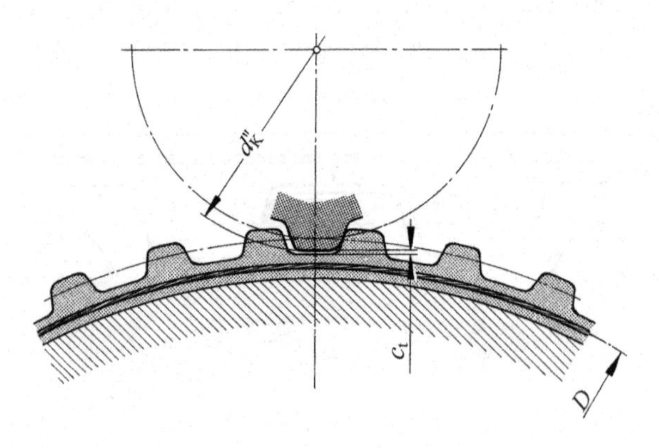

Fig. 2.32 Timing belt and pulley engaging on an exterior curve

If a belt back is curved over the cylindrical surface of a pulley as shown in Fig. 2.32, it can be compared to the tangential engagement of an elongated belt with a pitch increase at the pitch point for the outward facing belt teeth. In applying this belt layout (see also Fig. 3.17 in Chapter 3.1), an additional profile alteration is needed for mating to the outside diameter:

$$d''_K = d'_K \cdot \left[1 \pm \frac{2(u + h_f)}{D}\right]. \tag{2.58}$$

This sign "+" is used for an outward curved belt and "−" should be used for an inward curved belt.

Example 1: When using a T10 standard profile belt (10 mm metric pitch) with a toothed pulley with $z = 30$, the outside diameter and the head clearance for tangential engagement can be calculated as follows:

The tooth height is found in Chapter 2.3.2, $h_t = 2.5$ mm.
from Eq. 2.56, $h_a = h_t/2.25 = (2.5/2.25)$ mm $= 1.11$ mm,
from Eq. 2.57, $h_f = h_t/1.8 = (2.5/1.8)$ mm $= 1.39$ mm,
from Eq. 2.51, $d'_k = z \cdot p/\pi + 2 \cdot h_a = (30 \cdot 10/\pi + 2 \cdot 1.11)$ mm $= 97.72$ mm,
from Eq. 2.57, $c_t = h_f - h_a = (1.39 - 1.11)$ mm $= 0.28$ mm.

Example 2: A high performance timing belt type HTD8 M with an endless length of 1,200 mm is "gear" mounted (i.e. teeth outwards) over a cylindrical surface. For the pulley meshing at the pitch line of the outwardly curved belt teeth to the mating outside diameter, the tooth gap clearance of a pulley with 22 teeth is determined as follows:

Due to the wrap on the cylindrical surface, the middle of the tension member diameter of the belt is $(1{,}200/\pi)$ mm $= 382$ mm.
According to Chapter 2.3.4, the tooth height $h_t = 3.38$ mm and the U-value $u = 0.686$ mm.

from Eq. 2.56, $h_a = h_t/2.25 = (3.38/2.25)$ mm $= 1.50$ mm,
from Eq. 2.57, $h_f = h_t/1.8 = (3.38/1.8)$ mm $= 1.88$ mm,
from Eq. 2.51, $d'_k = z \cdot p/\pi + 2 \cdot h_a = (22 \cdot 8/\pi + 2 \cdot 1.5)$ mm $= 59.02$ mm
from Eq. 2.58, $d''_k = d'_k \cdot [1 + 2(u + h_f)/D] = 59.02 \cdot [1 + 2(0.686 + 1.88)/382]$ mm $= 59.81$ mm,
from Eq. 2.52, $c_t = h_f - h_a = (1.88 - 1.50)$ mm $= 0.38$ mm.

Summary

All timing belt profiles and their respective pulley tooth gaps are systems whose performance is essentially attributable to their meshing in the arc of contact and load balancing across multiple engaging teeth. The drive geometries for tangential meshing are something of a "solution" as they were not originally intended for toothed belts; however, practical tests have confirmed that this type of application is able to work.

For the expected life span, these drives have the same restrictions as with normal applications. In power calculations, however, only a *single* driving tooth (or a *single* tooth group, see next paragraph) can be used. If necessary, prototype drives should be evaluated before volume manufacture. Recommendations and calculations to tooth geometry and profile alterations apply to both the trapezoidal and curvilinear belt profiles.

Number of Meshing Teeth in Tangential Engagement

For small pulleys only a single driving tooth is taken into account in the performance calculations. Due to the resilience of the elastomer belt teeth, the adjacent teeth take up part-loads so that more teeth may be used in the drive calculation, see Fig. 2.33. Table 2.8 indicates the number of driving teeth z_e for tangential engagement depending upon number of teeth in the pulley z.

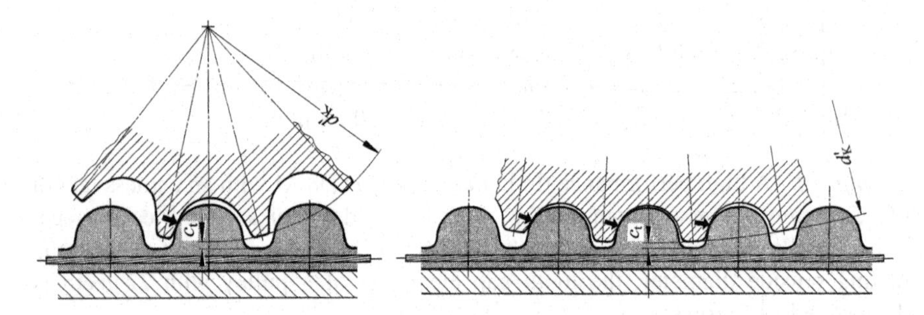

Fig. 2.33 *left*: HTD timing pulley $z = 15$ *right*: HTD timing pulley $z = 60$

With increasing numbers of teeth in the tangentially meshing pulley, the adjacent teeth carry portions of the transmitted tangential force.

Table 2.8 Number of teeth in mesh z_e in a tangential drive

No. of teeth in pulley z	No. of teeth in mesh z_e
<30	1
≥ 30	2
≥ 60	3
≥ 100	4

To calculate the tangential force transmitted, the number of teeth in mesh is used in a similar way as the arc of wrap function discussed in Chapter 2.9. A multiplier of z_e results in a load distribution and thus, real engagement conditions are used for the basis of calculation. The data in Table 2.8 for teeth in mesh figures are averages and are valid for both trapezoidal and curvilinear profiles for standard belt elastomers. With the use of an increased Shore-hardness belt material then tests should be made, where appropriate, due to the lower material compliance.

2.17 Belt Installation and the Adjustment of Pre-tension

A timing belt should be installed into a drive without the use of force and without being levered onto the pulleys. The pre-tension force shall be applied when the belt is installed and with a maximum of only one rotationally locked pulley.

Installation usually requires that the drive construction allows an adjustment of the centre distance. If the distance between driving and driven pulley is fixed then an additional pivoting or sliding tensioner is needed to adjust the pre-tension. The tensioner can either be on the toothed side or on the belt back and is positioned after belt installation to give the correct pre-tension and then secured against movement when the belt is working under load. A tensioner should ideally act on the slack side of the belt; however, use on the loaded side is also possible without affecting the belt function.

Toothed belt drives without any pre-tension adjustment should only be used in exceptional cases. Such designs and the resulting operational limitations are explained below in this chapter.

Tensioning the belt, either with a centre distance adjuster or a tensioner requires minimum design effort. They are almost indispensable components in the drive structure for both function and assembly. The importance of correctly applied pre-tension on drive refinement and performance of the belt is described in Chapter 2.6 and 2.7. Application examples of clamping options are dealt with in Chapter 3.7.

The variations that affect the length of a standard belt are compensated through this adjustment ability. These variations include the outside diameters of the pulleys, the installation accuracy of the pulley shafts and the length of the belt itself as well as the diameter tolerances of tensioners on the back of the belt and the manufacturing variation of the back height of the belt. Of the variances mentioned, the belt length tolerance is normally the largest to be taken into account. Depending on the make and the length of the belt, the tolerance range is about ± 1 to ± 5‰ based on the nominal belt length l_B. It is easy to understand that the desired belt pre-tension cannot be achieved without adjustment and moreover that drive assembly under this preload would be extremely difficult.

Assembling an un-tensioned belt requires extra free length to be able to pass over flanges and possible additional component interference as in Fig. 2.34. The wider and shorter a belt is, the more difficult this diagonal assembly is, which requires even more length. If necessary, the excess length can be minimized by planning so that a toothed pulley or a tensioner, instead of the belt, is the last

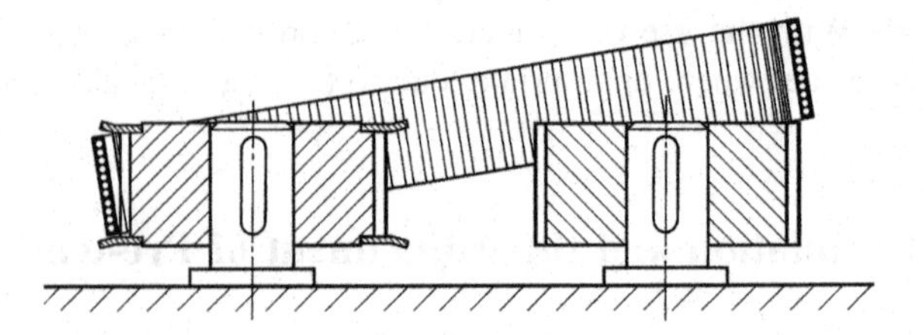

Fig. 2.34 Length restriction with diagonal assembly

component to be installed. Another solution is removable flanges as in Fig. 2.16c in Chapter 2.10.

The designer will necessarily have to consider the size of all the components in order to take into account the possible minimum and maximum installation dimensions, under initial pre-tension, as in Fig. 2.35. The difference gives the minimum adjustment dimension. For long belt lengths, where appropriate, elongation due to the elastic behaviour of the tension members should be considered separately. The elongation length Δl in the belt, due to the applied pre-tension force F_V, is calculated from the relationship

$$\Delta l = \frac{F_V}{c_{Bspez}} \cdot l_B, \tag{2.59}$$

where l_B is the nominal belt length, and the specific stiffness c_{Bspez} from Eq. 2.36 should be used. The value of Δl to be determined, relates to the elongation in the belt. This elongation is reduced by half when mounting between the pulley shafts.

Since belt elongation values in timing belt drives are always smaller due to the initial pre-tension, which is <2‰ or 2 mm for each metre belt length, this leads to an adoption of an estimated value for short belts in place of the calculation in Eq. 2.59.

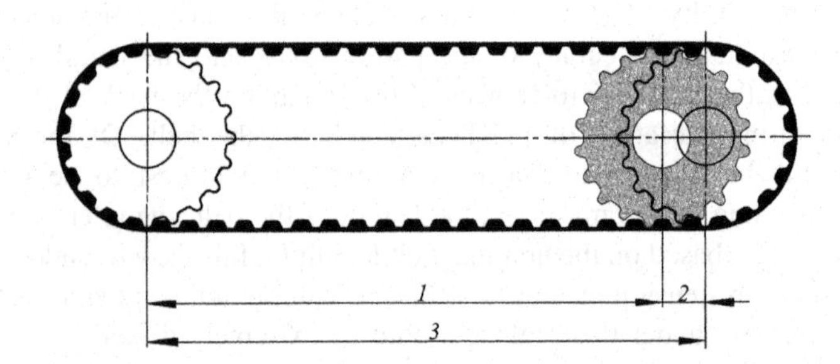

Fig. 2.35 Mounting dimensions: *1* minimum required assembly dimension, *2* minimum adjustment dimensions, *3* mounting dimension under maximum possible pre-tension

The calculation of the minimum pre-tension force F_V for two-pulley drives can be found in Eq. 2.12 and for multiple pulley drives in Eq. 2.17. In practice, the drive configuration is subject to calculations using catalogue data or a belt manufacturer's computer program. The results obtained from these sources point to the recommended belt width and the belt span forces, along with the recommended pre-tension and the natural frequency of vibration of the associated belt sections.

Measuring the Pre-tension Force

Today, the usual method to measure the initial pretension is to determine the natural frequency of vibration of one of the belt spans. The section of the belt to be tested is vibrated by striking it. The vibrations are then sensed by acoustic, optical or inductive sensors. The frequency measurement is usually given by a battery-powered, handheld device with a digital display. The accuracy of the actual measured frequency depends on the equipment quality, usually ±1 to ±5%.

The arithmetic relationship between natural frequency f_e and pre-tension force F_V is given by

$$f_e = \sqrt{\frac{F_V}{4 \cdot m_{spez} \cdot b \cdot l_T^2}}, \tag{2.60}$$

where pre-tension F_V in N, specific belt mass m_{spez} in kg/mm belt width and per metre belt length, belt width b in mm and span length l_T in m.

To adjust and check the initial tension, the fitter should use approximation steps, as necessary, until the given desired value is reached by changing the tensioner position. The pre-tension is set correctly by repeated measurement at different belt positions on the pulleys, due to irregularities in the belt and pulley eccentricities. The drive design provides a minimum necessary pre-tension force F_V which is labelled by default with "≥". Any customer produced assembly manuals should give the allowable setting values of $F_{V\,min}$ and $F_{V\,max}$. A re-check and correction of pre-tension loss is recommended, where appropriate, because of the settling behaviour of the tension member after initial drive setup.

Fixed Centre Distances

Timing belt drives that have no possibility of pre-tension force adjustment are subject to significant limitations in both the function of the drive, the possible geometric layout and also service and installation.

It is not possible to manufacture belts to an absolute length. Consequently, a nominal size is specified for the effective length which can be between a permissible maximum and minimum size. The tolerances of the length measuring instrumentation also should be considered. It is therefore possible, that a belt drive without adjustment will either operate with significantly high tension or be run entirely without tension. High pre-tension forces will result in negative effects on the drive function as well as very high bearing loads. On the other hand a loose belt will show torque and angular variations as well as uncontrolled flutter and tooth jump. Drives of this type should not be operated up to the maximum power limit.

During assembly and servicing, the installation and removal of the timing belt is complicated by the fact that there is no free length available to surmount the flanges. The fixed centre distances means that the belts and pulleys are installed or removed together as a preassembled unit on the shafts. Such an operation is relatively difficult because when a belt is near the bottom of its tolerance band then high pre-tension forces will occur.

Furthermore, it is desirable to limit such fixed centre timing belt drives to only "compact" dimensions, preferably with relatively short belt lengths. With increasing belt lengths the importance of properly adjusted pre-tension grows. Chapter 2.7 deals with drives with large centre distances and describes their tendency to tooth jump.

If fixed centre distances cannot be avoided, the supplier/manufacturer should be queried as to whether the belt length tolerances can be reduced.

Tensioners

Tensioners should be preferably fitted with roller bearings, whose dynamic and static load ratings must be considered to ensure sufficient life with the possible belt loads. As a precaution to prevent lubricant leakage, roller bearings for both tensioners as well as pulleys should be fitted with oil seals. A pre-lubricated bearing is sufficient for the lifetime of a drive in "normal use". Adjustment can be achieved by moving, pivoting or through an eccentric fixing. The tensioner and the surrounding machine construction must be rigid and unable to flex (see Chapter 3.7).

Spring Tensioners

A sprung tensioner in which the belt tension is applied by spring action should always remain the exception for timing belt drives and should only be used on the slack side of the belt. Due to the nature of this tension method it must be ensured that either a change of drive direction and/or a load reversal, for example during a braking operation, cannot happen.

Roll-Ring®

A new tensioning and anti-vibration device is now available for the user with the development of rotationally elastic Roll-Ring® system [28]. The installation between the two spans is simple and does not need a mounting base. The system is suitable for both original equipment and for retrofits, and no modifications are required to existing timing belt drives. Because of the simultaneous effect on both the slack and loaded span an unloaded state is avoided.

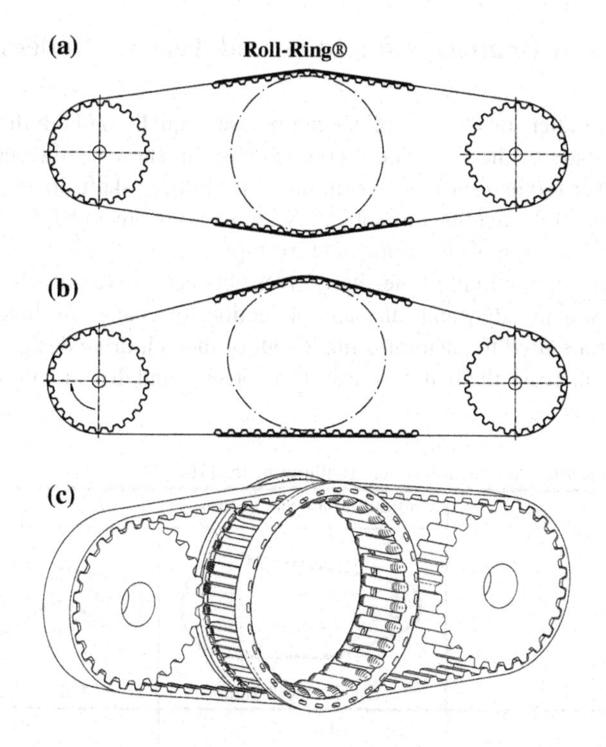

Fig. 2.36 Roll-Ring: **a** unloaded, **b** transmitting power, **c** 3D view

At rest and under ideal idling conditions, the Roll-Ring tension and anti-vibration element adopts a slightly elliptical initial shape, with minimal distortion, in a centre position between the upper and lower belt spans (Fig. 2.36a). If torque is transmitted, the forces in the belt spans will change and the ring centre shifts in the direction of the least loaded span (Fig. 2.36b). The new ring centre formed, lies at the force equilibrium of both belt spans and the increased deformation counteracting the force of the ring.

Properties according to the manufacturer [28]:

- The plastic Roll-Ring provides both tensioning and damping functions in one component.
- It is capable of reversing without needing to be mounted to the surrounding machine.
- It is self-adjusting to the real drive loads and works on both sides of the belt.
- On drives with intermittent and shock loads the Roll-Ring is very effective and it improves smoothness significantly. The deformation of the ring under heavy load and the associated deflection of the slack side of the belt mean a corresponding rotational angle difference has to be taken into account.

2.18 Minimum Diameters for Idler and Tension Rollers

Plain (toothless) idlers or tension rollers can be used equally on both the back and on the toothed side of the belt. These components can show significant potential cost savings over toothed pulleys. As timing belt multiple-shaft drives come in a wide range of possible layouts, this subject is addressed in other chapters providing numerous suggestions as well as practical examples.

There is a distinction in multiple shaft drives between active pulleys, which are involved in power transfer, and idler and deflection (or change of direction) pulleys. The latter are used to customize the layout of the belt drive design. They can also be used to increase the belt meshing at the driving and driven pulleys and can

Table 2.9 Smooth idler and tension rollers (without teeth) [16, 87]

Belt profile	Minimum diameter in mm	
	On the tooth side	On the belt back
XL	30	30
L	60	60
H	60	80
XH	150	180
T5	30	30
T10	60	60
T20	120	120
AT5	25	60
AT10	50	60
AT20	120	180
HTD5, RPP5, STD5	2.5 times larger than the outside diameter of the minimum allowable timing pulley	1.5 times larger than the outside diameter of the minimum allowable timing pulley
HTD8, RPP8, STD8		
HTD14, RPP14, STD14		
HTD20, RPP20, STD20		

The above figures are the specified minimum diameters related to standard timing belt designs. Depending on the make and any modification (e.g. increased tension member size) some manufacturers will recommend different values. For safety, the technical specifications of the specific manufacturer should be used be designed as stationary or as adjustable tensioning pulleys. In some manufacturer's product ranges [87] there are extensive standard stock programs of tension rollers that are adjustable via an eccentric shaft.

The belt, when in toothed contact with a smooth idler pulley and depending on the belt tooth profile geometry, can exhibit improved running properties, but also increased noise. If the cost saving potential with these kinds of idler pulleys is to be used, instead of toothed pulleys, the limit values of their minimum allowable diameters must be adhered to.

Trapezoidal belt tooth profiles are the most suitable types for supporting the belt teeth on smooth pulleys. These belts have linear tooth top surfaces which pass the support forces onto the outer surface of the pulley. A particularly good belt for this is the AT-profile because its broad tooth top surface results in very low compliances. Operating noise with these types of belts is lower than over toothed pulleys. In curvilinear belts, especially with the HTD-profile, the tooth profile reacts to the support loads by deformation in the form of flattening of the tooth section. These types of belts may exhibit conspicuous noise levels and thus the minimum diameter of a smooth pulley with this type of belt should be significantly larger.

The minimum allowable diameter for smooth idlers on the belt tooth side is determined mainly by the polygonal tooth surface contact. The smallest permitted belt back contraflexure diameter on the other hand will affect the properties of the tension member. The recommended minimum sizes for each profile type are shown in Table 2.9.

2.19 Measuring the Belt Working Length

Length measurement of standard endless timing belts [20, 24, 54] is achieved using length measuring equipment as shown in Fig. 2.37. This process checks the actual length in the active pitch line and assumes that the length tolerance is spread evenly across all tooth pitches. The measuring equipment consists of two equal-sized pulleys, which are chosen depending on the belt profile and pitch and have a predetermined number of teeth. These are known as measuring pulleys, because compared with standard pulleys their tolerances are significantly better. One measuring pulley is mounted on a stationary shaft and rotates freely and the other pulley is used to vary the centre distance and is mounted on an adjustable shaft. This shaft will have an appropriate system to adjust the measurement load. A measurement display is arranged to show the actual value of the centre distance between the two pulleys, which determines the belt length accuracy. Before data-collection starts, the belt needs to be turned at least two full revolutions, so that it sits properly on the pulleys and the total force between two belt spans is evenly distributed. Double-sided belts are measured on both sides. The pitch length is calculated from the measurement of the pitch circle diameter of the pulleys plus double the value of the measured centre distance.

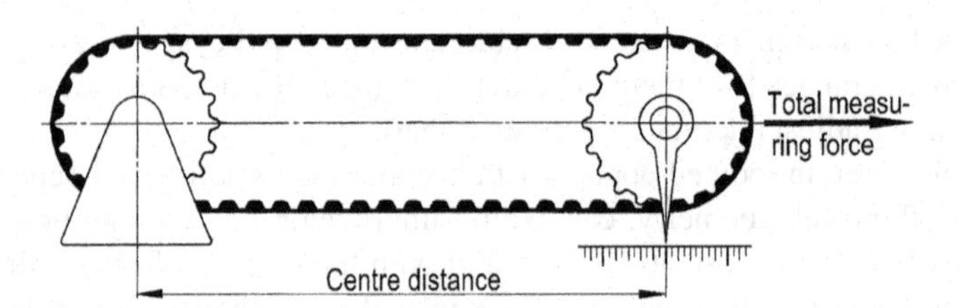

Fig. 2.37 Measurement setup for determining the pitch length

The measuring equipment can be pre-adjusted to the permitted minimum and maximum values of the belt to be checked. The deviations between the maximum and minimum amount to half the distance of the centre value of the pitch length deviation.

Length measuring equipment for endless belts is constructed depending upon the targeted application (e.g. laboratory or production applications), the kind of toothed belt and measuring load to be applied. For example, a design with vertically mounted belts allows a relatively simple method of applying the total measuring load by applying weights to the adjustable shaft. To minimize measurement differences between the belt manufacturer's quality control department and the end-user's goods inward inspection, it is recommended that both parties use identical types of equipment.

Tables 2.10, 2.11, 2.12, 2.13 and 2.14 shows the total measuring forces and the sizes of the associated measuring pulleys for standard timing belt profiles.

Table 2.10 Measuring loads for standard inch pitch belts to DIN ISO 5269

Pitch reference	**Total Measuring Load in N** for the belt width in mm										
	6.4	**7.9**	**9.5**	**12.7**	**19.1**	**25.4**	**38.1**	**50.8**	**76.2**	**101.6**	**127**
MXL	27										
XL		44	53								
L				105	180	245					
H					445	620	980	1,340	2,100		
XH								2,000	3,100	4,450	
XXH								2,500	3,900	5,600	7,100

No. of teeth in the measuring pulleys for MXL: 20, XL: 10, L: 16, H: 20, XH: 24, XXH: 24

For all other belt types, pitch length measurement is done in a similar manner using the same methods for the standard belt profiles above.

Pitch length measurement of open length belts is carried out on a straight section of belt. In determining the pitch accuracy it is assumed that any deviations

Table 2.11 Measuring loads for standard metric T-pitch belts to DIN 7721

Pitch reference	**Total Measuring Load in N** for the belt width in mm								
	4	**6**	**10**	**16**	**25**	**32**	**50**	**75**	**100**
T 2.5	6	10	20						
T 5		20	40	60	90				
T 10				90	140	170	270		
T 20						340	540	800	1,100

No. of teeth in the measuring pulleys: 20

Table 2.12 Measuring loads for high power profile HTD to ISO 13050

Pitch reference	**Total Measuring Load in N** for the belt width in mm							
	20	**30**	**40**	**50**	**55**	**85**	**115**	**170**
HTD8 M	470	750		1,320		2,310		
HTD14 M			1,350		2,130	3,660	5,180	7,960

No. of teeth in the measuring pulleys for HTD8 M: 34, HTD14 M: 40

Table 2.13 Measuring loads for high power profile RPP to ISO 13050

Pitch reference	**Total Measuring Load in N** for the belt width in mm							
	20	**30**	**40**	**50**	**55**	**85**	**115**	**170**
RPP8 M	470	750		1,320		2,310		
RPP14 M			1,350		2,130	3,660	5,180	7,960

No. of teeth in the measuring pulleys for RPP8 M: 34, RPP14 M: 40

Table 2.14 Measuring loads for high power profile STD to ISO 13050

Pitch reference	**Total Measuring load in N** for the belt width in mm						
	15	**25**	**40**	**60**	**80**	**100**	**120**
STD8 M	570	1,020	1,740	2,770			
STD14 M			2,420	3,840	5,340	6,880	8,470

No. of teeth in the measuring pulleys for STD8 M: 34, STD14 M: 40

are distributed evenly across all tooth pitches. To this end, the actual deviation from the target is measured in a test sample in millimetres of tooth position per 1,000 mm of belt length. The associated measuring load is 8% of the maximum tensile force F_{zul}. The belt in the end-clamps is at least the 10-times the stiffness value of the belt outside in the test length, so that strain effects from the clamping points can be excluded.

2.20 Efficiency

The efficiency η is the ratio of power delivered P_{ab} to power supplied P_{zu}. The resulting difference is the power dissipation P_{verl}:

$$\eta = \frac{P_{ab}}{P_{zu}} \tag{2.61}$$

$$P_{verl} = P_{zu} - P_{ab}. \tag{2.62}$$

Timing belt manufacturers suggest an efficiency range for their products between 97 and 99% where the values refer to the rated output. It is obvious, that under partial load conditions or if the drive is has an ample safety factor that this diminishes the efficiency accordingly. These values take into account the belt/ pulley interaction but they exclude the efficiency of any associated bearing systems and downstream components.

In a toothed belt drive, the frictional losses are the sum of the losses from tooth meshing plus the side contact losses on the flanges and the internal losses in the belt construction. Additionally, air pumping losses occur through the meshing of the moving parts. On the whole, timing belt efficiency is substantially higher than comparable flat, wedge or V-ribbed belts because there are no slip or creep losses and pre-tension is much lower.

The toothed belt drive delivers an overall positive balance of performance to power dissipation, better described as non-usable power, which is converted into heat. However, no applications are known where friction and hysteresis losses in the belt cause considerable heat generation, partly because of the ventilation and air recirculation generated by the continuous meshing of the belt and pulley teeth.

Today's requirements are increasing for high-efficiency drive components. A comparison of a tooth-belt driven camshaft against a timing chain drive will show up to 30% less frictional losses (see Chapter 3.11). The power losses in a vehicle, through this single change, can drop by 1–2%, thereby reducing CO_2 emissions by 2–3 g/km.

Chapter 3
Timing Belt Drive Technology

Abstract This chapter contains self-explanatory information for the user of timing belt applications, depicting a variety of solutions with integrated belt/drive layouts. It is often the case that real-life proven models provide the basic elements for new concepts, which can then be transferred and adapted to the task in hand. In particular, this allows the technical and creative potential of designers to be addressed, as they think in images and spatial structures which are then implemented. The combination of two or more known designs can be developed into innovative new solutions and existing examples borrowed from other sources can reduce the possible work load substantially.

3.1 Belt Drive Geometries

This chapter deals with continuous timing belt drives, which can be distinguished from linear drives, where the drive is rotational or applies a rotational force. Here, the designer can find a variety of actual drive examples plus helpful hints and detailed solutions to the drive geometry layout.

The diagrams are based on experience of sound drive-train design. Recommendations for manufacturer specific products are not offered, as this handbook is meant to be a neutral guide. Occasionally, there may be the indication of preferred belt types but without limiting or excluding other types. As a number of applications cannot be clearly assigned to one of belt type or another, then each case must be judged individually.

It is recommended that the preferred belt choice for each new design should always have the highest power density profile, thus allowing the drive to be dimensioned with a narrower width. Even with the more expensive high-quality timing belts as original equipment, the cost of the overall transmission due to narrower pulleys is often cheaper. The reduced installation space of the design is a positive benefit and will show improvements in the smoothness and the noise behaviour.

R. Perneder and I. Osborne, *Handbook Timing Belts*,
DOI: 10.1007/978-3-642-17755-2_3,

Fig. 3.1 Ratio $i = 1$

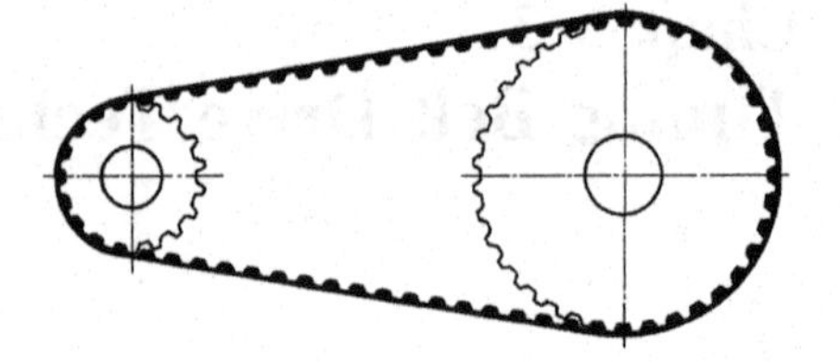

Fig. 3.2 Ratio $i \neq 1$

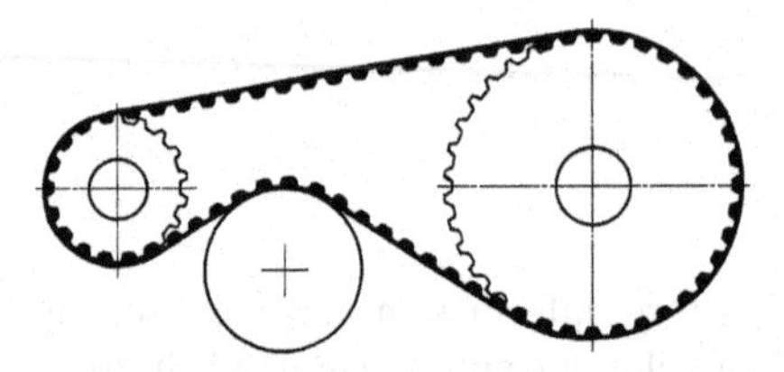

Fig. 3.3 Tensioner outside

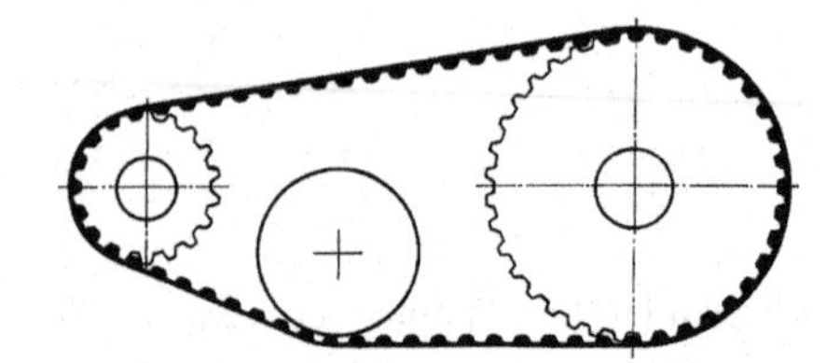

Fig. 3.4 Tensioner inside

A major property of belt drives is that the distance between driving and driven pulleys can be highly variable. Timing belts offer users additional options for drive solutions. The layouts shown are from the precision equipment manufacturing sectors as well as "tough" engineering solutions for heavy duty applications.

A belt tensioner on the outside of the belt improves the number of teeth in mesh and an internal tensioner has the benefit of only a single tension member flexing direction. Since both versions have advantages and disadvantages, it is up to the user to decide on the final geometric layout, depending on the design requirements. Tensioners without teeth on the tooth side, which do not transmit torque, are known as "plain" idlers (see Figs. 3.1–3.9).

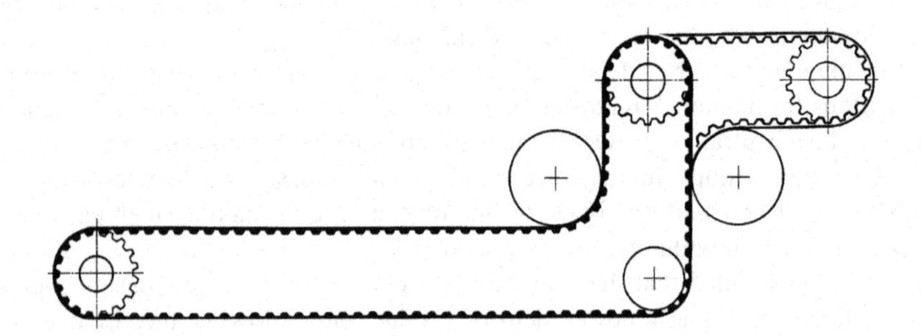

Fig. 3.5 Diversion drive

The belt layout can be planned with multiple diversions if the construction area is occupied with other machine elements.

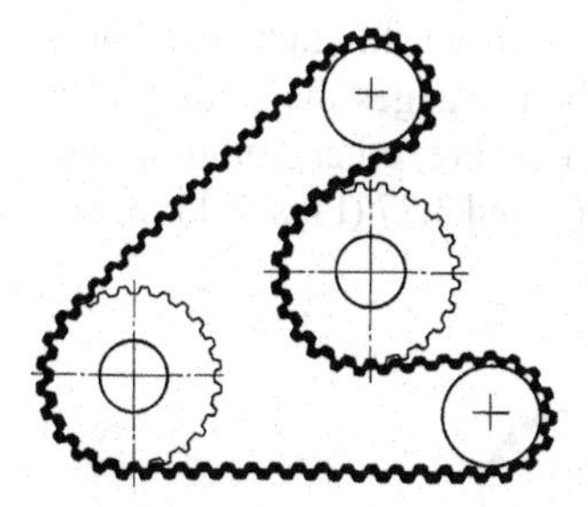

Fig. 3.6 Reversing drive with double-sided belt

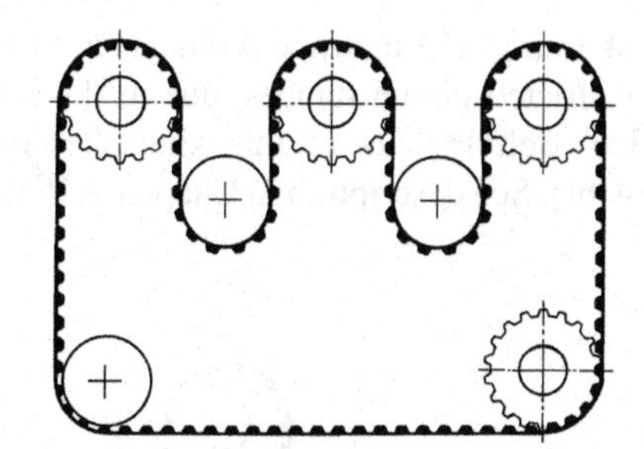

Fig. 3.7 Multiple shaft drive

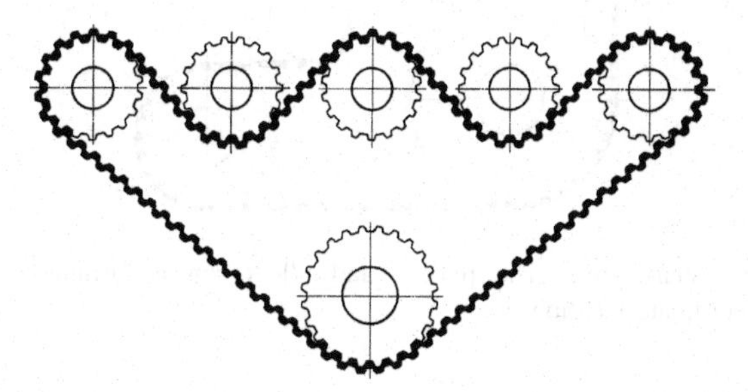

Fig. 3.8 Multiple shaft drive

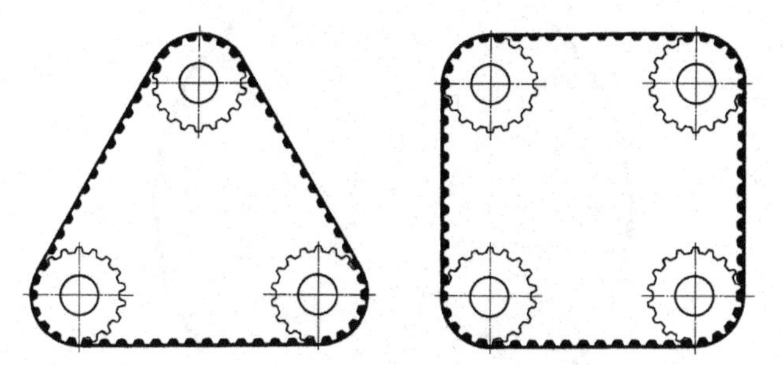

Fig. 3.9 Triangular and rectangular drive

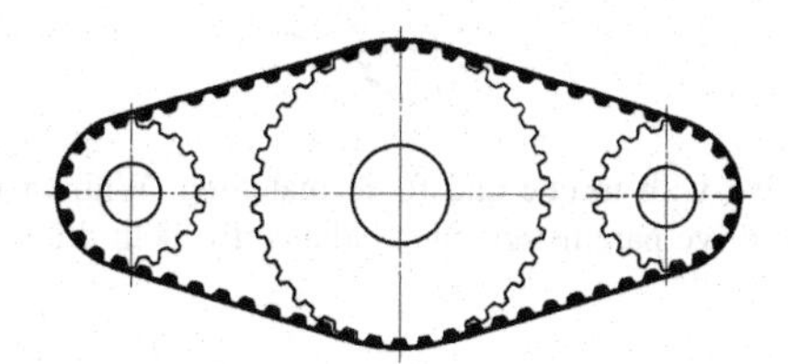

Fig. 3.10 Central drive with two small driven pulleys. Because of geometric redundancy this drive arrangement is to be avoided

Warning: Geometrical redundancy in Fig. 3.10! This drive should be used only in restricted circumstances, due to the fact that the belt engages the same pulley twice. Belt layouts of this kind should be avoided where other solutions are possible. See description in Chapter 3.15 with Figs. 3.66 and 3.77 (Figs. 3.11–3.13).

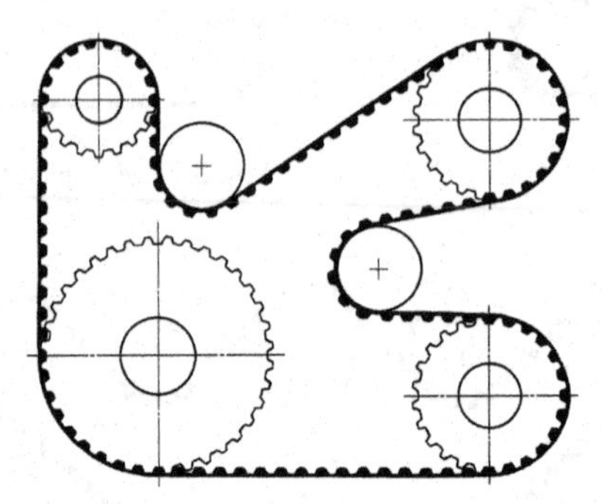

Fig. 3.11 Multiple shaft drive. The pulleys and idlers can be arranged in any number, combination or coordinate location

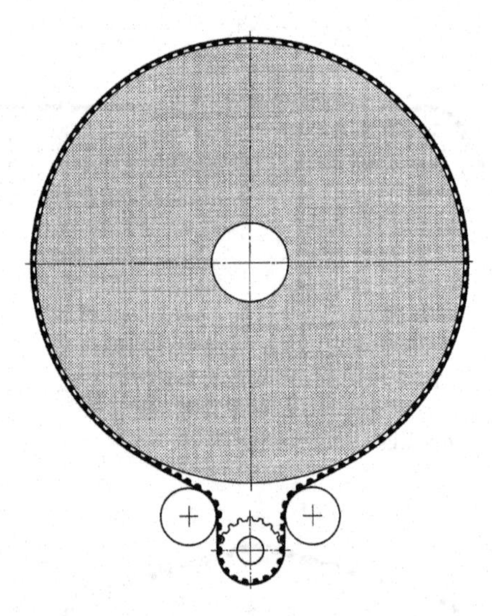

Fig. 3.12 Timing belt drive with torque and force matching. With toothed drive and smooth friction driven pulleys the drive parameters must balance for both drive methods

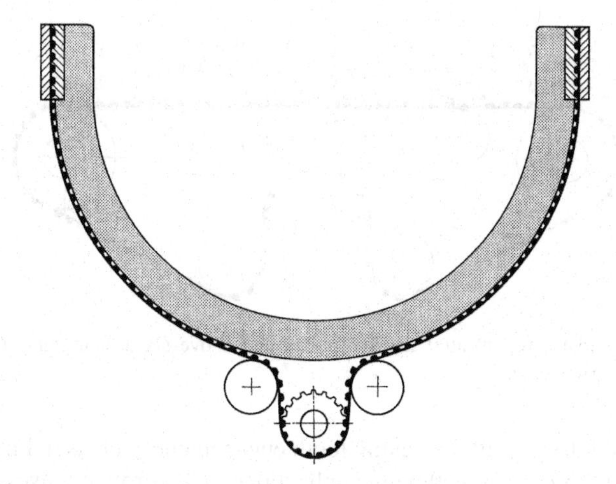

Fig. 3.13 180° actuator. It is possible to achieve approximately 360° of motion with this drive layout. The timing belt is clamped to the large wheel

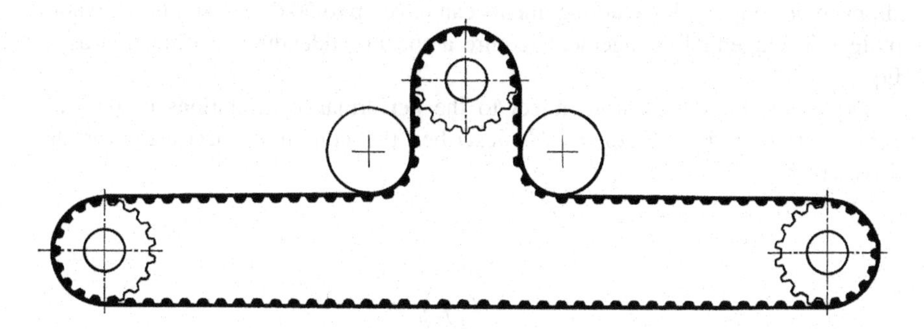

Fig. 3.14 Ω-drive. This drive is named after its geometrical shape, hence Omega drive

The Omega drive in Fig. 3.14 is a preferred drive design, where the motor pulley achieves advantageous numbers of teeth in mesh. The total power is transmitted by the belt under the most favourable of conditions. The transmitted power is shared between two or more driven pulleys (see Fig. 3.15 and 3.16).

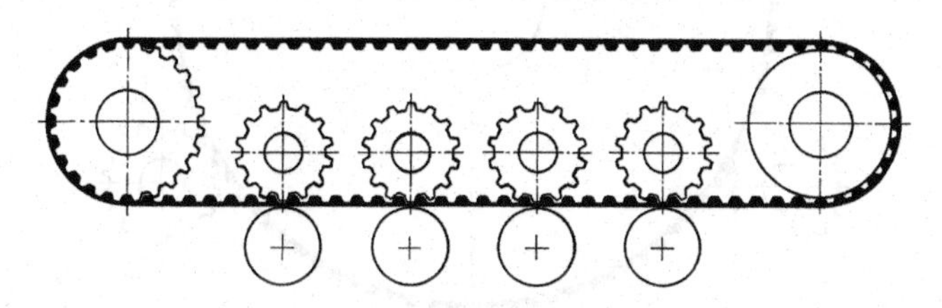

Fig. 3.15 Tangential drive. Left hand driver pulley, right hand plain idler and in the middle of the lower four pairs of pulleys in a tangential drive arrangement

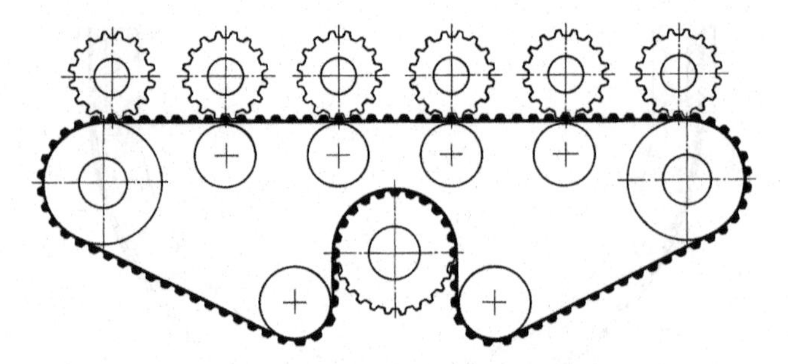

Fig. 3.16 Tangential drive. Central Ω-drive pulley and above six driven pairs of pulleys in a tangential drive arrangement

Toothed belt drives with tangential tooth engagement, such as in Figs. 3.15 and 3.16, are a preferred drive method in applications such as roller conveyors, spindle and winding drives. The total power is transmitted from the Omega drive pulley and the power is taken off as low part loads at the driven shafts. These consist of tangentially meshing toothed pulleys and opposed supporting idlers to prevent belt slippage or jumping. Such arrangements can have up to 50 driven shafts. The tooth pulleys in tangential engagement require a special addendum modification as in Eq. 2.51.

Drives of this design are subject to the performance limitations of the tangential drive design. Chapter 2.16 describes the conditions necessary for the drives to run.

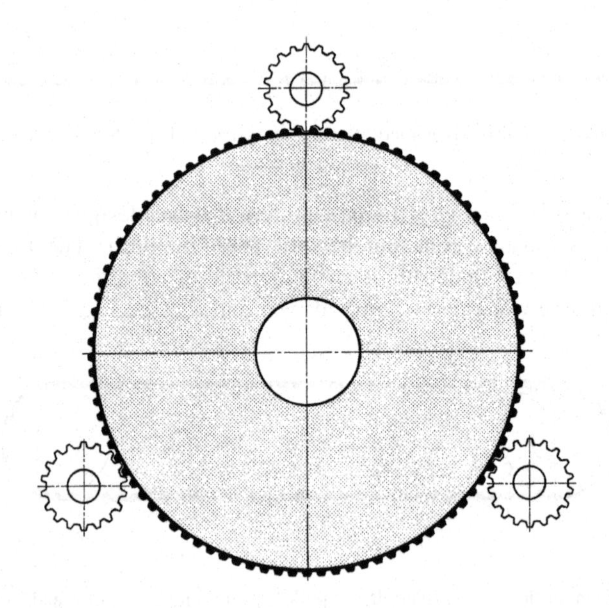

Fig. 3.17 Spur gear drive

The timing belt is "reversed" and fitted by its back to a central flat pulley. It adheres to this either by pre-tension or it can also be bonded separately. The toothed pulleys have a special tooth gap geometry, according to the type of belt, and a corrected addendum modification as in Chapter 2.16, Eq. 2.58 (see Fig. 3.17).

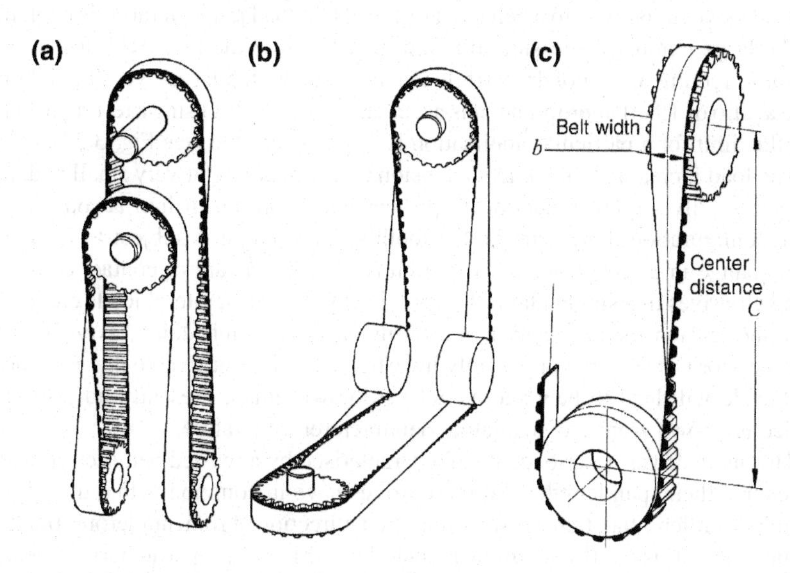

Fig. 3.18 **a–c** angle drives

Toothed pulley drives arranged at a Fig. 3.18 angle are also called unrestricted or spatial drives. They open up new possibilities of design for the user with many additional geometrical variations possible. Specific conditions must be met for angle drives: Fig. 3.18c shows the profile of the middle of the belt and the centre planes of the successive pulleys. They are arranged in a line! The belt exiting from one pulley may be twisted only to feed into the following pulley, as timing belts cannot function with lateral deflections.

The centre tension member(s) will seek the shortest route between the two pulleys.[1] All other belt tension members pass between the two pulleys through a spatial curve. The outer tension members will experience the largest loads and under this increasing load will build lateral forces on the belt centre. Under the inward directed forces, the belt tends to bow (bowl-shaped deformation of the belt

[1] A special feature of angle drives is that the belt always tries to take the shortest possible route within its drive layout and thus it positions itself independently on the pulleys without run-off. However, deviations can arise from the theoretical line in practical applications as a result of tolerances in the belt and pulleys as well as by tolerances when assembling. Thus, flanged pulley belt guidance for drives of this kind is quite usual, see Chapter 2.10. Side guides ultimately facilitate the installation and tensioning of belt for use and handling.

back). Limiting the lateral forces means that the minimum ratios of the belt width b and the centre distance C must be maintained.

$$\frac{b}{C} \geq \frac{1}{15} \text{ to } \frac{1}{5} \tag{3.1}$$

The above ratio refers to a belt twist of 90°. The broad range of ratios is explained by the large number of possible different belt types. Wide belts with low overall height h_s require a ratio of 1:15 while narrow belts with a large overall height can have a ratio of 1:5. If appropriate, it is necessary to verify this using a test rig to check that the limit for a particular application is actually feasible (see Fig. 3.23).

The load increases in the outside tension members are relatively small and need not necessarily require reduction in pre-tension or are taken into account in the drive configuration. Furthermore, the tooth shear strength is subject to no significant change in performance on transition to or from the arc of contact compared with belt drives in a single plane. The previously described lateral loads emanating from the tension member and acting on its elastomer embedding, as well as the tendency for bowing are not yet fully investigated. It is conceivable that this sort of load cycle will lead to new and as not yet known limits. Presently no restrictive advice is known from any available manufacturer's literature.

The angle drives described, are all characterised by a twisted belt section within the course their span lengths. All these drives have in common is that they do not exhibit a preferential transversal oscillation direction. From numerous practical applications it is well-known that such belt drives are particularly vibration resistant and furthermore the drive refinement stands out with remarkably lower noise behaviour in comparison with belt drives in a single plane. When meshing, the belt tooth surface has a slight tilt to the pulley teeth and thus, the pulsating air displacement escapes across the enlarged cross-section of the belt section which is not fully meshed and thus limiting the compression.

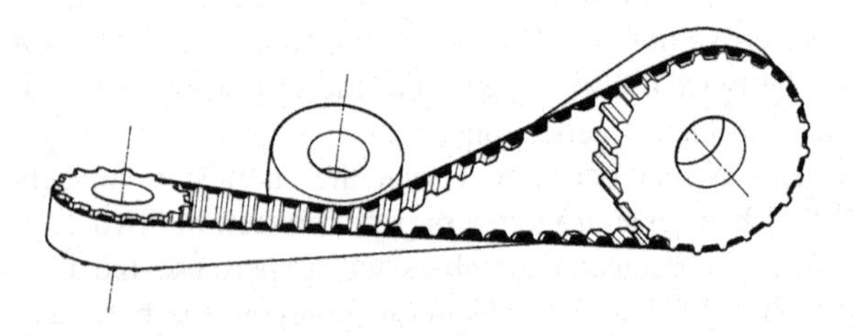

Fig. 3.19 Angle drive with three deflections. For spatial drives at least three pulleys are required. Angle drives with two deflections will not function

Angle drives require at least three pulleys. The belt layout in Fig. 3.19 is functional but is problematical in practical applications. Firstly, the usable tension from the belt back tensioner is very limited. Secondly, the geometric coordinates and axis positions can only be determined with considerable mathematical effort. Even if a calculation of the coordinates of a fixed back tensioner were successful, then its production as well as assembly would be subject to certain difficulties. A peculiarity is that the tensioner should be moved on a curved path while pivoting around its axis. It is therefore recommended that spatial angle drives should have a minimum of four deflection pulleys. The above design has not seen any practical use.

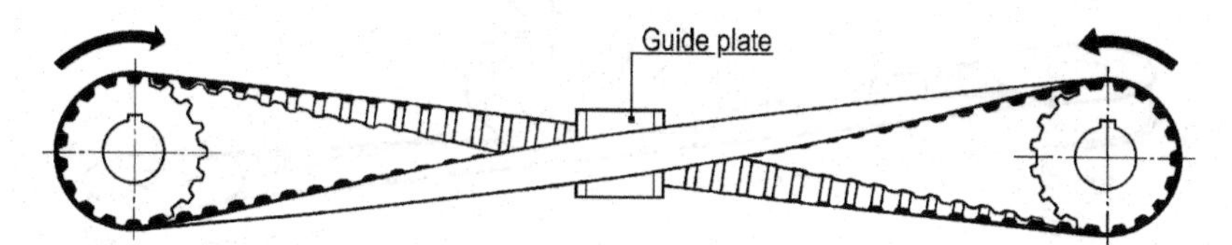

Fig. 3.20 Crossed belts

The design in Fig. 3.20 is of crossed belt sections which allow rotational direction reversal at the pulleys. The twist angle in this example is 180° and the width/centre distance ratio from Eq. 3.1 is doubled. The crossing point of the belt sections must be separated by a guide plate (see Fig. 3.21).

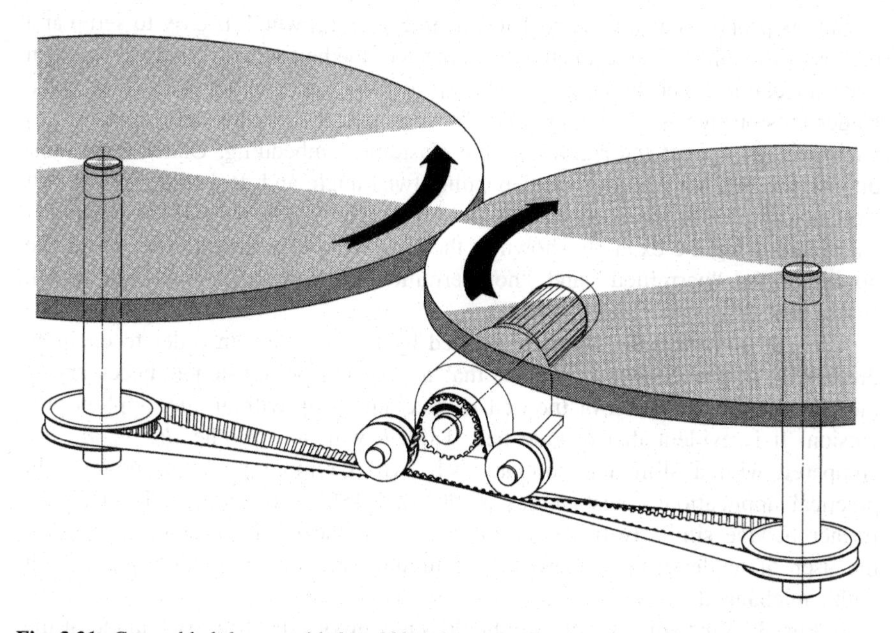

Fig. 3.21 Crossed belt layout with $2 \times 90°$ and $1 \times 180°$ twists. The belt sections do not touch as they pass in the gap created by the Omega drive

For angular twisted drives or crossing layouts it is not possible to secure evidence that the minimum ratio of "belt width spacing to centre distance" will function without damage (see Eq. 3.1). To solve this problem, the following

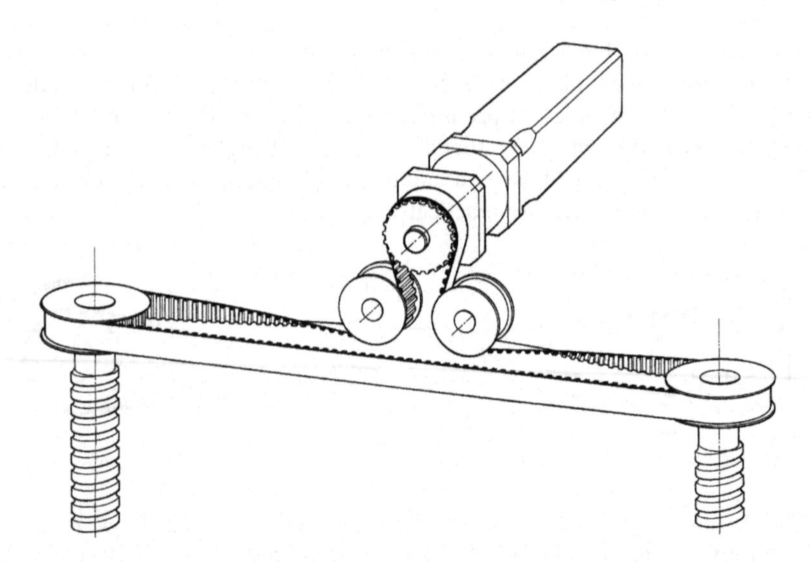

Fig. 3.22 Ω-drive in a spatial layout over five pulleys. This belt drive design is chosen where there are vertical driven axes and a horizontal drive motor

method is proposed as a static self-test as in Fig. 3.22, which is easy to setup and requires little effort. The test is to twist a piece of the belt section clamped between two end clamps. The outer cords take up a plane curve about the central axis. Under tension they seek the shortest path between the clamps and build inward lateral forces against the rigidity of the elastomer embedding. Under these conditions, the belt tends to deform depending twist angle and depending on the belt structure and width. The belt will begin to bow. The visible start of the bowing is calculated using the example shown in the Fig. 3.23a experiment. This forms the border of the "permitted" and "not permitted" regions of twisted belt use in spatial belt drives.

The test procedure is further simplified by the fact that, in order to examine the border of the "bowing" region, that no defined pre-tension is necessary in the belt. The deformation of the belt surface shows up without any applied pre-tension. It is evident that a wide belt width b with low profile (h_d, h_r, h_s, h_t), compared with a slim and compact belt, will be more prone to bowing. In practical applications, twist angles of 90° and 180° are common. Both angles further behave similarly by becoming bowed at the border region. It is up to the user of angle drives to use the minimum ratio of centre distance to belt width unchanged or with an additional safety factor.

In Fig. 3.24 the spindle nuts are the driven components. The slope angle of the threads will impart linear motion to the fixed spindles. The threaded spindles both absorb the static axial loads as well as guide the table. See also the reversal of this principle in Fig. 3.26 in which the spindle has the rotational movement.

For increased positioning accuracy and parallelism, the lift table pulleys are manufactured with reduced backlash c_{m1}.

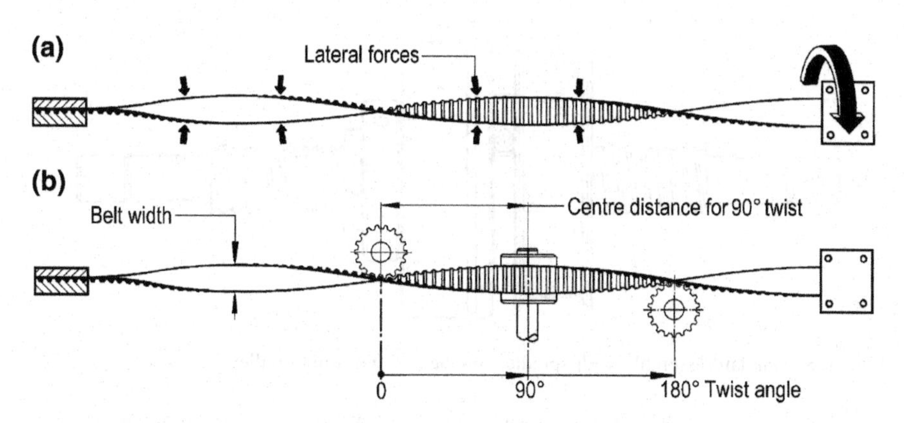

Fig. 3.23 Twisted timing belt example. **a** Experimental set-up "warping". **b** Twist angle

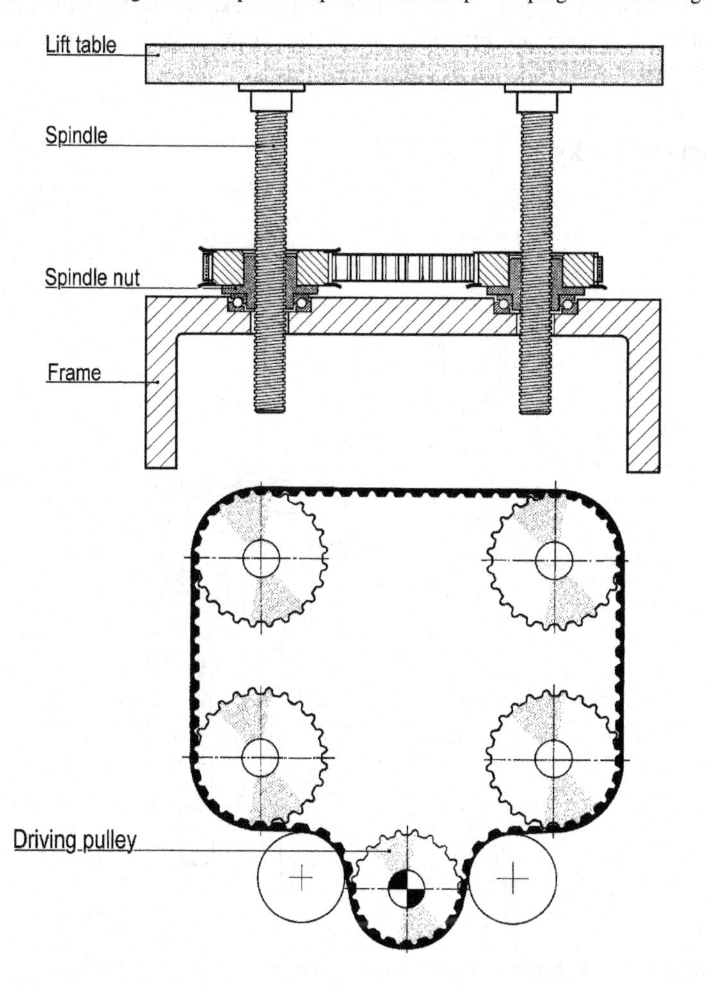

Fig. 3.24 4-spindle lift table

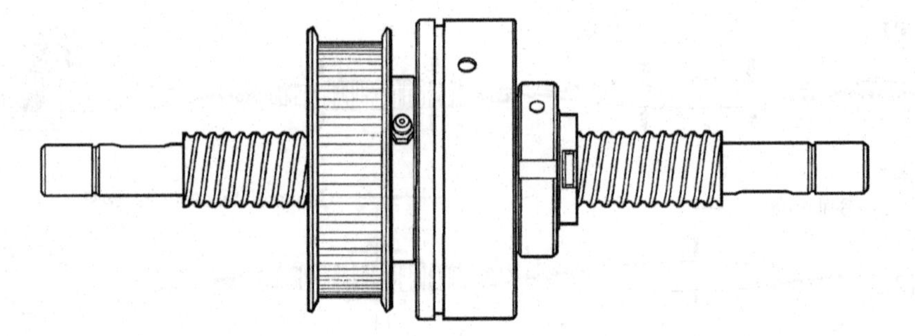

Fig. 3.25 Standard assembly with spindle, spindle nut and timing pulley

Many suppliers provide standard mechanical components in connection with the timing belt drives which often simplify the design effort significantly. Such assemblies as shown in Fig. 3.25 allow the user to buy certified safe parts (See references [102]).

3.2 Screw Jacks

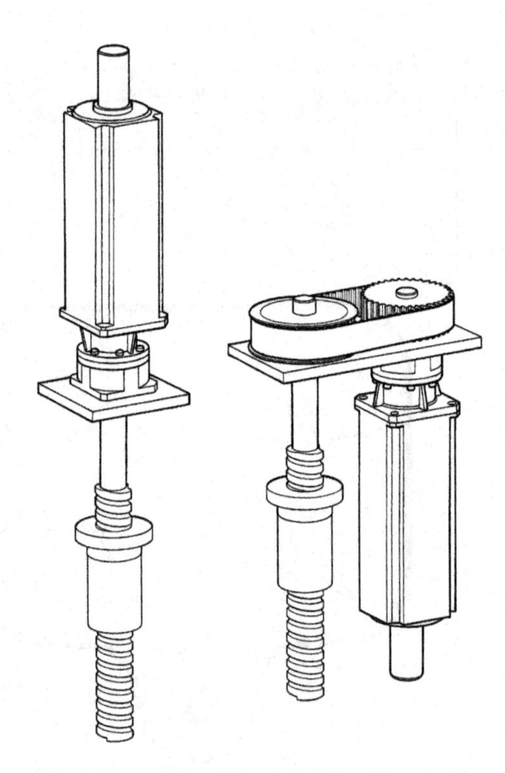

Fig. 3.26 Screw jack drive. *Figure left*: without belt drive. *Figure right*: A compact layout due to the timing belt drive [99]

The modular design of screw jacks can be seen above with both in-line (left) and parallel types (see Fig. 3.26, picture right). Drives of this type are associated with lifting, applying pressure or used as feed modules. They serve as units in mechanical handling operations for pressing, joining, riveting and crimping. These operations require flexible application, programmable feed speeds and reliable and accurate travel limit positional tolerances of less than 1/100 mm.

The design of the timing belt drive is most important in this type of application. The drive design, due to the parallel axial arrangement, is compact and limited space conditions can be easily handled and accounted for. The toothed belt drive design is characterized by large diameter timing pulleys with wide belt widths and is optimized according to the recommendations in Chapter 2.13 for high positional accuracy. The pulley tooth profiles have reduced backlash c_{m1} to allow for the screw jacks working under various start-stop conditions.

3.3 Packaging Machine Mechanism

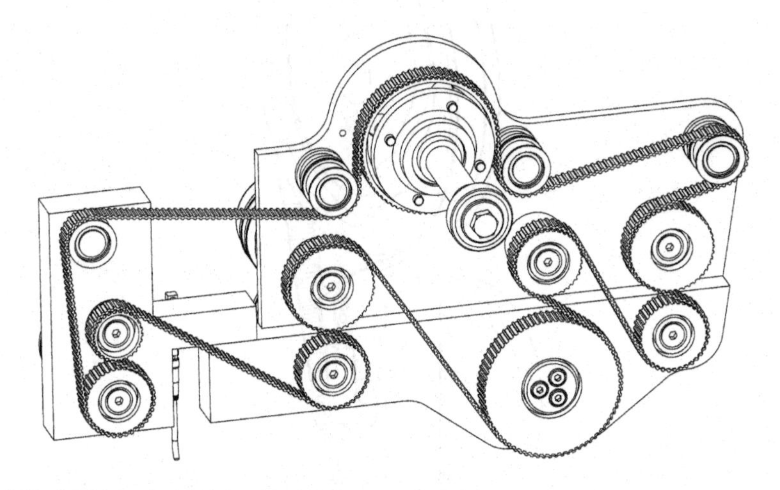

Fig. 3.27 Multiple shaft drive with double-side belt [35]

In this packaging machine mechanism, the designer uses the reverse direction properties of double-sided belts across a total of 13 pulleys, including nine timing pulleys all of which directly transmit power (see Fig. 3.27). The other four pulleys are used for optimal belt layout as well as tensioning. The machine frame separates the mechanical drive section from the production side of the machine. In this mechanism, for the food and beverage industry, laminated and printed cardboard blanks for the manufacture of small packets or cigarette packs are folded, glued and transported to the next station. Its functioning inherently requires a synchronous drive system. The high target productivity is achieved by operating at 500–1300 rpm, corresponding to a rate of about 3 m/s or equivalent to 500 packs per minute.

This multi-shaft timing belt drive example has replaced a proven solution using spur and change gears, which running in oil, necessitated elaborate oil seals fitted to each shaft. The particular advantages of this new timing belt solution are that it is low-noise and lubrication-free. This allows the production environment to remain clean and hygienic for the food industry, thus achieving customer satisfaction with less effort. Thanks to the timing belt drive, the shaft centre distances can be chosen freely with positive benefits for the surrounding mechanism. The technical data for the drive is: 25T5/2500-DL polyurethane timing belt, drive pulley $z = 72$, $n = 500\ \text{min}^{-1}$, $P = 1.5$ kW.

3.4 Press Drive

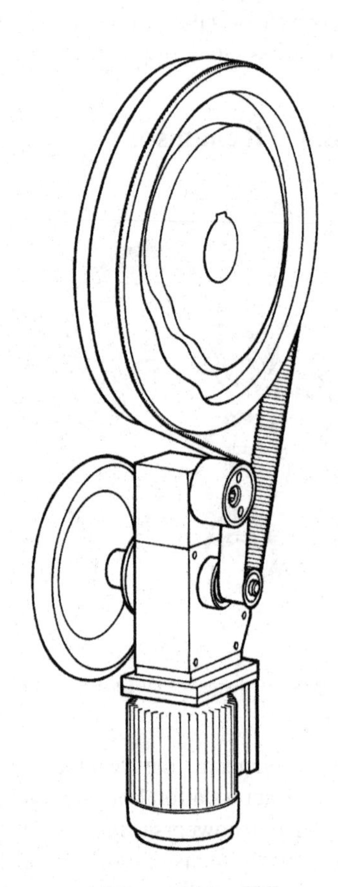

Fig. 3.28 Press drive with torque and force matching [66]

Presses are available in many types, depending on the operational requirements. The common features relate mainly to relaying the movement of the flywheel to the ram stroke, using the dynamic principles of inertia. The illustrated machine is a laboratory press for the pharmaceutical industry. The powder mixture is compressed by a punch and die into tablets. The stroke is cam initiated (Fig. 3.28).

The chosen solution is a rotary actuator using the cylindrical outer surface of the flywheel as frictional contact surface for the timing belt. Since the technical requirements of the drive train require a large drive ratio, the motor pulley is necessarily dimensioned with a relatively small diameter. Given the choice of belt types, only timing belts can provide the necessary transmission characteristics through such small bending radii.

The selected timing belt transmission thus works on both the toothed small pulley and frictionally around the large flywheel. The cost reduction due to the elimination of teeth on the flywheel is significant. Such combined toothed and frictional performance drives are examined separately in the appropriate Eytelwein's equation:

$$F_1 = F_2 \cdot e^{\mu \cdot \alpha_1}, \tag{3.2a}$$

and

$$F_t = F_1 - F_2. \tag{3.2b}$$

Where F_1 force in belt span under load, F_2 force in the slack span, F_t tangential force, μ coefficient of friction between belt and outside surface of the flywheel, α_1 belt wrap.

3.5 Cable Ferries

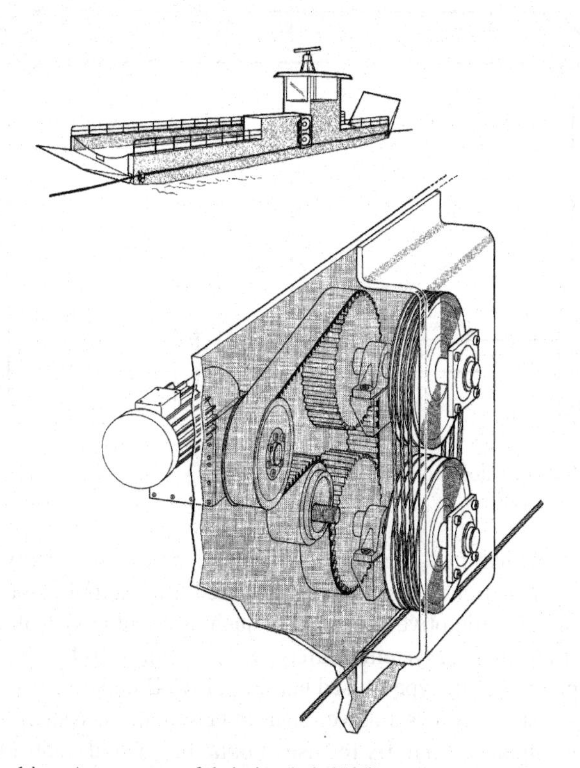

Fig. 3.29 Cable drive. A very powerful timing belt [105]

Cable ferries run between the banks of lakes and rivers. The friction on the cable pulleys transmits the load with the three-stage design giving an active contact angle of $3 \times 360° = 1{,}080°$. The calculation of the coefficient of friction in the cable drive follows the familiar Eytelwein equation, see Eq. 3.2a, b in the drive example on the previous page. The requirement of load transmission for this wrap drive is that both cable drive timing pulleys are powered as a pair. The chosen timing belt has the following specification: (see Fig. 3.29).

PUR-Timing belt 150 AT20 / 3900 with V2A special tension members, Timing pulleys $z = 70$ with pitch circle diameter $d_W = 445.6$ mm, allowable belt tension $F_{zul} = 30{,}000$ N, drive torque $M = 500$ N·m.

3.6 Test Bed

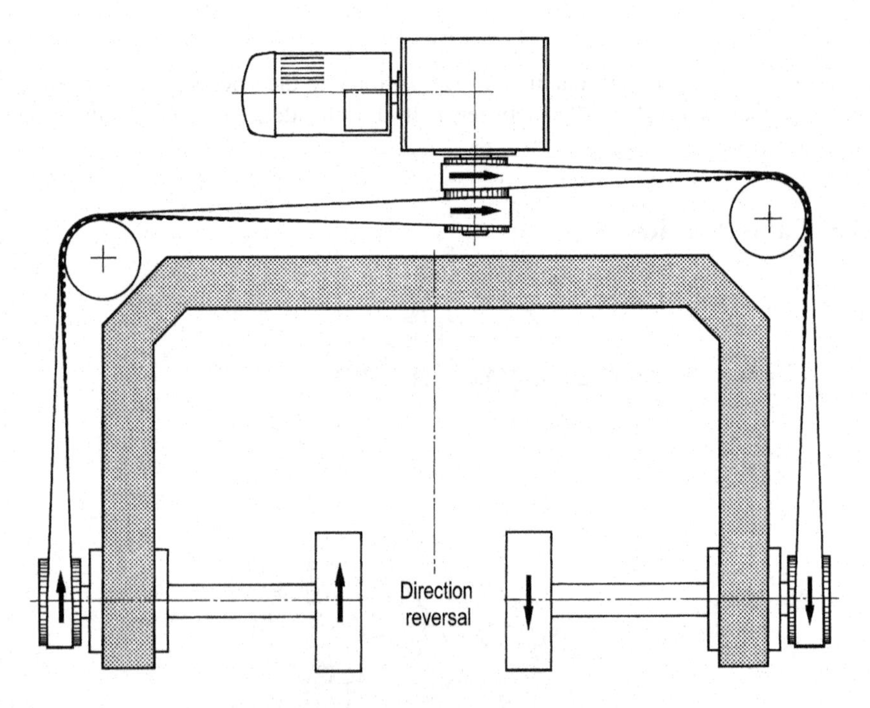

Fig. 3.30 Test bed for automotive components

Automated product test facilities ensure the functionality and quality control standards for mass production products. They are used, for example, to test the performance standards of automotive components such as couplings, cooling pumps and also to qualify transmission systems. Safety devices require 100% quality control. In addition, equipment of this type is used in areas of R&D departments. The principle structure of the test bed shows a power-sharing central drive system using two separate belts in an angular layout. By the use of both the left and right sides together, a doubling of speed can be achieved for the appropriate the test sample (see Fig. 3.30).

This power splitting drive layout offers distinct advantages when high speeds are required especially as the timing belt drive components themselves are only working half as fast. Other possibilities for use include friction welding production equipment.

3.7 Adjustable Centre Distance Solutions

Providing adjustable centre distances in the drive design, whilst having to take the drive geometry into account, is always a challenge for the designer. A good solution must be equally suitable for assembly, disassembly and the operational processes. Cost-effectiveness must also be considered. Other desired characteristics can be fine tuning and rigidity of the drive. The applied pre-tension and the minimum adjustment were previously discussed in Chapter 2.17.

Bearing housings are proven solutions from all producers in the bearing industry and are available in cast iron, steel or plastic construction, see Fig. 3.31. The complete assemblies, rather than standard bearings, are widely accepted in both chain and belt tensioning devices. Depending on the version used, they allow linear or angular adjustment movements. Fine adjustment is partly done through tie-rods or turn-screws. The tolerance of the bearing seat in the housing is fixed so as to compensate for any deviations in the shaft.

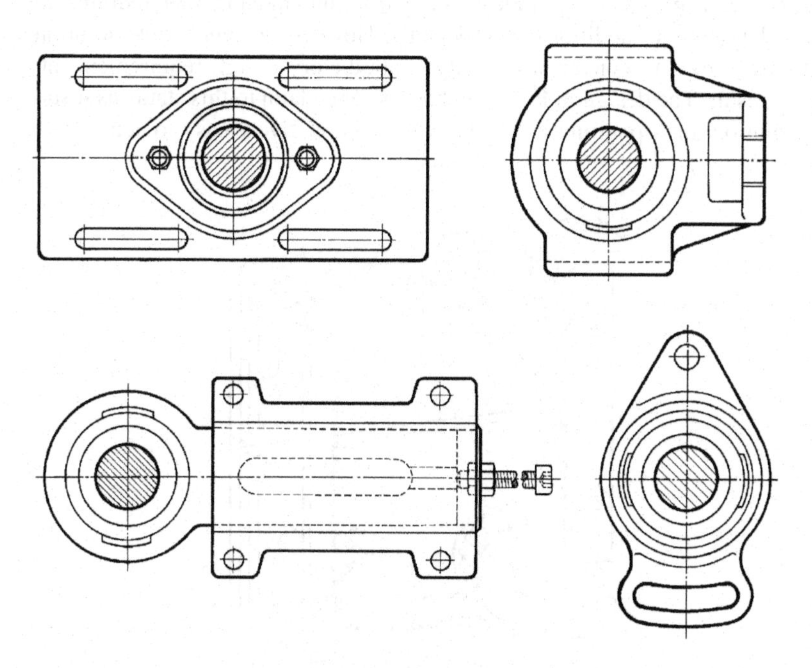

Fig. 3.31 Tension bearings from INA [52]

Further adjustments to the belt preload can be achieved by the addition of one or more axes in the form of adjustable idlers. They can also assist in the additional job of laying-out the belt run and selectively help to optimize the angle of contact on individual pulleys. Good solutions for pivoting idlers with eccentric adjustment are available from many belt manufacturers. The illustrated tension pulley in Fig. 3.32 supports the cantilevered bearing on an amply proportioned base. Only a single socket head screw is required for mounting. Adjustment is achieved by a pin wrench.

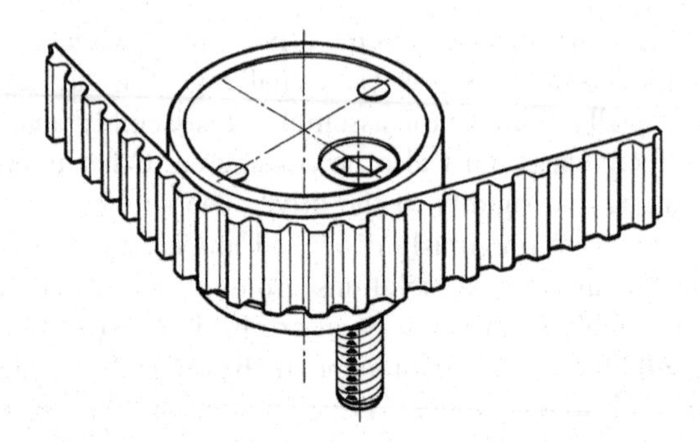

Fig. 3.32 Eccentric tension idler from Mulco [85]

Adjustable axes are the simplest belt tensioning method that requires little effort and few or no additional components. However an *innovative* solution is sometimes needed to convert a partially contradictory design requirement into a practical result. The design in Fig. 3.33 has a strong claim to this status as it shows that a solution, incorporating a split backplate, is relatively easy to achieve.

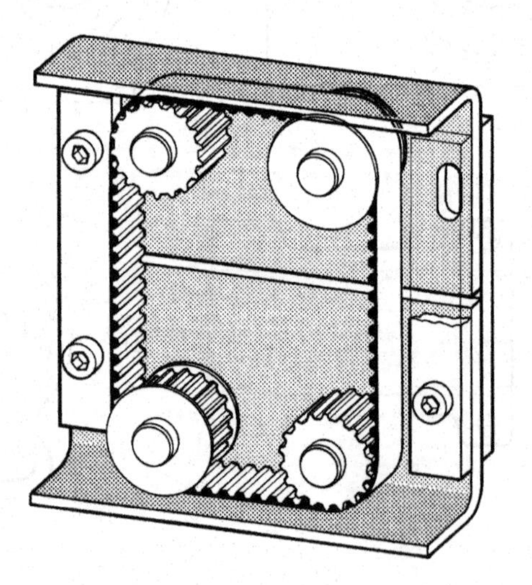

Fig. 3.33 Adjustable centre distance through split backplate

Fig. 3.34 Clamp plate with adjustment screw from Breco [85]

At least one pulley must be adjustable for tensioning. Once in position, at the correct belt pre-tension, it must be rigidly mounted (not sprung) and be secured against further movement.

There are additional possibilities to achieve the belt pre-tension with linear timing belt drives by the use of adjustable clamp-plates. With wide belts in the higher load ranges, the belt pre-tension is set by an adjusting screw. Generally, assemblies of this type are supplied by timing belt manufacturers or technical distributors (see Fig. 3.34).

3.8 Injection Moulded Plastic Timing Pulleys

For cost-effectiveness, the choice of timing pulley materials is a priority. For volume applications, from around 2,000 units plus, injection moulded pulleys in PA and POM have cost advantages over conventional machined pulleys. For zinc and aluminium die-cast pulleys the economic start point is from around 4,000 units because tooling costs are typically so much higher. When injected/die-cast manufacturing solutions are chosen, the geometry of the pulley should be adapted to the different requirements of the production processes (see Fig. 3.35). These include: uniform wall thicknesses with a draft angle and an integral side mounted flange.

The timing pulley shown in Fig. 3.36 consists of two identical halves which have a mutually interlocking geometry. The d-shaped drive shaft aligns the left and right halves of the pulley with the locking washer applying axial pressure against the shaft shoulder. Drive torque is transmitted by the d-shaped shaft. The split line

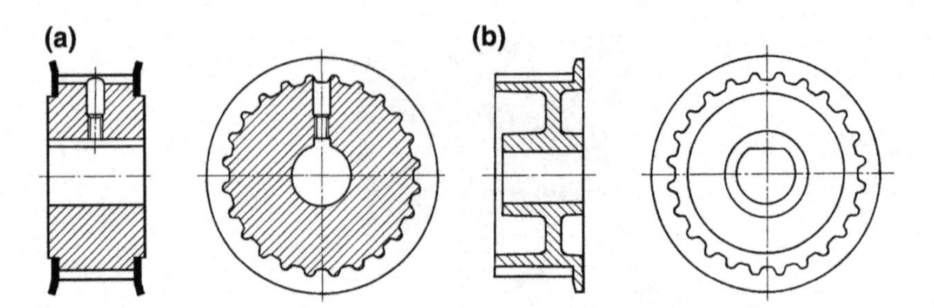

Fig. 3.35 Timing pulley comparison. a Machined. b Injection moulded or die-cast

within the timing pulley does not affect the function of the timing belt. Such types of pulley are only ever feasible with even numbers of teeth.

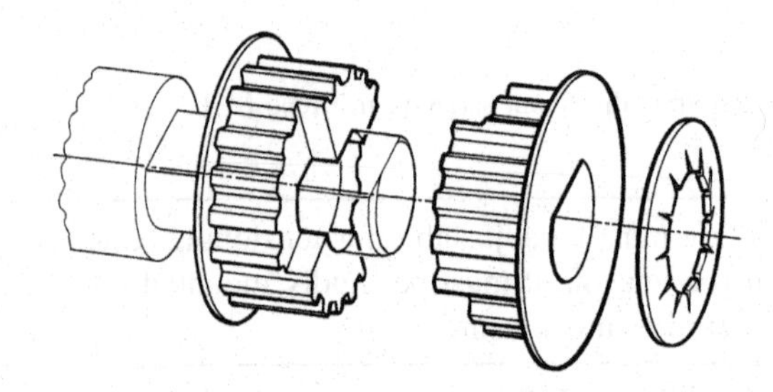

Fig. 3.36 Special pulley design for injection-moulding/die-casting

3.9 Motor Glider Propeller Drive

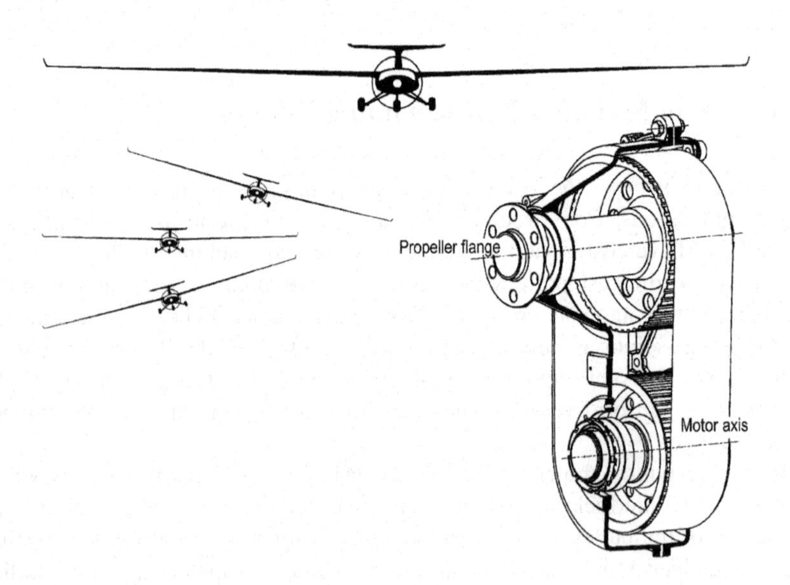

Fig. 3.37 Propeller drive for a super lightweight plane [108]

A motor glider combines both powered and unpowered modes in one aircraft. The drive must allow self-launching capabilities, thus making the machine independent of its home landing strip. All the characteristics of the aerodynamic design must conform to that of a glider, where the motor is relocated centrally into the fuselage, meaning the air-flow over the outer structure remains unaffected. The relatively large distance between the engine and propeller is bridged by an extended drive shaft. A toothed belt drive is used between the propeller and the drive shaft coupling. The drive housing is of lightweight construction and the timing pulleys are made from a hard anodised high-strength aluminium alloy (see Fig. 3.37).

Nominal power $P = 85$ kW
Small pulley $z_1 = 48$
Large pulley $z_2 = 54$
Number of revolutions $n_1 = 1934$ min^{-1}
Synchroflex belt 75 ATP 10 / 920 Gen. III

By any measure it is a very compact transmission and the short centre distance gives high values of stiffness with no tendency to belt-jump. For such drive geometries the pre-tension has been reduced to one-half or one-third of the normal value (see Chapter 2.7).

3.10 Industrial Robot Arms

"An industrial robot arm, robot for short, is a universal motion automaton, consisting of several movable axis, whose functions, with regard to a sequence of paths or angles, are freely programmable and possibly sensor controlled". This is the definition given in the VDI guideline 2860 [122]. A rough guide divides the robot into

- Mechanics
    - Sensors
        - Control

The mechanism consists of the components.

- Kinematics
    - Structure
        - Drives
            - Gripper

Kinematics refers to the mechanical transmission elements of all the individual axes. They have the task of bringing the final axis attached to the gripper to a specific position or move it on a predetermined path. Kinematics and control create a spatial–temporal correlation between the robot base and the object to be handled. Generally each axis consists of one degree of freedom. A maximum of six degrees of freedom are available, including the three main dimensions for controlling the spatial coordinates and three secondary movements for position control.

Robot Axis

A robot axis is a rotationally or translationally controlled structure comprising of one degree of freedom. Articulated arm robots consist exclusively of rotational structure elements and we can distinguish the spatial location of their centres of rotation. Horizontal joints have additional loads due to extra torques on the drive links whereas vertical joints experience no torque due to gravity on the drive links (see Fig. 3.38). Increased length of the arm is coupled with greater compliances and variations. Structures connected by joints to a series of members result in a kinematic chain. Whatever is the current movement of the gripper, it always involves all the members in a force-action combination. With control commands, it is possible to arrange a smooth, shock-free motion path in order to avoid peak loads and to optimise the motion quality of the handling device.

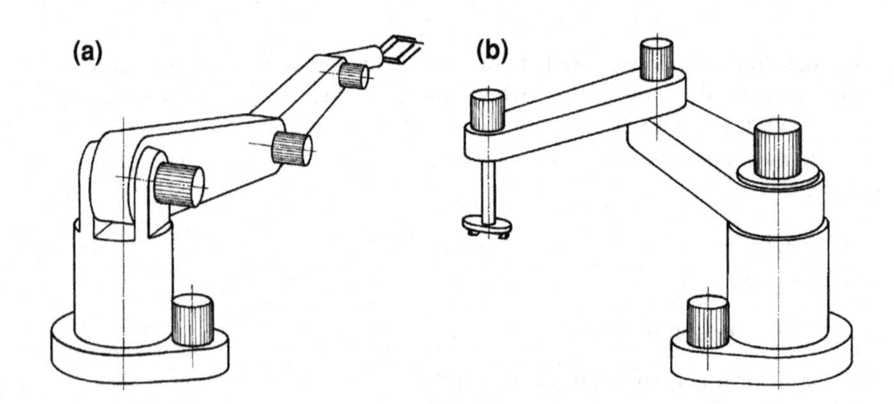

Fig. 3.38 Articulated arm robots. **a** With horizontal axis. **b** With vertical axis

Timing belts dominate this application through their ability to bridge the centre distances between the joints of the arm. Belts are low mass, positionally accurate and are suitable for operating conditions with changing loads and rotations, and for implementing the motion commands from the base to the gripper. In organising the

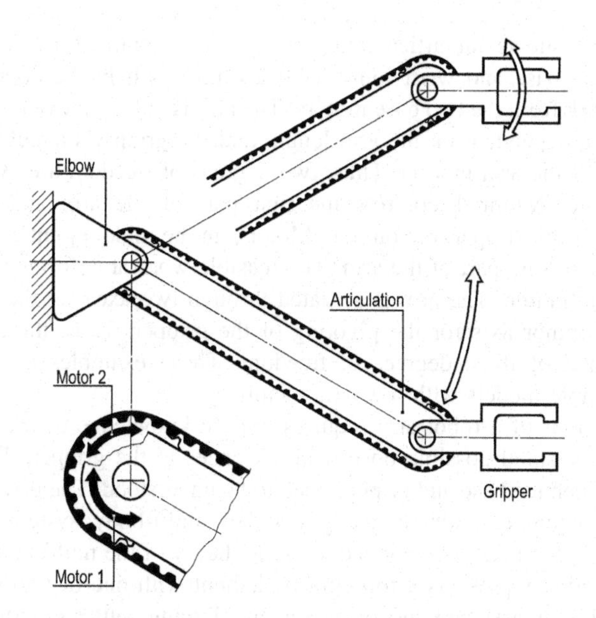

Fig. 3.39 Motion model* with single part arm. Two independent motors control the arm from the pivot in the base. Motor 1 is coupled to the arm and thereby moves the gripper in an arc. Motor 2 controls the angular position of the gripper (*The term robot is avoided here because the device has only one axis of main motion.)

kinematics, the motors are optimised for low mass at the extremities and are thus rarely in the gripper arm but usually close to the base. The work area should have an undisturbed field of reach. Each joint can use up to two motors, where a Harmonic-Drive® [45] or a Cyklo-Drive® [110] is used to transmit the torque.

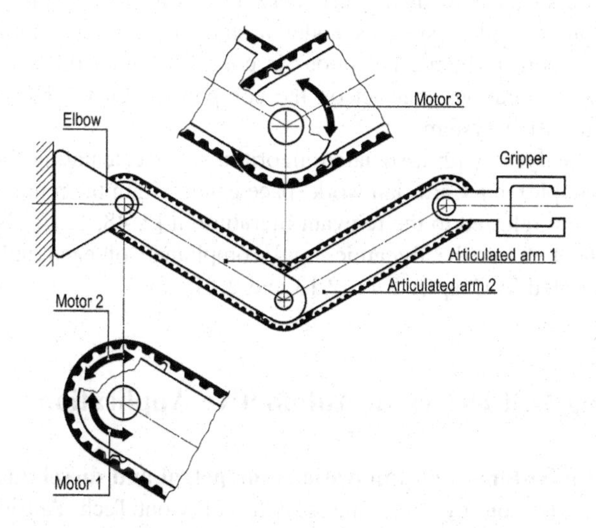

Fig. 3.40 Motion model with a two part arm

Figure 3.39 shows an articulated arm with a simplified motion model. It consists of the base and wrist joints with a timing belt in the connecting arm. The joints work from the two base motors. The arm is able to move the gripper in a circumference dependent on the arm length and the gripper's angular position is independent of the arm motion. Thus, two degrees of freedom are available.

In Fig. 3.40, Motor 1 controls the first part of the arm and Motor 2 is responsible for the gripper motion with Motor 3 in the middle joint controlling the movement of second part of the arm. The feasible work area forms a surface and the gripper is tiltable. The arm is actuated through two axes for the main movements and a minor axis for the pivoting of the gripper. Thus, the arm operates through a total of three degrees of freedom. These examples are deliberately related to simple models with few axis members.

The behaviour of a robot arm requires precise knowledge of the whole drive system to understand possible positional variances at the gripper. Thus, all arm sections between the base and gripper need to be understood. The drive system is a complex, dynamic and kinematically coupled multi-body system (MBS) and systems of this kind can be modelled with the theory of the multi-body dynamics. Each arm section represents a force/form element with one or two-way connections which has an assigned task of movement. Timing belts are normally used to solve those arm section drives which have larger centre distances. Thus, there are tension member elongations and belt/pulley compliances to take into account. The structure of the arm members affects the dynamics as well as the rigidity, inertia and the vibration behaviour. You have to know all the factors that act on the joint, including the forces caused by inertial masses. The inertia, in turn, depends on the position of the arm and that is always changing during motion.

It is relatively simple to determine the effects for each individual arm section. However, the mathematical treatment of a kinematic chain with many joints is complex. We also have to distinguish between static and dynamic behaviour. Static behaviour considers systems under a state of constant equilibrium after cessation of motion. Dynamic behaviour is concerned with the analysis of torques, forces and vibrations that act on the components on the basis of accelerations from the drive system.

A detailed analysis, with associated algorithms, for calculating the motion of robots (for example) in a Cartesian work space is not within the remit of this book. You are therefore referred to the relevant literature in [2, 48, 127]. The influences exerted by the belt on the dynamics and compliance of each individual arm member are treated in Chapter 2.11, 2.13 and 2.14.

3.11 Timing Belt Drives In Automotive Applications

Development milestones and innovations for petrol and diesel engines

A special contribution by Hermann Schulte of ContiTech Antriebselemente, Hannover, Germany [15].

3.11.1 Introduction

Since the introduction of overhead camshafts in internal combustion engines in the 1970s, timing belts have been mainly used for crankshaft/camshaft synchronization [44]. Timing belts have been able to win a 75% market share in Europe because of their inherent advantages. Modern timing belt drives are now designed and guaranteed for the engine lifespan. The realisation of the objective of a lifespan for the engine lifetime, i.e. 240,000 km for a timing belt drive, has been achieved through a series of innovations. Amongst those innovations are:

- The optimization of the abrasion resistance of PTFE-coated tooth reinforcing fabrics.
- Improving the dimensional stability of the new high-strength, hydrolysis and bending resistant K-glass tension member types.
- Increasing cold and heat resistance due to the use of peroxide cross-linked HNBR rubbers covering a continuous temperature range from −40 to +150°C and short-term peaks of +170°C.

Oil resistant HNBR timing belts are currently being developed as an alternative to the often extremely noisy chain camshaft drives. With innovative oval timing pulley technology, belt widths for 4-cylinder engines can be reduced by approximately 30%. Oval timing pulley solutions, with lifespan improvements, are no wider than similar chain drives but provide benefits of lower noise and reduced friction.

3.11.2 Evolution of Camshaft Belt Drives

The first generation of camshaft belt drives were constructed from CR-elastomers (Polychloroprene), however CR-elastomers have only a relatively limited temperature range of −28 up to +100°C. In 1985, an optimized CR-elastomer, CHP (Conti High Power) was introduced with a temperature limit of +110°C. In the early 90s, better engine encapsulation and closely coupled catalytic converters resulted in increased engine bay temperatures of up to +120°C. Timing belt manufacturers reacted to the rising ambient temperatures with significantly better thermally stable material variants like the heat-resistant high-performance HNBR (Hydrogenated Nitrile Butadiene Rubber) elastomer from Bayer. The first generation of HNBR timing belts, HSN-VHT, tolerated high temperature applications from −25 to +120°C. However, up to the 1990s, these were rigidly tensioned drives and thus came with a maximum lifespan of 120,000 km (see Fig. 3.41).

Early 2000 was the first time that the full life of the automotive timing belt could be achieved, with a lifespan of up to 240,000 km, through new material

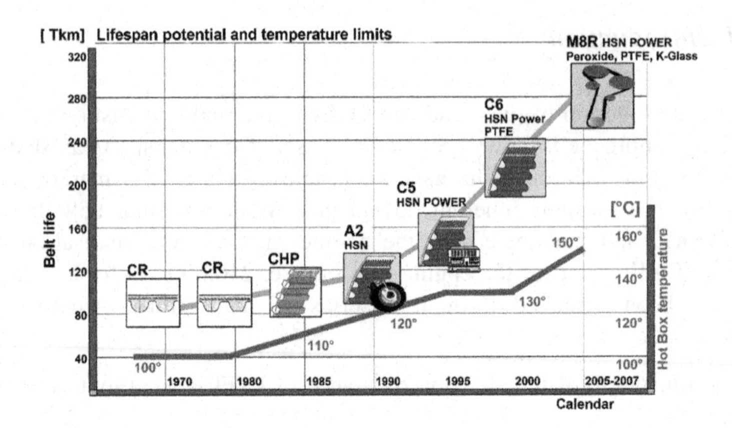

Fig. 3.41 Automotive timing belt milestones

developments. Also, a temperature increase, from +130 to +150°C for long-term hotbox tests, could now also be achieved through the cross-linking of sulphur types in peroxide cured HNBR types.

Belts of the latest generation consist of high-performance, aging-resistant, peroxide cured HNBR elastomer. Aramid fibres are mixed into the HNBR elastomer to give additional stability and increased tooth shear strength [79].

The glass cord tension member (E-glass and the more reliable, but also much more expensive K-glass) gives high tensile strength, hydrolysis resistance and dimensional stability. The improvement of the polyamide tooth facing, with optimized fabric structures and advanced patented finishes, increases the resistance to wear and provides optimum adhesion to the belt body, and reduces the friction between the fabric and pulley surface.

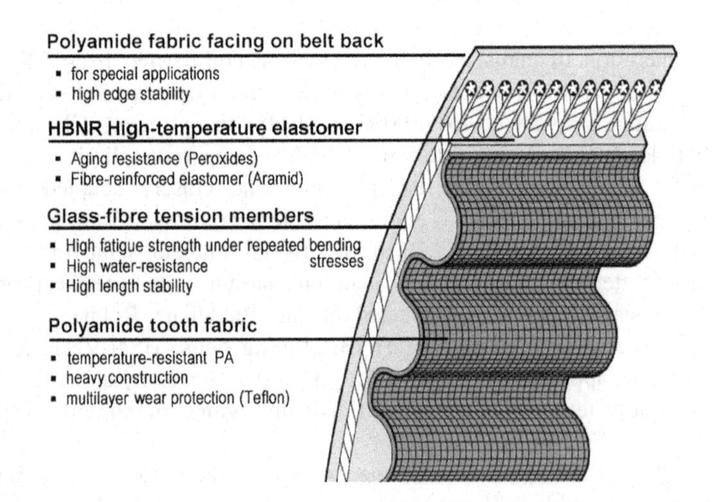

Fig. 3.42 Timing belt construction

Tooth profile development is still on-going and while the first belts used simple trapezoidal teeth, almost all of the current automotive timing belts are either the curvilinear HTD belts (High Torque Drive), or alternatively, the involute STD belt (Super Torque Drive, see Fig. 3.42).

Further improvements in timing belt materials allowed the use of automatic belt tensioning idlers to ensure a constant belt tension either by spring loading or by hydraulic pressure. Such systems minimize dynamic power peaks and belt span vibrations, which like temperature, were the main contributors to belt wear and tear. Tensioning idlers, manufactured by companies such as INA Schaeffler KG [52], match the temperature requirements by including heat-stabilized lubricants for the ball bearings, thereby achieving a crucial lifespan increase. High temperature resistant elastomers such as FPM also ensure that the bearing seals are flexible to the end of their life and thus retain the bearing lubricant. Additional seals are used, such as a contact seal at the rear of the tensioner with an additional shield, to keep contaminants away from the plain bearing (see Fig. 3.43).

Belt development testing was initially limited to test rigs with unrealistic constant torque levels but drives are now realistically tested with alternating torques and the transmission dynamics analyzed on a live engine by means of force and vibration sensors. Manufacturers now use a variety of durability test rigs for the development and quality assurance of the belts.

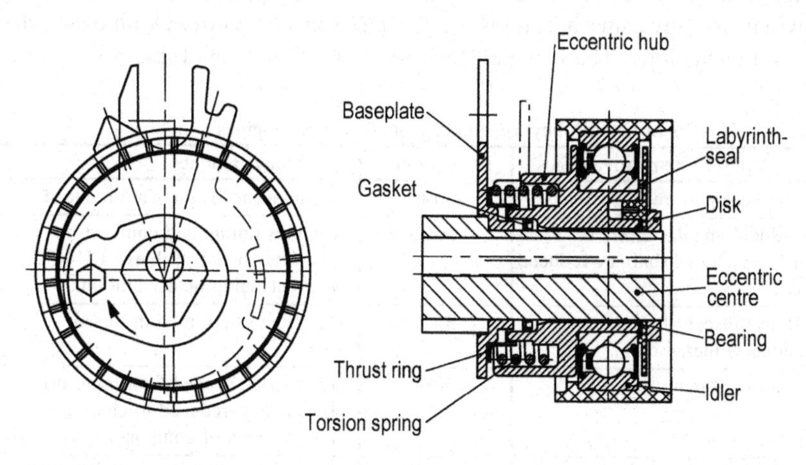

Fig. 3.43 Mechanical timing belt tensioner (Source: INA Schaeffler KG [52])

In product life testing, the belt drive is investigated and checked for geometry, forces and materials compounding. For the best possible integration of the individual components with each other, the development of the camshaft control system is increasingly the responsibility of system developers. Today, vehicle validation tests are offered with different driving profiles covering highway, rural and city driving, combined with hot and cold climate testing under extreme climatic conditions, to ensure the required engine lifespan. All this taken together means, that since early 2000, it has been able to be proven that timing belt drives would last the motor lifespan. This is still valid for all the ongoing development projects and improvements today.

3.11.3 Oval Pulley Vibration Reduction Technology

Oval crankshaft pulleys for 4-cylinder internal combustion engines are one of the best inventions since the timing belt. With oval pulley technology, applied to a 4-cylinder camshaft drive, the belt load capability increases by up to 30% and angular rotational errors are reduced by up to 50%.

The original oval pulley vibration reduction solution was honoured in September 2004 with an Innovation Award from the Association of German Automotive Industry Suppliers. The Litens Automotive Group of Canada and ContiTech Power Transmission Systems of Germany [15] received this prestigious award for the oval pulley system on the Audi 2.0 L TFSI engine, a 200-horsepower 4-cylinder engine.

Although the oval pulley vibration reduction system for camshaft drives is still relatively new, it has already been implemented in the shortest possible development time as a solution in various automotive applications. The dynamic behaviour of many camshaft drives can be significantly improved with oval pulley vibration technology. The benefits are listed in the following Table 3.1.

Table 3.1 Benefits of Oval Pulley Technology on 4-cylinder engines

Oval Pulley System Advantages	Improvement	Customer Benefits
Power peaks are reduced	30%	Component loadings are reduced
Rotational angular errors of camshaft to crankshaft are reduced	50%	Enables timing and combustion characteristics to achieve EU4/EU5. Prevents additional vibration
System lifespan (belts and tensioners) increased	30%	Lifetime safety margin raised
System packaging reduced	30%	Less space needed for timing drive. Belt widths reduced to chain drive levels. Mass of components reduced
Noise reduction	3–5 dB	Comfort levels increased

Oval pulley technology basically consists of the camshaft, the valve train, the belt spring effect and excitation and the oval, unevenly rotating, crankshaft pulley. The inertia of all parts acts on the dynamic behaviour of the powertrain. The ovality of the pulley momentarily raises and lowers the belt speed and this causes an acceleration and deceleration of the camshaft drive system and thus a different torque curve can be superimposed on the system.

With optimal phasing, the oval pulley can be used to reduce the vibrational loads on the belt. Figure 3.44 shows the operating principle of the oval pulley system.

A graph of the camshaft drive torque of a 4-cylinder engine can be represented by a simplified model of a sine wave. High positive camshaft torque occurs when the valves are opened against the force of both the valve springs and the gas pressure in the combustion chamber. If the cam has passed its kinematic dead centre then the torque load is reversed and from this moment the camshaft is driven by the valve spring forces.

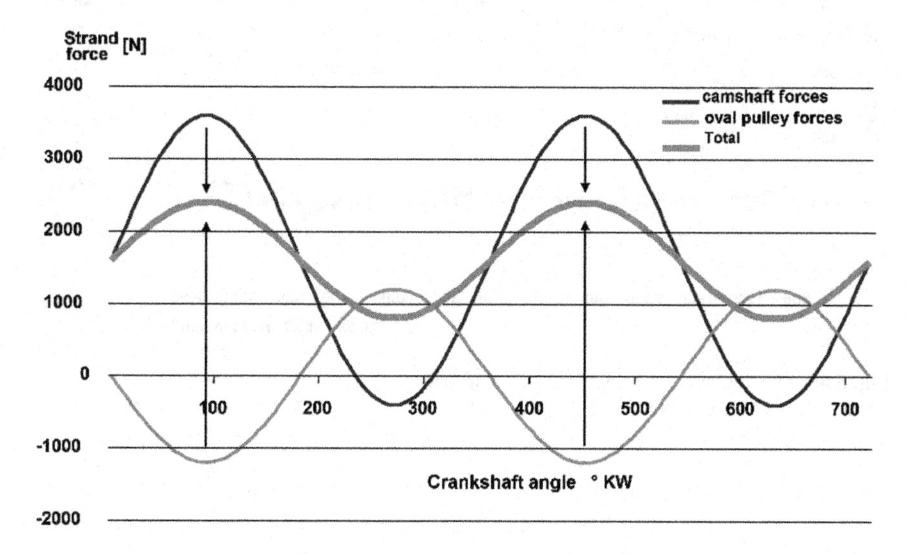

Fig. 3.44 Principle of the Oval Pulley Vibration Reduction System

The oval pulley vibration reduction system opposes the dynamic forces acting on the loaded belt span. The accelerations and decelerations caused by these forces contribute their torque at specific phase relationships to the drive components.

When the belt runs at the oval pulley minimum effective diameter then the belt speed is slightly delayed. If the belt runs at the oval pulley maximum effective diameter then the belt speed is slightly increased.

The optimally designed oval pulley vibration reduction system reduces peak forces and avoids the unfavourable zero axis crossings (see Fig. 3.44) which can form in camshaft drive.

With an investment in affordable oval crankshaft pulleys, some engines would not find it necessary to use special and costly camshaft vibration dampers.

As a practical example, the associated reduction in belt forces and the improvement of the synchronisation characteristics are shown with diagrams in Figs. 3.45 and 3.46. These are the belt span load measurements of a 4-cylinder petrol engine.

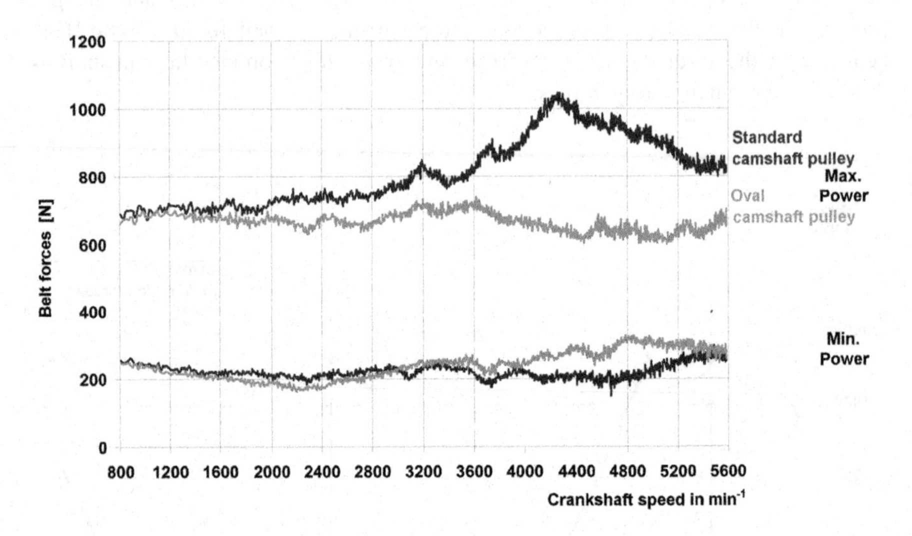

Fig. 3.45 Oval pulley technology reduces belt loads

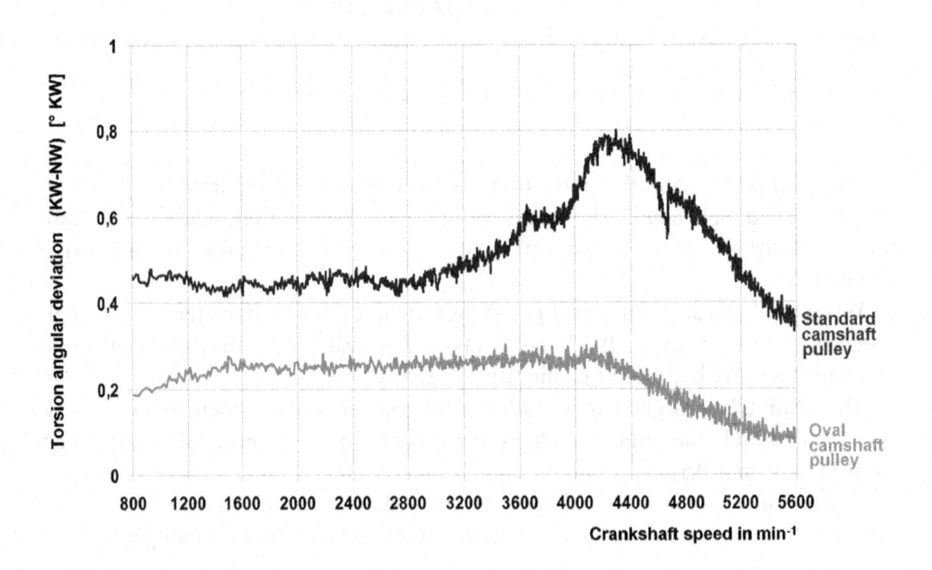

Fig. 3.46 Oval pulley technology reduces belt loads

3.11.4 Current Customer Requirements for Timing Belt Drives

In the following Table 3.2 customer requirements for automotive timing belt drives are noted.

Table 3.2 Customer requirements for Automotive Timing Belts

Customer Requirement	Target	Solution
Cold resistance	−32 or −40°C	Cold resistant HNBR
Heat resistance	Ambient temp.: 150°C Peak temp.: 170°C	Peroxide cross-linked compound
Reduced belt stretch	<0.1% every 150,000 km	E-glass-fibre or K-glass-fibre
Extreme power peak capability	Max. 3,000 N tensile load	K-glass-fibre, HNBR + Aramid fibre loading, Teflon coated fabric
Packaging reduction	Timing belt width reduced by up to 30%	Oval pulley technology
Noise reduction	Noise reduction 3–5 dB	Fabric tooth facing with noise-damping properties
Replacement of chain drives	Noise reduction Friction reduction	Oil resistant elastomer, fabric and tension member

3.11.5 OIL RUNNER Timing Belts

Oil-lubricated timing belts that can drive oil pumps, balance shafts, and camshafts can be used to reduce frictional losses by up to 30%, compared to a chain drive, thus helping to reduce fuel consumption by 1–2% and lower emissions of carbon dioxide (CO_2) by 2–3 g/km.

The OIL RUNNER timing belt is an oil resistant belt composed of a HNBR elastomers, PA fabric and innovative high-strength glass or hybrid tension members. In conjunction with the oval pulley technology, belt widths can be as narrow as the chain drives they replace.

Engine oil will swell CR (Chloroprene Rubber), the standard belt elastomer material of the 1980s, however ContiTech uses heat and oil resistant HNBR polymer (Hydrogenated Nitrile Butadiene Rubber) in OIL RUNNER timing belts (see Figs. 3.47 and 3.48).

HNBR elastomers are suitable for use in ambient temperatures from −40 to +150°C (up to +170°C peak). According to the elastomer manufacturer's material specifications, HNBR can be classified as oil-resistant and is therefore ideally suited for engine bay applications. As a measure of the oil resistance, the elastomer remains at a constant volume or does not swell while in an oil test.

The tension members used in OIL RUNNER timing belts are special oil and hydrolysis-resistant glass-fibre or carbon-fibre cords. The fine filaments of glass-fibre can be affected by oil if they are not sufficiently protected first from water and

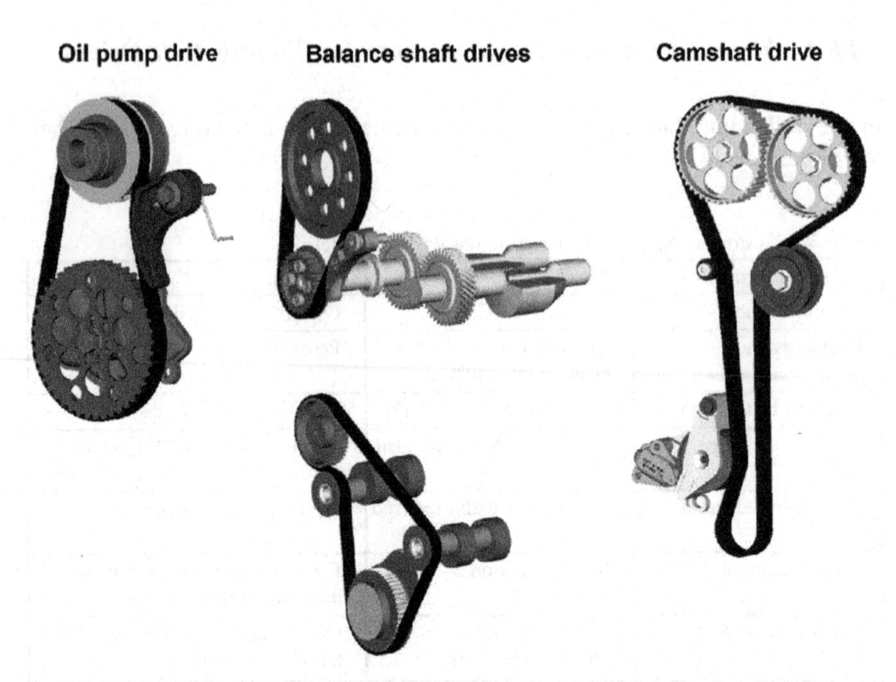

Fig. 3.47 Timing drives with OIL RUNNER timing belts substituting for chain

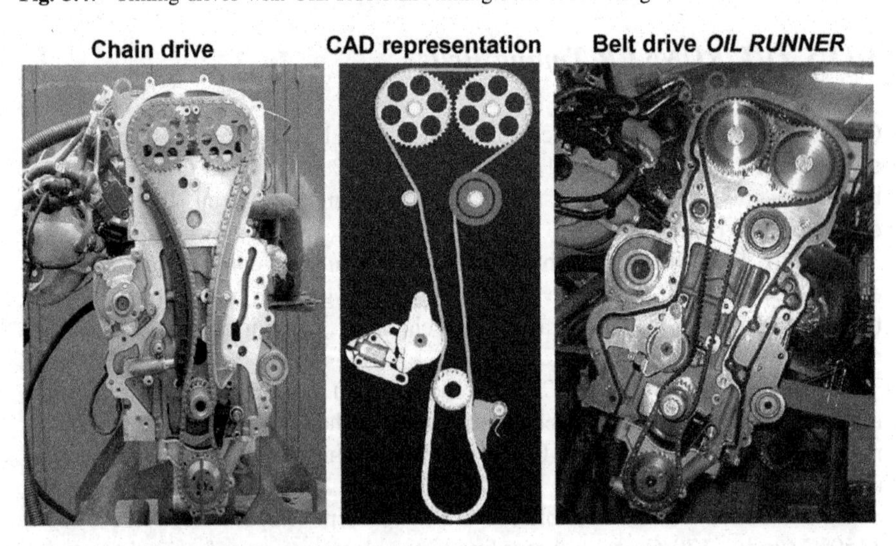

Fig. 3.48 Example of the redesign of a chain camshaft drive to timing belt camshaft drive

oil. For extreme applications, carbon-fibre cords or glass-carbon hybrid cords are being tested.

The fabric tooth facing for use in an oil-lubricated environment uses a specially prepared ContiTech PA 6.6 fabric. Even in an oil bath, the PTFE coating of the material has positive benefits, in particular, in permanent low-friction contact with a belt guides or tensioner pulleys. There can be problems with acid build-up in the oil, but this is not critical as long as the prescribed oil change intervals are complied with. To overcome this possible residual risk, existing fabric coatings and impregnations are being further optimized.

In contrast to chain drives, toothed belts are unaffected by engine oil contamination by combustion products. Especially in modern TDI direct-injection and common-rail engines where combustion products enter the engine oil and increasingly damage the chains used today by abrasive joint wear (see Fig. 3.48).

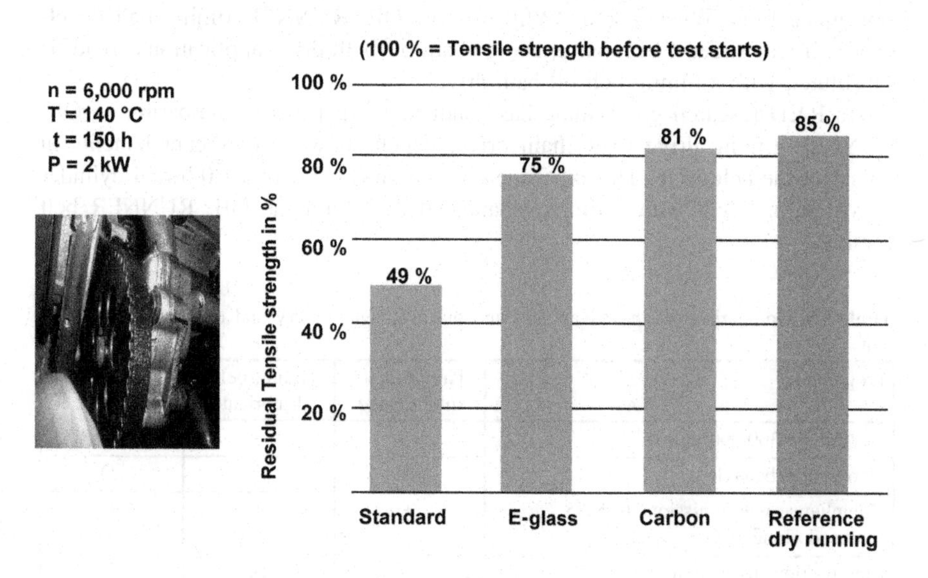

Fig. 3.49 Residual break load of OIL RUNNER belts after high temperature testing

Heavily polluted oil and oil quality altered by the "blow-by gases" is a particular challenge for materials development.

OIL RUNNER belts on ContiTech's own vehicle fleet have seen life spans of over 240,000 vehicle-kilometres, demonstrating their practicability, and showing the oil resistant belt exceeds an engine lifespan. First, the use in oil pumps and balance shafts drives is planned and later the belts will be modified for other applications to reduce CO_2 emissions and thus further protect the environment. With regard to life spans, there are hardly any differences now between belt and chain drives. Both systems are now designed with the same lifetime of 240,000 km validated to 300,000 km.

The main advantages of OIL RUNNER belts lies in their quiet operation and the frictional advantages of belt drives against chains where metal parts, i.e. the plates and rollers, continuously wear against the teeth of the sprockets.

When fabric-faced belt teeth mesh with the polymer-coated steel teeth of the pulleys then the noise due to tooth meshing, caused by the displacement of air as well as by the impact of the belt on the teeth on the pulleys, is significantly reduced. The comparisons in Table 3.3 show that in terms of elongation and timing accuracy over the engine lifetime there is an obvious advantage in the use of belts.

Belt stretch in vehicles over 100,000 km is around 0.1%, on the other hand chain stretches approximately 0.5% or more for the same mileage.

It should be also be mentioned that OIL RUNNER belts, if they are driven by toothed pulleys, can be used both in lubricated and dry operation. OIL RUNNER belts must be compulsory lubricated only when used in combination with back or side guide rails. When asked "Why was the OIL RUNNER timing belt developed", it was stated "Because existing chain oil-bath drive applications could be substituted with a timing belt oil-bath drive".

An R&D research programme has conducted experiments comparing an OIL RUNNER timing drive to a chain drive, which showed significant benefits in favour of the belt drive. The experiment was performed with a 1.0 litre 4-cylinder petrol engine, first with a timing chain and then with the OIL RUNNER belt version.

Table 3.3 Comparison of the advantages and disadvantages of dry and oil-lubricated timing belts

Properties	Timing belt unlubricated	Timing belt oil-lubricated	Chain oil-lubricated
Durability (300,000 km)	++	++	++
Good noise behaviour	+/O	++	O
Compliance with emissions to EU5 (reduced elongation)	+	+	O
Low friction, fuel saving	++	++	O
Warning of impending failure (Sensor)	– Sensor?	– Sensor?	+ Noise
DI diesel applications–oil contamination from blow-by gasses & particulates	+	+	O/– Chain pins/ Sprocket teeth.
Axial packaging requirement	O	O	+
Reduced system cost	PA Cover (+) VVT (–)	Metal Cover – VVT ÖL (+)	Metal Cover – VVT ÖL (+)

The experiments used a tensioning system constructed of a hydraulically actuated lever with tension rollers. In order to be able to use a timing belt that could be run in the axial space of the chain, an oval pulley vibration reduction system was used in parallel to the OIL RUNNER belt to reduce the crankshaft forces on the belt. Oval pulleys can, as mentioned above, reduce the belt loads by approximately 20–30% and synchronization errors by about 50%.

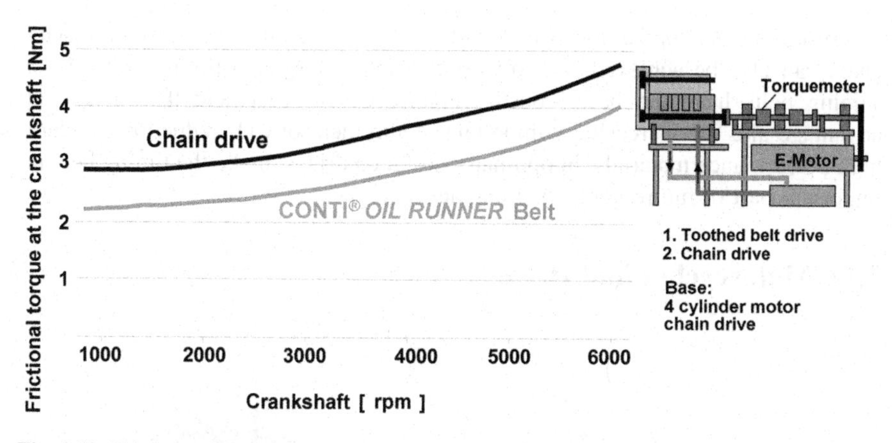

Fig. 3.50 Friction reduction comparison between the chain and belt drives

The frictional advantage of the timing belt is around 1 Nm. This corresponds to a frictional advantage of approximately 30% at idling speed and approximately 20% at a nominal running speed. For the entire vehicle this results in a fuel savings in the range between 1 and 2%, i.e. about 0.1–0.2 l / 100 km.

A major advantage of chain systems lies in their narrowness. Chains generally require an installation width of approximately 14 mm and, if including the guide rails, no more than 16 mm. Timing belts generally require 20–25 mm of installation width. However, through the use of oval pulleys, the crankshaft timing belt width can be reduced to 16–18 mm.

Chain has the advantage over timing belts that in a case of premature wear or partial failure, such as chain stretch, it emits such high levels of running noise that the vehicle will be returned to a garage to prevent impending engine failure. Chains could be considered to have almost a built-in sound sensor that warns of catastrophic failure. Timing belts, however, stretch so little that even with premature wear, no increased levels of running noise can be distinguished. To avoid unexpected timing belt failures it is necessary to strictly adhere to the recommendations of lifetime service intervals. Timing belts should be installed with a separate failure sensor able to diagnose the low levels of belt stretch over the drive lifetime.

In a cost comparison, simple roller chains in combination with tension guides show a favourable price advantage. Modern silent chains and chain sprockets are about the same price as modern belt solutions. However, the simple guide/tension systems compared with ball-bearing tensioners show a slight cost advantage in favour of chain systems.

When comparing costs for camshaft drives, the complete system costs must be used. For example, substitution of variable valve timing (VVT), which runs submerged in an oil bath, could show a significant cost advantage (see Fig. 3.50).

As a contrast, the simple plastic timing belt cover of a dry-running belt drive is considerably cheaper than the costly metal cover of chain a systems. In simple timing belt-driven oil pump drives, the OIL RUNNER solution has a favourable cost advantage because a tensioner can be dispensed with altogether.

To summarize, applications with small axial space requirements currently using chain can also be achieved with oval pulley technology and timing belt drives. Timing belts have the best acoustic running performance and their frictional advantage aids the increasingly important requirement of CO_2 reduction. Modern high-performance materials, in optimally designed drives, make the timing belt an important part of future generation engines.

3.12 Motorcycle Final Drive

Fig. 3.51 BMW Motorcycle, Model F 800 S/ST [5]

The unique advantage of the BMW Model F800 S/ST is that the swinging arm is mounted in conjunction with the belt final drive (Fig. 3.51). In comparison to a chain drive there is no slack in the drive. While pulling away, and under varying load conditions, the powertrain works positively and completely smoothly. It is distinguished by a high level of refinement and long life. For the rider, this form of drive is vibration-free, clean and virtually maintenance-free. The fitting and removal of the wheel and the belt tensioning of are quickly and easily done.

2-cylinder 4-stroke engine with engine capacity of 798 cc
Motor power 62.5 kW at 5,800 rpm
ContiTech timing belt type "Extreme" 1903-M11-34 (length–pitch–width) [15]
Small pulley $z_1 = 34$
Large pulley $z_2 = 80$

The large pulley is made in two parts from stainless steel. It is bolted to the rear wheel and incorporates a cush-drive. The cush-drive, using a four-block elastomeric system, distributes the driving torque evenly between the tyre and road and also absorbs shocks from uneven road surfaces.

3.13 Pulley Shaft and Hub Locking Assemblies

A timing belt drive needs other components to complete its expected function. Shaft-hub fasteners will be required to secure the pulleys to the shafts. These are usually standard components that transfer torque from the shaft to a rotating hub (timing pulley) or the other way around. They will also resist axial and transverse forces. Locking assemblies of this nature are divided by their mode of operation into

- *force-based* fastenings whose function is based on elastic–plastic press fits. These include the conical locking assemblies.
- *form-based* fastenings that have a modified geometry between shaft and hub such as a polygon or splines or which are linked by adding driver elements such as keys or dowel pins.
- a combination of *force* and *form-based* fastenings in, for example, safety modules (equipped in lifts).
- *materials-based* fastenings such as in welding, soldering or gluing, mostly brought about by the addition of extra materials.

The designer is able to choose from the wide range of possible alternative solutions which meets the best requirements of the task. The designer's focus is technical suitability along with the cost of installation and procurement. He has to consider, amongst other properties, torque transmission, rotational accuracy, spatial compatibility and service life. There are many manufacturers and technical specialists in the shaft-hub connections industry with a wide range of available solutions.

The following solutions for torque transmission between shaft and hub are presented with their associated technical standards. For existing products, the information and relevant bibliography can be found on the selected manufacturer's website. This listing of components, standards and manufacturers is not exhaustive, see Figs. 3.52–3.63.

The conical locking assemblies available in the marketplace generally have many detail variations. Their construction results in different dimensions for *length* to *width* and this ratio, with the absolute diameter of the shaft, significantly affect the transmissible torque, depending upon type, whether axial location is necessary or the ability to re-position at installation. At a cone angle of $<1:5$ the shaft-hub connections of this type are self-locking and in these cases extraction systems must be used for disassembly. It is common to all designs of these components that the axially arranged screws act on the cone to give the force required to achieve a radially acting secure fit. These are exclusively imposed internal stresses, which are free from external forces, giving contact pressures to the desired level on the

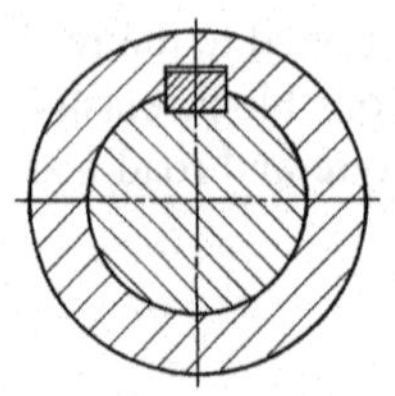

Fig. 3.52 Keyway (DIN 6885). Interlocking power transfer. Not appropriate for transmission of high alternating torques

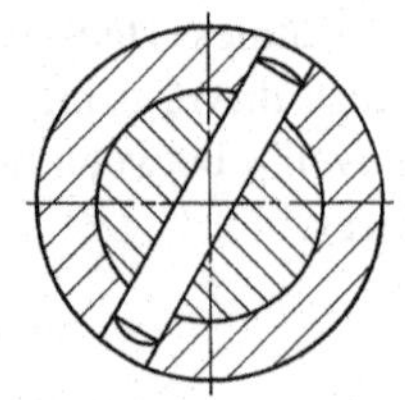

Fig. 3.53 Cross pin (DIN EN ISO 8734), Grooved pin (DIN EN ISO 8740), Roll pin (DIN 1481 und DIN EN ISO 13337 or Conical pin (DIN EN 22339)

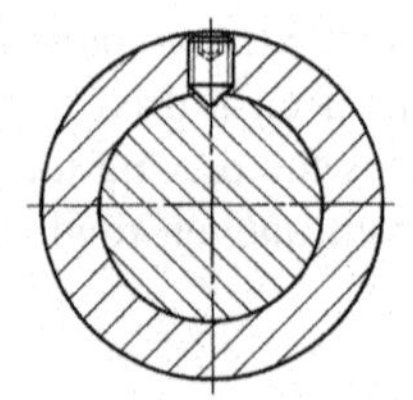

Fig. 3.54 Grub-screw with conical point (DIN 914). Only suitable for low power drives. The radially acting force of the screw locks the shaft in the centre

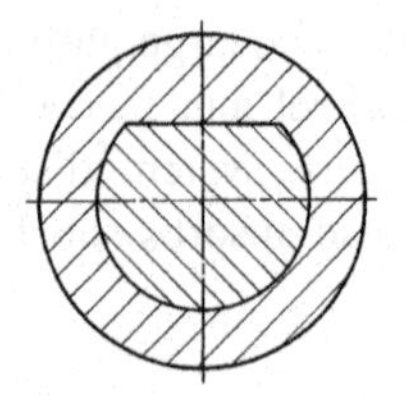

Fig. 3.55 Flats on shaft and hub. Suitable for timing pulleys manufactured by forming processes. See example on plastic timing pulleys in Chapter 3.8

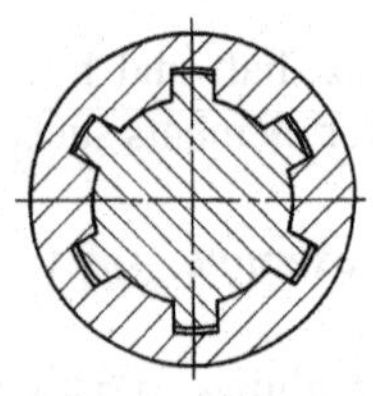

Fig. 3.56 Splined shaft with straight edges (DIN ISO 14). Centred by either internal or flank. Highly resilient. Also standardised with an axial sliding fit

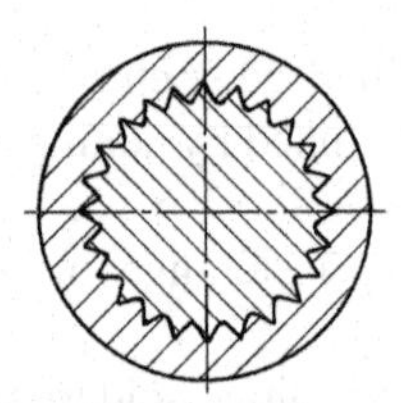

Fig. 3.57 Splined serrated shafts (DIN 5481) and involute profiles (DIN 5480). Highly resilient. Outside gear formed by hobbing and inside gear is broached. Sliding as well as interference fit possible

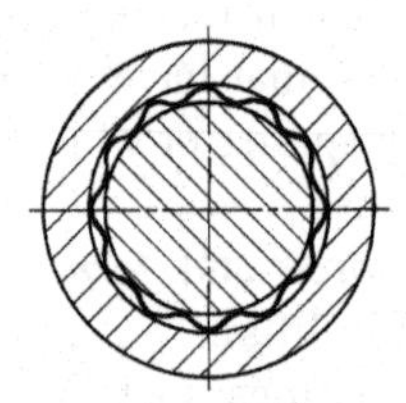

Fig. 3.58 Tolerance ring or shaft spring fastener. Frictional connection, which is based on surface pressure by radially acting spring stiffness. Simple and economic installation. Limited torque transmission. Bosch-Rexroth [7]

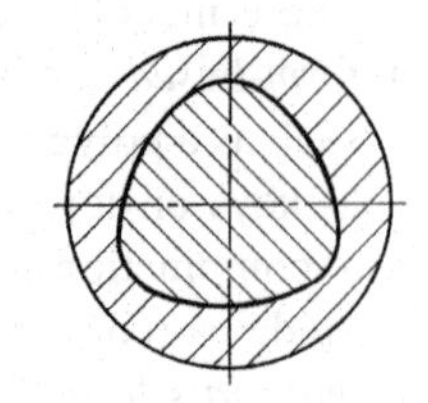

Fig. 3.59 Polygonal profiles. Usually P3G and P4C-profiles to DIN 32771 and DIN 32712. Manufactured to tolerance class 6. Good concentricity. High torque transfer without stress concentration. Polygon & Rund Schleiftechnik [98]

active surfaces. The suitability of "self-centring" is shown separately in the catalogues, see Figs. 3.46–3.74.

Figure 3.63 also relates to press fits, where fastening takes place by the longitudinal pressing together of the two parts at ambient temperature. The press fit is usually achieved by a pair of tolerances, e.g. H7/r6, where the manufacture of the bore on the outer component is to H7 and the inner component to an r6 tolerance, according to DIN ISO 268.

Calculations of frictional press fits can be referenced in the bibliography to the works of Franz G. Kollmann [61]. The actual use of conical shaft-hub fasteners

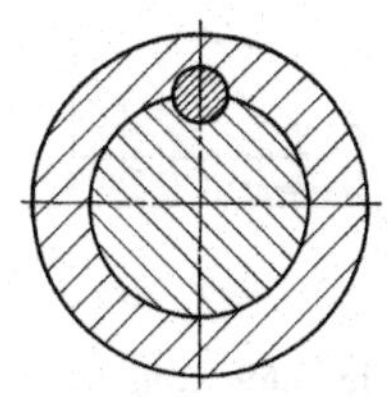

Fig. 3.60 Round key fitted axially using a cylindrical pin (DIN EN ISO 8734), Grooved dowel pin (DIN EN ISO 8740), Roll pin (DIN 1481 und DIN EN ISO 13337) Or Conical pin (DIN EN 22339)

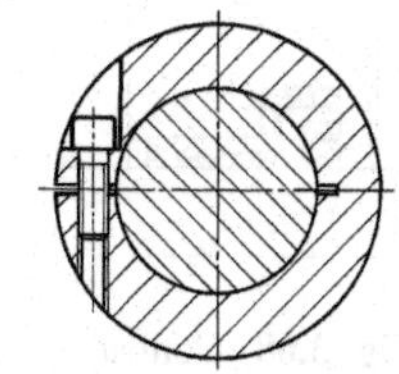

Fig. 3.61 Locking collar on slit or split hub. Friction grip for adjustment operations. Low torque

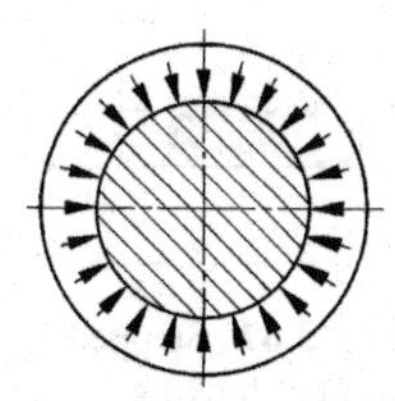

Fig. 3.62 Cylindrical press fit, effected by the heating of outer part and/or cooling of the inner part. Design and calculation according to DIN 7190 (withdrawn standard)

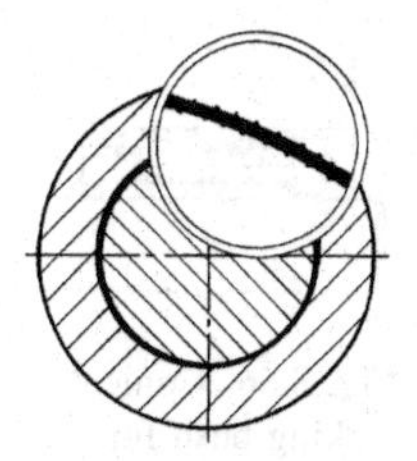

Fig. 3.63 Shaft-hub connection by bonding. Loctite 638 cures between close fitting metal surfaces. Surface roughness increases shear strength. Henkel [46]

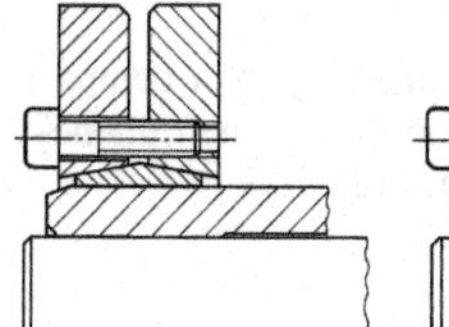

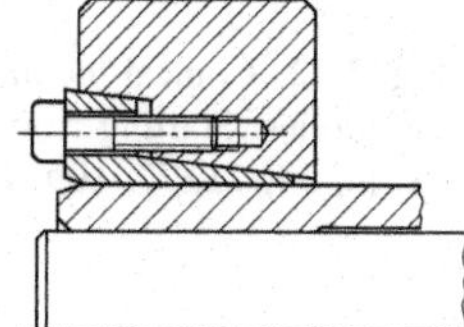

Fig. 3.64 Externally clamping locking bush. Self-centring. Suitable as a hollow shaft mounting or with timing pulley designed with a narrow hub. The locking element itself is not involved in the friction pair. Transmission of radial and axial forces on the cylindrical surface of the shaft. High torque. [4, 14, 73, 77, 109, 112, 117]

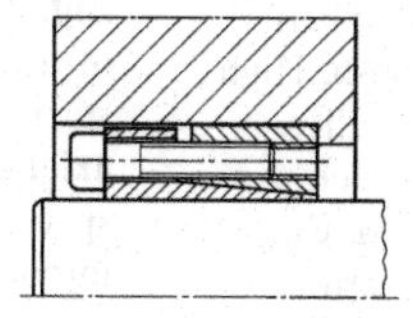

Fig. 3.65 Universal conical locking bush. Self-centring. Transmission of axial and radial forces by friction between the shaft and the hub through the locking element. High torque. [4, 14, 73, 77, 109, 112, 117]

will always use the same information. Manufacturer's catalogues contain tables listing the geometric dimensions, transmittable torque, axial force transmitted, contact pressure between the connected parts, number and dimensions of the screws and the required torque. Thus leaving the user only a simple comparison between the required and theoretical torques offered. Where appropriate, the torque transfer capability can be increased by using two or more fasteners in tandem. During design, the available space and mounting conditions should also be considered.

It should be noted whether the individual fasteners are self-centring or whether they require additional centring bushes in the hub. The same applies to observing the tolerances and surface roughness of the mating surfaces of the parts to be

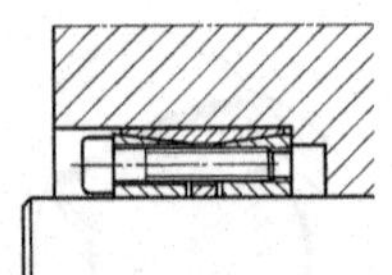

Fig. 3.66 Conical locking bush for maximum torque. Self-centring. No axial displacement during clamping. [4, 14, 73, 77, 109, 112, 117]

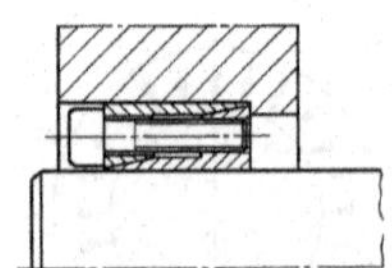

Fig. 3.67 Stepped conical locking bush. Self-centring. Assembly with two conical active surfaces. High torque. [4, 14, 73, 77, 109, 112, 117]

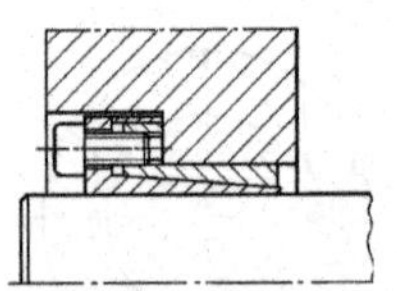

Fig. 3.68 Conical locking bush for reduced mounting space. Self-centring. Medium torques. [4, 14, 73, 77, 109, 112, 117]

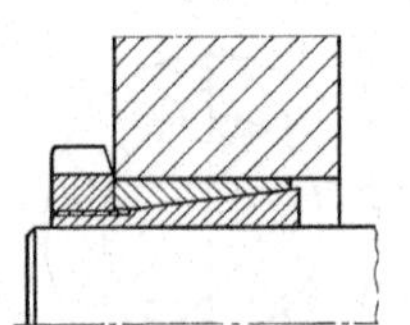

Fig. 3.69 Ring nut locking bush. Self-centring. Clamping force is applied by hexagonal nut. For medium torques. Easy installation and removal. [73, 101, 117]

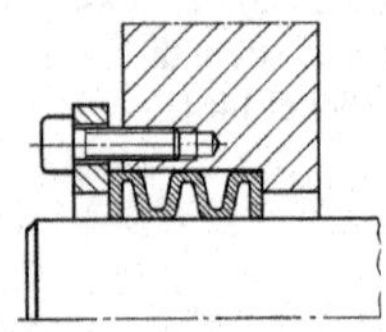

Fig. 3.70 Shaft locking bush. High-centring ability. Made in tolerance classes 5–6. Concentricity error <0.01 mm. For medium torques. Spieht [107]

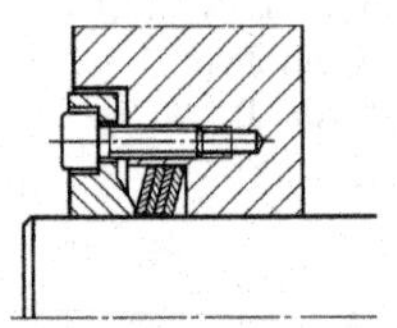

Fig. 3.71 Star disks of hardened spring steel. For restricted axial installation space. Medium torque. Good repeatability. Single or up to 25 pieces in an assembly. Ringspann [101]

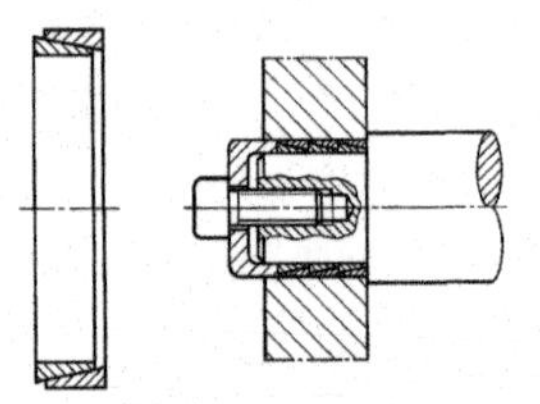

Fig. 3.72 Conical locking rings. For compact assemblies. Starting from 6 mm shaft diameter. [4, 14, 73, 77, 109, 112, 117]

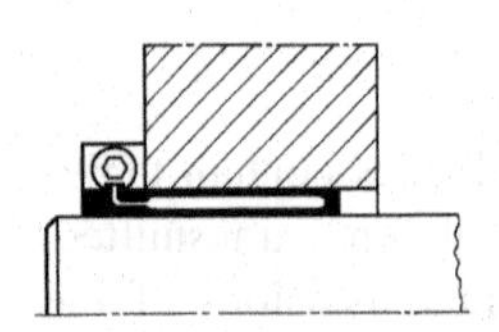

Fig. 3.73 Hydraulic locking bush. Run-out <0.01 mm. Easy on shafts. Good repeatability. [30]

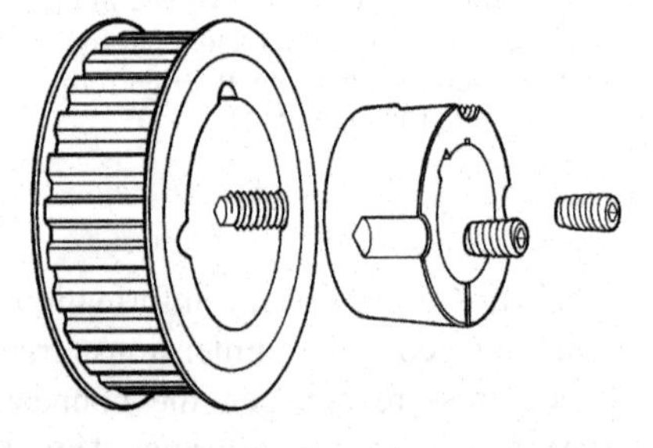

Fig. 3.74 Taper-lock bushes from Fenner [32]. Fastening system with the largest worldwide distribution. Concentricity <0.1 mm. Range includes both inch and metric dimensions. Suitable for shaft diameters from 12 to 120 mm. The matching taper is machined into the standard pulley. Further possibilities are available with the use of intermediate sleeves, bolt-on and weld-on hubs. Taper-lock bushes are also used for mounting v-pulleys, chain sprockets and flat belt pulleys. These products are generally available from technical distributors

connected. The information provided by the manufacturer must be strictly observed to ensure proper operation of the relevant shaft-hub connection. Since product ranges of this type are also subject to technical changes it is always necessary to check the current performance and quality data from the manufacturer.

Materials-based fastening uses an anaerobic (absence of oxygen) curable synthetic resin adhesive as shown in Fig. 3.63. These adhesives are capable of completely filling the assembly gap, including the surface roughness. According to the manufacturer [46] the optimal shear strength can be reached between the mating parts with a surface roughness of Rt 10–20 μm. Disassembly is only possible by heating.

Disassembly is generally not possible when using soldering or welding as a fastening method. When soldering, the solder melt temperature should always be lower than that of the parts to be fastened together. Welding is usually open arc welding with the use of a fusible electrode. Friction welding is suitable for rotationally symmetric components and is particularly used in mass production. Other special procedures with low heat input are plasma and laser welding, which are carried out in high vacuum chambers.

Summary

The choice of shaft-hub fasteners is an important and very common sub-task in drive designs. It is appropriate at the rough design stage, to first draw up a shortlist, which can be fine-tuned after review. If the timing belt drive works under frequently changing operating conditions, then a force-based press fit is clearly the most appropriate solution.

3.14 Drive Test Rig

Dynamic full-load tests on timing belts are possible on drive test rigs employing an energy cycle as described by Steinbrück [114] in the 1922 patent. The basic principle is the circulation of power in a closed circuit. Only the power losses are fed by the drive as an energy supply for the test transmission. The idle load stresses the test specimen in real time simulation. This system can be tensioned by applying torsion mechanically, electrically or hydraulically with the tensioning motor which has no test load pass-through. On equipment of this kind, drive components such as shafts, joints, couplings and belt drives up to complete rotating transmissions are tested over the short and long-term under continuous stress. The test specimen is part of the power circuit.

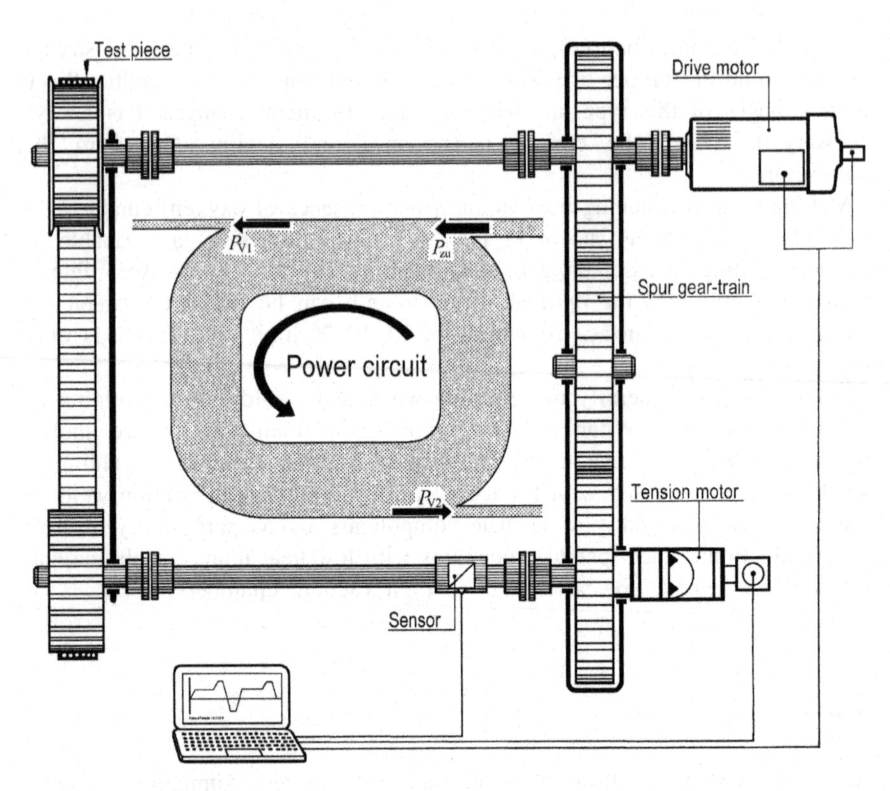

Fig. 3.75 Tension dynamometer. P_{zu} input power, P_{V1} test piece power loss, P_{V2} test rig component power loss

The test rig power circuit shown in Fig. 3.75 consists of two matched drive-lines, facing each other, coupled with a spur gearbox on one side and with a *single* timing belt on the other side (the test specimen). The coupling of the power circuit is also possible using two equal and mirror-image timing belts. With a belt test rig it is important that, due to the length range of the test specimens, the test rig centre distance is adjustable within this range. Furthermore, the sum of all the elasticities of test components must be known, so that the part elongation of the belt can be determined.

The core component of the test rig is the tensioning motor. To generate a tension angle or torsional moment a planetary or countershaft gearbox can be used. It can also be axially adjustable with the use of helical gears. According to [40] a high speed hydraulic rotary cylinder using variable-vane geometry and non-contact rotary sealing is particularly suitable. Tensioning motors of this type are available in several sizes and can be installed outside or within the drive train.

With this equipment, the drive can achieve the test speed without load. By activating the tension motor a control angle and torsional moment is introduced into the power circuit. Compared with brake dynamometers, energy savings of about 90% can be achieved. The separate regulation of torque and number of revolution results in favourable conditions for economical control systems allowing the simulation of any power spectrum over the entire speed range. Customized sensors with connectors for a PC data-logger allow data acquisition and program control.

3.15 Reactive Power

In electric, single or multi-phase equipment there is an additional energy in the circuit that does not contribute to the active power generated by the electric motor drive. This is called reactive power and it is caused by inductive loads. Extensive research and numerous publications are related to reactive power in electrical systems. On the other hand, one rarely finds technical reports in publications of mechanical reactive power and the analysis of its causes. Here we consider reactive power in the context of timing belt drives in more detail. What mechanical and electrical reactive power has in common is that they usually generate undesirable side loads in the drive train and affect the efficiency negatively. The reactive power is always a part of higher internal power that does not flow outward and it should be avoided.

The only possible use of reactive power is related to applications in tension test rigs as is described in the previous chapter. Through the specific torsions of two parallel transmissions one produces reactive powers in the power circuit in order to test drive components or complete epicyclic transmissions.

Drive Loading

A common cause of accidentally producing reactive power is by the "loading" of multiple drive systems. A drive system that is first divided by a power split into parallel sub-drives and then reassembled into a common output can develop reactive power, see Fig. 3.76. The parallel drive system may use different or the same types of gears, for example, or two adjacent belt drives. If the sub-drives work at a torsion angle to one another and have an unequal tooth alignment then a tension is generated. While the flanks of one strand of the timing belt transmission transmits power positively those on the other strand may lie on the opposite flanks, see Fig. 3.77. This is caused by a negative power flow, the reactive power. This can stress the drive components concerned with a multiple of the nominal power. It is also possible that a belt will operate in idle mode without flank contact and in this case, the total power is transmitted by the remaining belt.

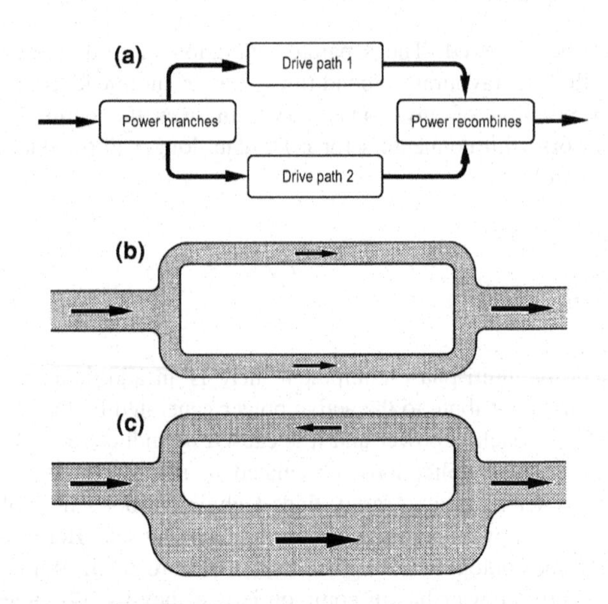

Fig. 3.76 Power split. **a** Drive system with parallel drive strands. **b** Even load transfer. **c** Uneven load transfer with reactive power

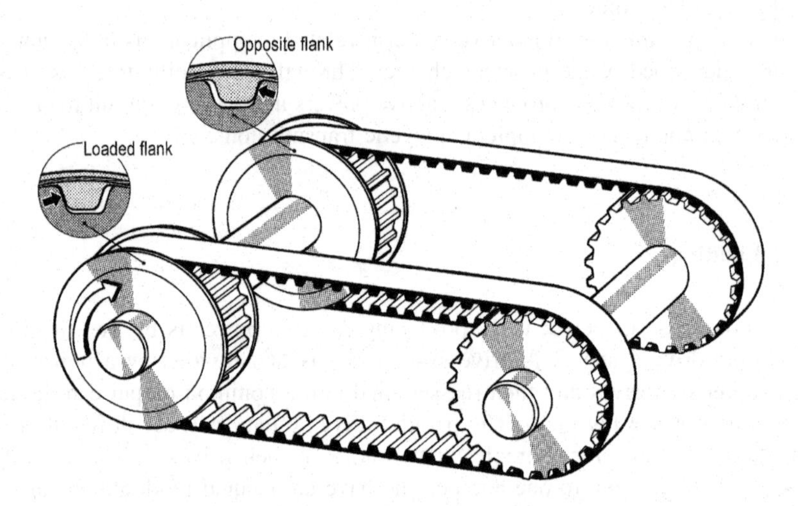

Fig. 3.77 Two belt parallel drive

If you want to use two or multi-strand belt drives, where the parallel single belts share the loads equally, then all the pulleys must have their tooth positions exactly aligned to each other. With more complex assemblies, the subsequent measurement of uniform power distribution is not possible. With increasing shaft lengths, the drive torque influences the tooth positions by twist. Further causes are tolerances, in both the pulleys and the belt, allowing changes to the force values through the flank contact. In a two-strand belt drive, a load ratio of 1/3–2/3 can occur together.

Geometric Redundancy

A further reason that reactive power develops is forces due to geometric redundancy. Such an example is shown in Chapter 3.1, Fig. 3.10. If the centre pulley runs-out by as little as 0.1 mm from its symmetric centre then different tension forces are formed in all the belt spans. A design with an odd number of teeth in the centre pulley and an even number of teeth on the belt should not be used (and vice versa). In a symmetric drive layout, it can already be seen during assembly that the belt teeth don't mesh well with the pulley tooth gaps, giving a build-up of significant distortion forces. This example shows there is a geometric redundancy if you try to make a belt engage two or more times in the same pulley. Belt designs of this kind should be avoided. As a necessary counter-measure a sectional correction of the centre pulley with an adjustment range of at least half a tooth pitch would need to be provided.

Feedback in Grinding Mills

Mill drives always lead to reactive power flows and Leo Hopf analyzed this in his work "Power Flow in Cylinder Mills" [49]. Main mill drives have used many different types of transmission such as v-belts, spur gears, chains (rare) and timing belts in this application (see Fig. 3.78). Consider the power flow during grinding between two rolls, the faster roll does the grain grinding while the

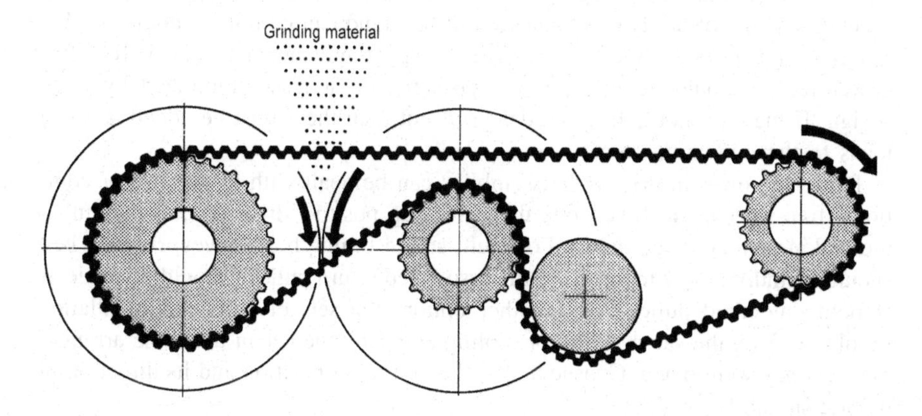

Fig. 3.78 Mill drive

slower counter-roll holds up the grain (grinding material). The rollers do not touch and grain mills are normally unloaded when run-up to speed. In addition, the grinding material builds-up due to the cutting operation (the grinding) and counter-friction forces in the slow roll and as a result the counter-roll tries to match the faster rate of the grinding roll. Thus, the belt acts as a brake on the slower roll to obtain and maintain the desired difference in grinding speed. The generated braking power corresponds to a reactive power as an additional load on the drive. Through this feedback the internal power generated roughly doubles the total load in the drive belt.

Recognition and Avoidance of Reactive Power

In *series* gear trains the output motion of an upstream gear is used to transfer the motion to the following gear. Generally, large ratios are achieved with multiple stages. With *parallel* gear trains the output motion is divided, at a branch, into parallel partial drives. The movements of which, can be used individually or combined again into a common output. The combination possibilities are endless. A functionally related mechatronic unit is usually made of several identical and/or different types of drive. The emergence of possible reactive power is not obvious at first glance as the causes are often geometrically induced forces and operational feedback. There are always additional internal forces not taken into account.

Force analysis offers security of function as well as the basis for cost-effective dimensioning in the design stage. One differentiates between the actual outside forces and the internal forces of inertia produced by motion changes (for example, acceleration). Both kinds of force can be treated in the same way, because they can be referred to as a *quasi-stationary condition*. What is looked for, for example, is the maximum force that a transmission component experiences or the torque in a certain part of the drive train which generates the given acceleration condition. How-to analyses of this kind can be found, amongst other things in Johannes Volmer, "Gear Technology Fundamentals" [124]. Reactive power forces should be ruled out, if possible, or at least minimized by good design. If they are not able to be designed out then they must be added as extra loads to the remaining forces.

Reactive power in drive sub-assemblies can be dealt with by the use of additional free-wheels. However, one then gives up positive drive and, in particular, the braking effect of the motor. For multi-strand timing belt conveyors, each belt should be individually mounted or the associated return pulleys should be made as smooth untoothed pulleys. As another option, the sets of belts for installation should be from the same sleeve or tooling and run together in the same arrangement as they were made. Consequently, the belt tooth position and its direction of motion should be marked.

3.16 High Accuracy Timing Belt Drives

There is no secret to the implementation of high accuracy timing belt drives. While they are not easy to achieve, they can be produced fairly simply by following best practice timing belt drive design and by careful component selection and matching. High accuracy timing belt drives have all the advantages inherent in the product such as high speed, low mass, clean, maintenance-free and above all, cost-effective against other accurate drive systems. Due to the elongation properties of timing belts under load they are not normally suitable for high accuracy drives where large masses have to be accelerated/decelerated in short periods of time.

Design of a high accuracy system must be done holistically, e.g. as an integrated system and not a collection of parts, as every component in the drive structure can either add or subtract to the base accuracy and with good design we always wish to increase the possible accuracy and repeatability.

Machine Frame/Support Structure and Environment

As all drive components interact to give the final accuracy then the machine frame/support structure is one of the most overlooked parts as all effort is normally put into the pulley/belt design.

The machine frame/backplate/sub-assembly on which the belts and pulleys are mounted must be as rigid as possible. Designing a high stiffness belt drive is pointless if the supporting frame is able to flex as the drive comes under load. Such flexing alters the centre distance and thereby affects the pre-tension, so the structure that supports the belt drive must be as stiff as practicable. Two or three orders of magnitude greater than the belt stiffness would not be unreasonable.

The pulley shafts are connected to the machine structure and they are the next component to consider along with whatever bearing systems are being used. Pulley shafts and their bearing assemblies must be as rigid as possible to eliminate another point of flexure giving centre distance and pre-tension problems. Small diameter shafts and overhung pulley designs should not be chosen. Bearing selection is a lesser problem but good quality, low friction-stiction bearings are a benefit to the final accuracy and repeatability levels. Generally there is a small trade-off in the required properties between ball, cylindrical roller and needle roller bearings. As this is a timing belt handbook we leave the user to make their own choices.

Another relatively unconsidered element in designing for accuracy is the environment in which the drive runs. Factors that affect accuracy include contamination and temperature. If the drive is in a dusty, oily, or damp environment then there can be a build-up of matter in the pulley teeth, and as the pulleys will almost certainly be of Zero toothform, then this build-up will push the belt teeth up from the bottom of the pulley teeth affecting both belt pitch and pre-tension. Over a period of time such drives exhibit a drift in the accuracy figures and eventually fail. Temperature variations can also affect the drive if the support structure material does not match the tension member material. An aluminium support structure coupled with a drive with steel tension members would have a differential coefficient of expansion/contraction as the temperature changes and would thus affect the pre-tension. If the drive is a wide gantry system with two drives moving the gantry there should not be a temperature differential from side to side.

Drive Component Selection

It is obvious that all components in the actual drive should be of the highest quality as parts manufactured to wide tolerances have no place in accurate timing drives.

Pulleys should be manufactured to the highest specification possible for outside diameter, toothform and concentricity of bore to outside diameter. Zero toothform is the one chosen for high accuracy as it eliminates belt to pulley teeth backlash. Pulley manufacturers should be chosen for their ability to produce an accurate toothform as this will guarantee smooth meshing and accurate pitching of belt to pulley. As all the high power/stiffness belts available for this sort of application have pulleys of a non-involute toothform then the actual toothform changes slightly from one pulley size to the next, see Chapter 2.15. This means that pulley manufacturers using a restricted range of cutters to produce the timing pulleys will have much wider tolerances on the toothforms of their pulleys. Ideally each size of pulley for a high accuracy drive should be made with a specific cutter for that number of teeth and not a range cutter. In the real world this is obviously not cost-effective.

The pulley/shaft connection should also be considered carefully as another potential problem area. Many high accuracy drives use taper locking elements as the pulley/shaft connection for their self-centring properties, see Chapter 3.13.

Belt selection could be considered the most critical choice of the drive design. The ideal belt will have a high stiffness and large tooth section and this, almost automatically leads us to the AT/ATL family of belts. Perusal of manufacturer catalogues will allow a simple comparison of belt properties between brands and the stiffest, least compliant belts should be chosen. Some manufacturers have optimised their production methods for higher than normal accuracy and these belts could be prime candidates for selection. Such belts have achieved precision within the range of ±0.03 to ±0.05 mm/m of travel and also exhibit improved tooth to tooth pitch accuracy with a reduction from 0.7% on a typical timing belt to less than 0.2% on a 3000 mm length belt (see Fig. 3.79).

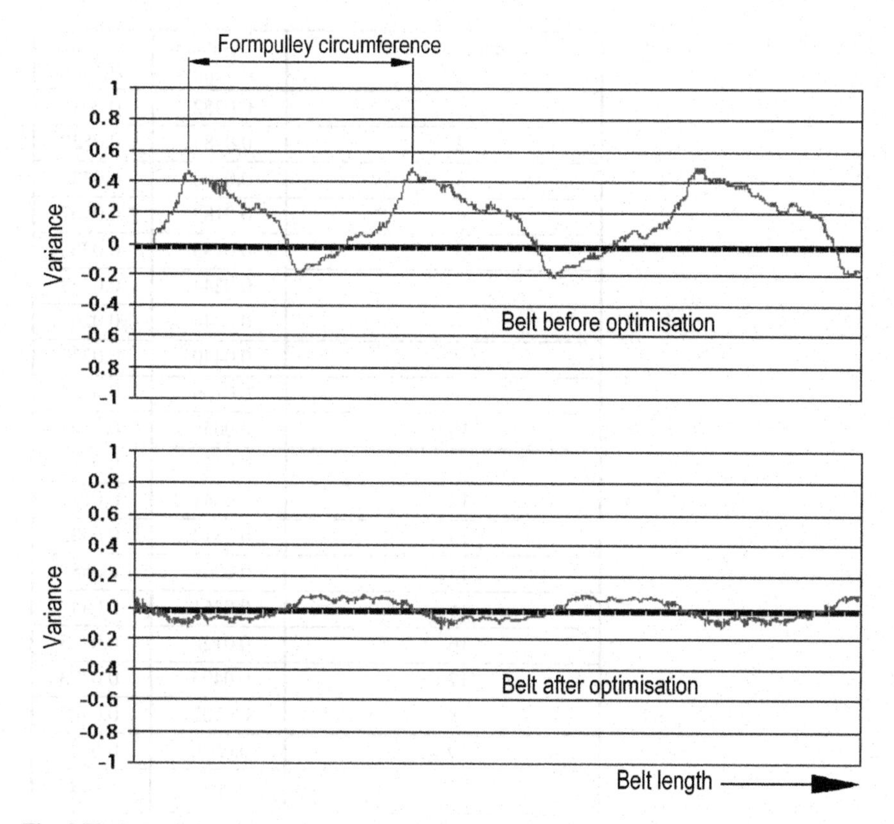

Fig. 3.79 Belt pitch accuracy

Belts with bifilar S and Z wound steel tension members would also be an advantage as the multiple steel cords are arranged in pairs that are twisted in opposing directions helping to reduce lateral reaction forces and minimise belt run-off. Belt tracking forces and friction will be reduced and will thus improve accuracy.

If a pair of belts is being used on a gantry system, for example, then the belts should be consecutively cut from the same reel stock and kept together for assembly. If the belts are of a closed mould type then they should cut side-by-side from the same sleeve. Their teeth should be marked to retain their relative position and they should be kept together for assembly.

Drive Assembly and Maintenance

Apart from the normal assembly practices of cleanliness and rejection of visually sub-standard components, the most important parts of the process are alignment of the drive and setting the pre-tension. Alignment is critical to the proper running of the drive and its long-term effectiveness. High accuracy drives should ideally be

Table 3.4 An investigation

Machine No	X-axis P_a[a] mm	Y-axis P_a[a] mm
1	0.0252	0.0745
2	0.0384	0.0520
3	0.0290	0.0402
4	0.0183	0.0567
5	0.0135	0.0366
6	0.0244	0.0258
7	0.0644	0.0266
8	0.0440	0.0233
9	0.0169	0.0304
10	0.0855	0.0206
11	0.0151	0.0384
12	0.0541	0.0179
13	0.0314	0.0425
14	0.0706	0.0587
15	0.0361	0.0443
16	0.0128	0.0223
17	0.0495	0.0473
18	0.0787	0.0427
19	0.0600	0.0221
20	0.0649	0.0297
Evaluation of 20 Results		
Minimum	0.0128	0.0179
Maximum	0.0855	0.0745
Sum	0.8328	0.7526
Average	0.0416	0.0376
Standard Deviation, "Sigma"	0.0225	0.0147
Range 6 Sigma	0.1349	0.0880

[a] P_a Positional variance

aligned by laser, especially if the centre distances are large or if multiple pulley drives are involved. Setting the pre-tension is also critical for the drive to be both accurate and repeatable and therefore the only acceptable method of setting the pre-tension is with an electronic tension gauge. No other process is accurate enough. All adjustable centres and idler pulleys must now be secured against movement that would affect the pre-tension. After the drive has run for a few hours there may be some 'settling' of the belt in the pulleys and this could change the pre-tension. A check at this point is advisable before the machine is released to the user.

Maintenance checks should be scheduled to check the pre-tension on a regular basis with an electronic tension gauge. If the original setting figure has changed then there is a problem with the drive. This needs to be investigated.

Example In 2005 DEK International, a business of DTG International GmbH [9, 51] wished to implement a cost-effective X–Y camera gantry drive using timing belt drives instead of ballscrews. The X-axis is equipped with a 16AT5/1965 open length belt and the Y-axis with a 16AT5/2590 open length belt and both running on 22 tooth Zero toothform anti-backlash pulleys. This corresponds to a maximum X-axis traverse of 1150 mm and Y-axis traverse of 850 mm. The user's maximum acceptable error was 90 microns with the drives travelling to a target position 200 mm away then back to the start position; the initial testing was over 1 million cycles. After careful design and assembly the following was measured: (see Table 3.4).

It can be seen from the results that the user requirement of 90 microns was met and the drives went into production. A question raised is how is there such a disparity between the highest and the lowest positional variance? The answer is that the disparities are almost all due to tolerances in the shaft centre distances, the belt lengths and the pulley diameters. The effects of such tolerances can be reduced by component matching, where pulleys and belts at the top or bottom of their tolerance range are matched together to reduce the effect.

Conclusion

To achieve high accuracy timing belt drives, the process can be time consuming and is sometimes not cost-effective but it does show that with good design, intelligent component selection, and scrupulous assembly techniques that a relatively simple and inaccurate set of drive components can produce outstanding results.

Chapter 4
Timing Belts in Linear Drives

Abstract Linear belt drives are a branch of power transmission technology and they deal with mechanical systems for linear motion. Timing belts ideally couple both rotary and linear motion whereas the most common electromechanical actuators have only rotary outputs. The user can find all the relevant kinematic equations in this chapter to convert rotation into linear values. Torque and angular motion in the drive pulley arc of contact translates to a force value and distance in the linear section of the drive. By careful calculation it is possible to make definitive statements about positional accuracy. The essential content of this chapter refers to application illustrations with explanations of linear timing belt drives. There are numerous practical examples and references to help the user find innovative linear drive solutions.

4.1 Conversion of Motion

A toothed belt moves rotationally around pulleys and in a linear manner across the spans between pulleys. It is able to turn linear motion into rotary motion and vice versa.

Each linear axis has a limited travel in a closed system and thus, starting and braking loads are always part of the linear drive. For these situations it is an advantage that the timing belt is of low mass and that the load distribution across the teeth in mesh on the pulley gives an ideal force/torque conversion between the linear span and pulley. The belt teeth, which always change position, both with rotation reversal and with the change from acceleration to braking on the loaded flanks of the pulleys, allow smooth force direction reversals due to their elastomeric resilience (see also description in Chapter 2.1). The explanation for the outstanding service life of toothed belt linear drives within many areas of production engineering and automation arises from this absorbed and cushioned movement dynamic.

R. Perneder and I. Osborne, *Handbook Timing Belts*,
DOI: 10.1007/978-3-642-17755-2_4, © Springer-Verlag Berlin Heidelberg 2012

The discussion in this chapter focuses primarily on the slip-free transmission behaviour of toothed belt drives as converters of rotary motion to linear movement. Such systems are used, for example, in the main axes of gantry robots, the carriage drive in plotters and for positioning grippers in assembly and handling equipment. In the context of the optimization of power relationships in timing belt linear drives, as well as the prediction of positional accuracy, there are various research projects and publications available such as [97, 124].

4.2 Dimensioning Linear Drives

The initial choice of belt type, pitch and width for a rotary drive, is usually based on the dynamic properties of the drive motor and the dimensioning of the linear timing belt drive is no different. When high accuracy positioning is a requirement it is useful to increase the rigidity of the belt. This technique is described in the next chapter.

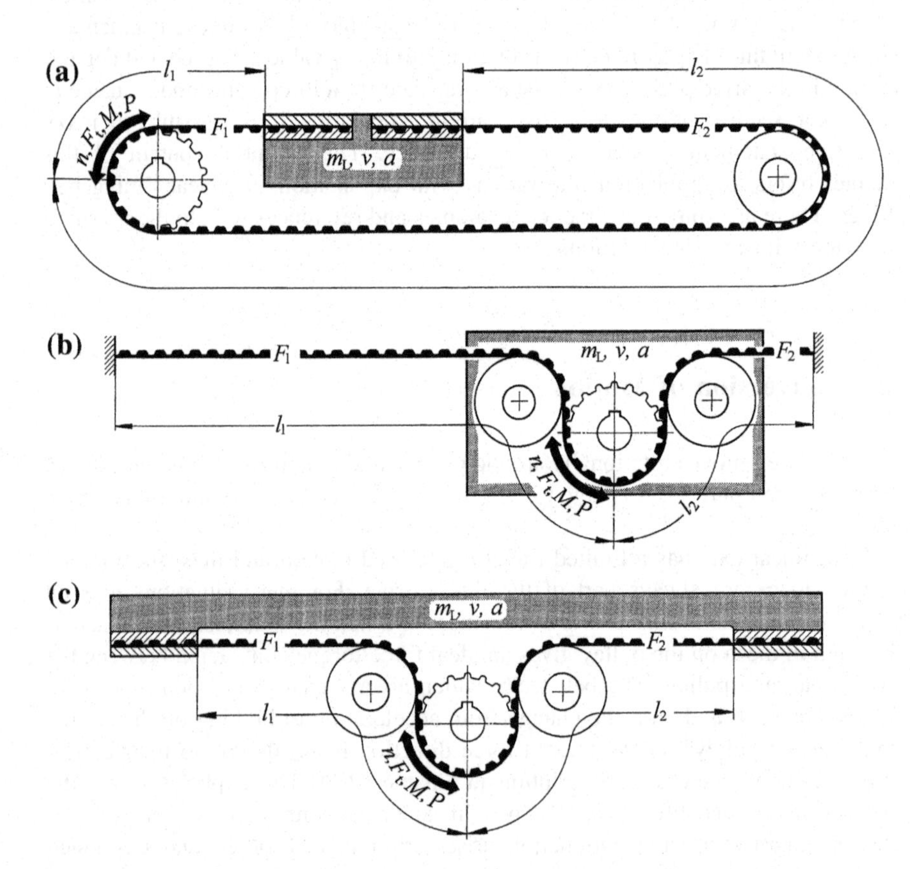

Fig. 4.1 Linear drive variants. **a** Linear slide. **b** Linear traveller. **c** Linear table

There are many possible drive geometries. Figure 4.1 shows the basic variants. The drive motor can be fixed or travelling.

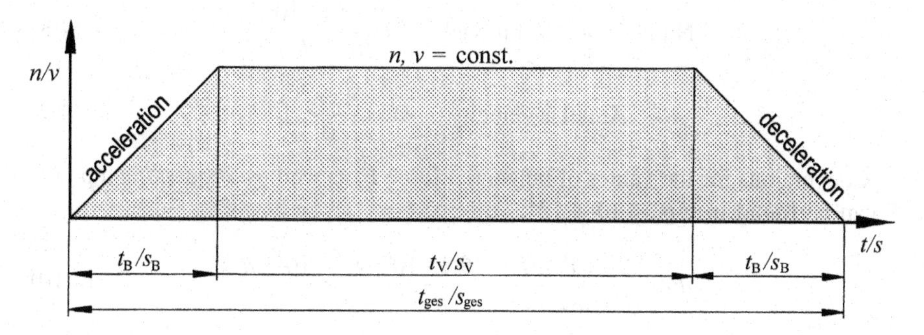

Fig. 4.2 Motion profile

The calculation process needed for sizing linear drives is difficult to standardize due to widely differing initial design requirements. Where travel time is often optimised in linear automation applications, the positional accuracy is most important for handling applications. Furthermore, different conditions can result from where only the acceleration is already known; while for another application only the starting torque is known. Linear drive design is therefore tailored to each individual application. The illustrated motion profile in Fig. 4.2 and the equations below are based on the simplified assumption that the conditions for starting and braking have the same acceleration a.

If the motion of the linear drive is to be time optimised then the Eqs. 4.1 to 4.12 apply:

$$\text{Total distance} \quad s_{ges} = s_B + s_v + s_B \tag{4.1}$$

$$\text{Total time} \quad t_{ges} = t_B + t_V + t_B \tag{4.2}$$

$$\text{Travel distance where } v = \text{const.} \quad s_V = v \cdot t_V \tag{4.3}$$

$$\text{Travel time where } v = \text{const.} \quad t_V = \frac{s_V}{v} \tag{4.4}$$

$$\text{Acceleration/braking distance} \quad s_B = \frac{a \cdot t_B^2}{2} = \frac{v}{2 \cdot a} \tag{4.5}$$

$$\text{Acceleration/braking time} \quad t_B = \frac{v}{a} = \sqrt{\frac{2 \cdot s_B}{a}} \tag{4.6}$$

Peripheral speed[1] $$v = \frac{d_W \cdot n}{19.1 \cdot 10^3} = \sqrt{\frac{2 \cdot s_B \cdot a}{10^3}} \tag{4.7}$$

Number of revolutions[1] $$n = \frac{19.1 \cdot 10^3 \cdot v}{d_W} \tag{4.8}$$

Angular velocity[1] $$\omega = \frac{n \cdot \pi}{30}. \tag{4.9}$$

Conversions for the values of tangential force F_t as well as torque M and power P can be found in Eqs. 4.10–4.12[1]:

$$F_t = \frac{2 \cdot 10^3 \cdot M}{d_W} = \frac{19.1 \cdot 10^6 \cdot P}{n \cdot d_W} = \frac{10^3 \cdot P}{v} \tag{4.10}$$

$$M = \frac{d_W \cdot F_t}{2 \cdot 10^3} = \frac{9.55 \cdot 10^3 \cdot P}{n} = \frac{d_W \cdot P}{2 \cdot v} \tag{4.11}$$

$$P = \frac{M \cdot n}{9.55 \cdot 10^3} = \frac{F_t \cdot d_W \cdot n}{19.1 \cdot 10^6} = \frac{F_t \cdot v}{10^3}. \tag{4.12}$$

If a mass is to be accelerated in a linear manner then the driver pulley of the linear drive has to apply a tangential force F_t of

$$F_t = F_B + F_H + F_R, \tag{4.13a}$$

$$F_t = m \cdot a + m \cdot g + \mu \cdot m \cdot g, \tag{4.13b}$$

wherein the acceleration force F_B is necessary to ensure the mass can be moved from rest and accelerated to terminal velocity v. The lifting force F_H has to be applied if the direction of acceleration is opposite to that of gravity. $F_H = 0$ in horizontal linear movement. The friction force F_R acts against the direction of motion. If the frictional resistance can be ignored then $F_R = 0$.

The mass m to be moved consists of

$$m = m_L + m_B + m_{Zred} + m_{Sred}. \tag{4.14}$$

Where: m_L mass of the linear carriage, m_B the mass of the belt, m_{Zred} reduced mass of the pulley(s) and m_{Sred} reduced mass of the tensioner pulley(s).

The mass m_Z of a toothed wheel and a tension pulley m_S is calculated from

$$m_Z = \frac{(d_K^2 - d^2)}{4} \cdot \pi \cdot B \cdot \rho \quad m_S = \frac{(d_S^2 - d^2)}{4} \cdot \pi \cdot B \cdot \rho \tag{4.15}$$

[1] Equations 4.7 to 4.12 are numerical value equations. The specific units applied are: force F in N, torque M in N·m, power P in kW, acceleration a in m/s^2, pitch circle diameter d_w in mm, number of revolutions n in min^{-1}, acceleration distance and stopping distance s_B in mm, velocity v in m s^{-1}, angular velocity ω in s^{-1}.

The reduced mass m_{Zred} of a toothed pulley and/or tension pulley m_{Sred} is a replacement mass with the same mass inertia to the neutral line of the timing belt as the rotation body on the rotation axis.

$$m_{Zred} = \frac{m_Z}{2}\left[1 + \left(\frac{d}{d_K}\right)^2\right] \quad m_{Sred} = \frac{m_S}{2}\left[1 + \left(\frac{d}{d_S}\right)^2\right] \tag{4.16}$$

The maximum force in the belt span F_{max} is achieved if the pre-tension force F_V (static) and the tangential force F_t (dynamic) work together.

$$F_{max} = F_t + F_V. \tag{4.17}$$

A linear actuator is properly pre-tensioned if the slack side of the belt, acting under the tangential force F_t, is always taut. A suggested minimum pre-tension load is:

$$F_V \geq 1 \cdot F_t. \tag{4.18}$$

The static shaft load F_{Asta} is valid for stopped or no-load conditions. F_{Adyn} is dependent on the effective tangential force value:

$$F_{Asta} = 2 \cdot F_V, \tag{4.19a}$$

$$F_{Adyn} \approx 2 \cdot F_V + F_t. \tag{4.19b}$$

In general, the dominant variables for the calculation of linear drives are the mass of the linear carriage m_L and the acceleration a. The optimisation process for determining the individual size depends on the pitch circle diameter of the drive pulley, the belt width and the belt type. Rigidity, safety and positional accuracy also need to be considered where appropriate. It is also necessary to determine whether the desired speed/torque curve is available from the drive motor and its control system. Thus design can only start after the selection of the prime mover, whose characteristics form the basis for the detailed design of belt and pulley(s). The technical data of the selected belt must comply with the maximum loads from acceleration and braking plus a safety factor. The designer is thus dependent on reliable information from manufacturer catalogues.

Tooth Load Capacity

The maximum torque of the selected drive motor is used to calculate the size of the motor pulley. The rule is to use the greatest motor torque available during start-up, however, braking loads should also be considered. As a result, the drive is designed in the same manner as the rotary drives in Chapter 2.9. The motor pulley is therefore calculated from the tooth load capacity of the teeth in mesh in Eq. 2.19.

Tension Member Loads

The force curve in the tension member is not only dependent on the introduced torque. Displacement of the linear slide always results in new geometric relationships between loaded and unloaded part of the belt which affects the current tensile loads.

In a timing belt linear drive, the motor power is rotational over the drive pulley and then translates to a linear motion force $P = F_t{\cdot}v$. In comparison with purely rotational applications there is *no driven pulley* and in the example "Linear Slide" (Fig. 4.1a) it is recommended that the associated idler or return pulley is an untoothed (smooth) pulley. Depending upon position of the linear slide and also depending upon direction of rotation, the loaded and unloaded belt part lengths l_1 and l_2 change, and with them the associated stiffness values change, as well as the forces F_1 and F_2. Therefore, the linear unit to be moved is regarded as being clamped between two coupled belt ends. In the dynamic state, the motor drive pulley transfers the torques from the teeth in mesh into the force differences $F_1 - F_2$ over both partial lengths in the belt (see Fig. 4.3). The developing force difference corresponds to the tangential force F_t, which is the mass m accelerated by a (see Eqs. 4.13a and 4.13b). It is especially important where the ratio of the drive geometry of the loaded to unloaded belt part lengths are very large. Because of the force build-up in the loaded side of the belt, the elongation in a very short part of the belt can create a force reduction. The force difference $F_1 - F_2$ can only be developed in the aforementioned belt sections if the minimum pre-tension force from Eq. 4.18 is observed. For fail-safe function, there must always remain a residual tension force in the unloaded belt section so that no slack is created. Where no tension exists there is a tendency to a "whip effect". This borrowed colloquial term very aptly describes a loose cord which reacts to the transition to stretch to give a crack of the whip.

In the case of a linear slide drive, the sudden change would be accompanied by a sagging of the elongated belt section due to the oversized acceleration forces. The latter condition also generates significant noise and the belt could be possibly damaged or torn.

Furthermore, linear drives with the slide position having equal belt part lengths l_1 and l_2 need to be particularly considered. This situation results in the largest compliance. The operating conditions of the different lengths of loaded and unloaded belt sections are also discussed in Chapter 2.8, "Pre-tension in multiple shaft drives". The special feature of linear drives is that the section length differences depend on the position of the moving mass during operation of the drive and, as a result, their associated stiffnesses and the actual force relationships change. The maximum belt tension given in Eq. 4.19b is determined from the slide position where there is a very short loaded span length opposing a long unloaded span length. These geometric relationships are found only in the type of drives that correspond to the linear traveller drive in Fig. 4.3b. This position develops because the maximum tension member load that builds the force difference $F_1 - F_2$ is not, or only slightly, relieved as the pre-tension on the

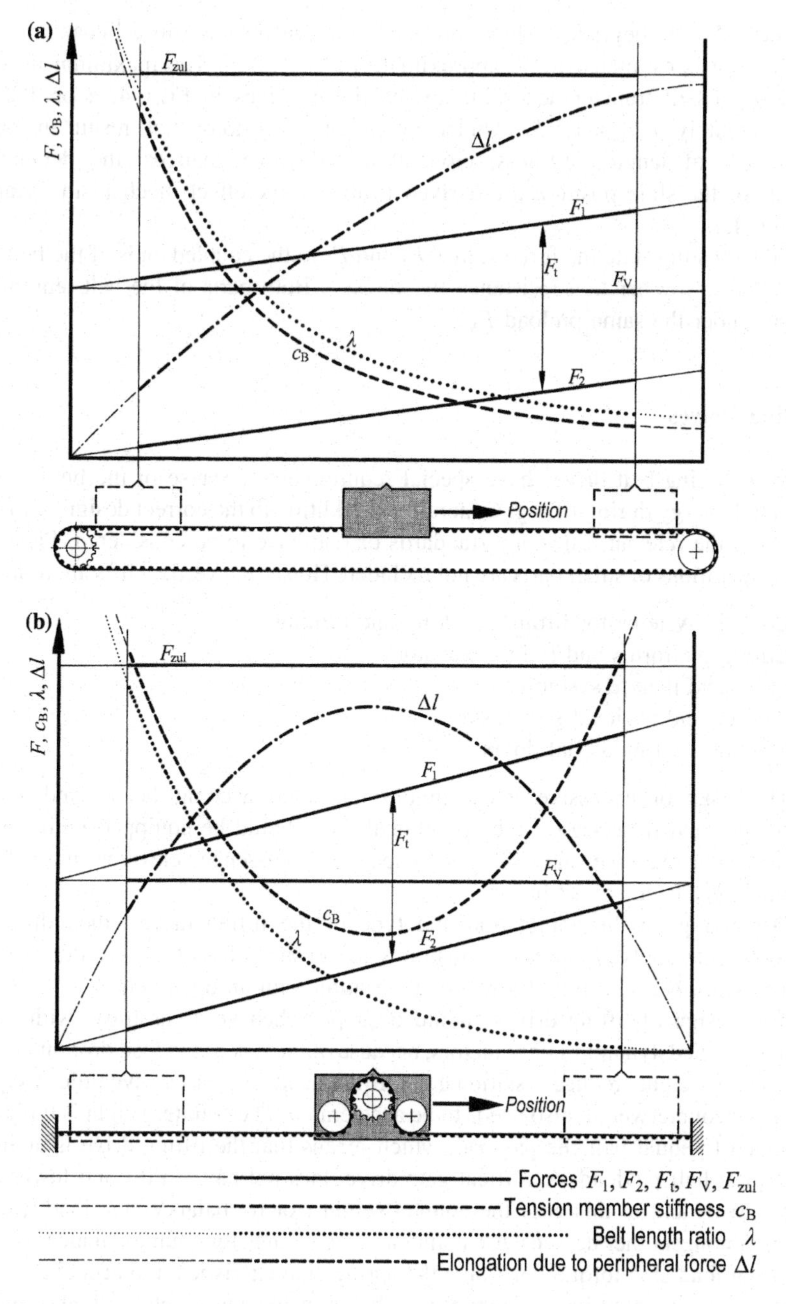

Fig. 4.3 Force graphs for **a** linear slide, **b** linear table and linear traveller

loaded belt side degrades. Thus, the maximum tensile force in a linear drive of this design is estimated to be approximately $F_{max} \approx 2 \cdot F_t$. The maximum tension load for a belt transfer according to the linear drives in Figs. 4.1a or 4.3a is approximately $F_{max} \approx 1{,}5 \cdot F_t$. Under operating conditions that result in rapid sequences of length, stiffness, elongation and force changes, the dependent values of the slide position are derived in the "force-effect mechanism" graphs in Fig. 4.3.

In the resting state the force values F_1 and F_2 in the coupled ends of the belt are subject to a low friction resistance (hysteresis). Both parts of the belt length are nearly under the same preload F_V.

Lifting Drives

Vertical timing belt drives have special requirements because of the build-up of potential energy during lifting. Generally, in addition to the correct design of motor and belt, the relevant rules and standards of risk have to be considered. Pick and place operations of small parts are not included. Hoists and vertical lifts are found in

- Logistic systems for lifting, stacking and turning
- Lifting platforms and freight elevators
- Stage and theatre systems
- Storage and retrieval systems
- Machine and industrial doors

The risks of excessive loads should take into account health and safety guidelines and if necessary the operational area should be equipped with access systems. For example, the cable load capacity of passenger elevators has a minimum safety factor of 12 [64].

Depending on the application and type of the lifting drives, then different requirements apply. Figure 4.4 illustrates the design of a closed vertical power transmission belt and Fig. 4.5 shows a variation with an open belt.

The design of lifting drives should be approached very carefully, with sufficient static and dynamic safety factors, because of the risk of the load dropping [64]. A counterweight reduces stationary torques and increases dynamic torques since the counterweight also needs to be accelerated. The counterweight is typically designed to equal half the payload, which means that the lifting drive is at equilibrium at half payload. A vertical gate drive does not have to lift and lower different loads and in this case the counterweight exactly balances the load. Roller doors, amongst other devices, use compensation springs for counterbalance and the resulting load-side torque must take the spring characteristics into account.

With a vertical drive (lifting drive) the movable mass acts with its normal force $F_H = m \cdot g$ on the belt but more on the loaded section and less on the slack side.

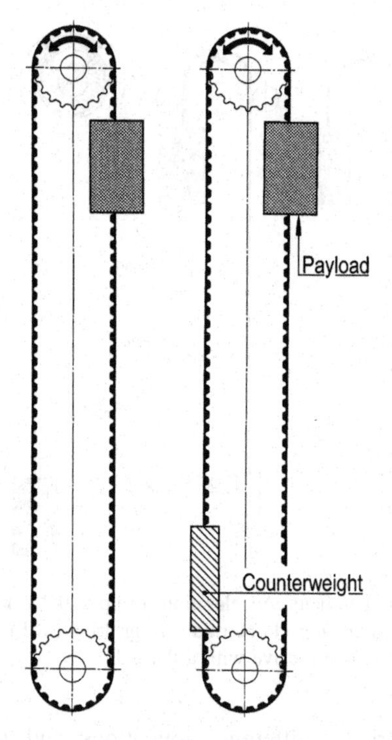

Fig. 4.4 Lifting drives with closed belts. Accelerations of over 10 g are possible. The possible designs are: (1) Motor above, without counterweight (2) Motor below, without counterweight (3) Motor above, with counterweight (4) Motor below, with counterweight

For a state of rest and/or for holding in the desired position, the toothed motor pulley must be able to apply, at a minimum, the appropriate holding torque without a counterweight during operation. For a lifting drive with a transfer belt, as in the linear slide shown in Fig. 4.1a, there are, depending on whether the drive unit is positioned at the *top* or at the *bottom*, different length ratios and elongation ratios for l_1 and l_2. It follows that the minimum pre-tension is adjusted to the design used, and

$$\text{Motor at the top}: \quad F_V \geq \frac{1}{2} \cdot F_t \tag{2.12}$$

$$\text{Motor at the bottom}: \quad F_V \geq 1 \cdot F_t. \tag{2.18}$$

It is easy to understand that with the motor *at the top* that the drive belt and the components of the surrounding structure (bearings, shafts) experience significantly lower loads. However for easy access and maintenance the drive motor is often situated *at the bottom*. In a lifting drive with a counterweight, the necessary holding torque required from the motor is reduced, depending on the balance of the masses achieved. With increasing acceleration, the counterweight also increases the dynamic forces. Equations 4.13a and/or 4.13b should be used for both for the motor and the belt calculations.

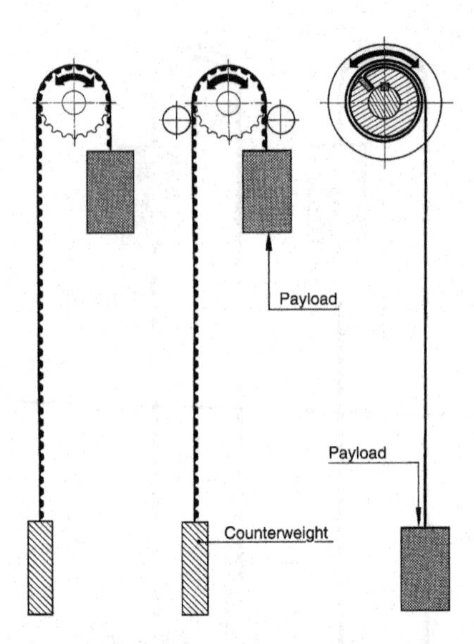

Fig. 4.5 Lifting drives with open belts. Accelerations of up to 0.2 g are normal, in special cases up to 0.5 g possible. The designs are: (1) Normal lifting drive (2) Lifting drive with tangential rollers (3) Lifting drive as a winding drive with a flat belt

In vertical linear drives, different elongations and power balances arise depending on the motor assembly and belt lengths between the drive pulley, lifting weight and counterweight. It is not possible to take into account the variety of load conditions in detail since if starting with accelerations of over 1 g then cases of force reversal would be needed to be considered during downward motion. It is recommended that the coupled drive movements are analysed for the stress/strain behaviour with two or more masses and then, from this information, calculate the minimum preload.

> The minimum pre-tension force F_V of the belt drive should be set such that, under worst case conditions, the unloaded belt section will always remain stretched under residual tension.

Open belt lifting drives as in Fig. 4.5 are used with or without lateral support for raising and lowering loads in factory and theatre scenery automation. Additional tangential rollers ensure the safe feeding of the belt into the pulley. Open belts can only transmit force in a single direction, therefore only applications with accelerations which are significantly below the acceleration of gravity are possible.

Cable winches are widely used in cranes and in handling and theatre equipment. In typical applications, winding techniques are used to drop the cable into the shaped groove of the drum. Belts are able to be used in this sort of application but only for multilayer drum winding and unwinding which results in a position-dependent variable diameter. Winding drives for winches of this type offer interesting kinematic alternatives. The belt attachment is problem-free to implement, since after one complete drum revolution a self-retaining effect starts, see Fig. 4.5.

Figure 4.6 shows the operational area of a Lenze SE [75] hoist using a closed two-strand belt drive which can be configured with or without a counterweight. Hoists of this type can be operated over several floors, for example, for lifting pallets in shafts. To move to a specifically defined position an operator is needed, who commands the hoist by sight, using button or joystick control. This sort of manual control is standard on cranes and hoists. Once a lifting device is operated automatically, positioning systems are required. Fixed stopping positions for floors in buildings or shelf levels can be initiated by changing the operating speed to a creep speed and then controlled via a limit switch to the desired position. For higher requirements, a relative or absolute position detection system is used on the motor pulley or a linear measurement system (e.g., laser measurement) is used with or without reference sensors on hoist. There are multiplicities of different control types depending upon safety requirements as well as double redundancy sensor systems. Recommendation for optimal control may be possible from [75].

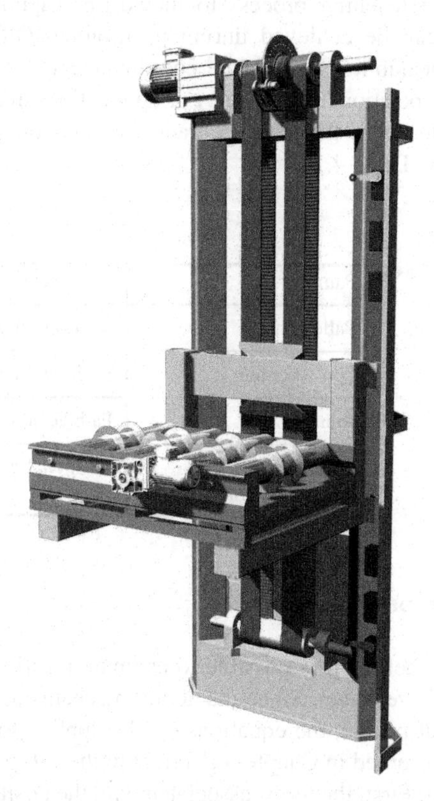

Fig. 4.6 Two belt lifting drive with closed belts. A Lenze [75] hoist

Timing Belt Selection

When deciding to use a particular brand of belt in a drive, the technical design parameters must be reconciled with the selected manufacturer catalogue values. This handbook does not recommend specific product data due to the need to keep its neutrality and both the belts and their performance are often subject to continuous improvement. The timing belt is a relatively new drive element and its technical properties are still evolving (see Chapter 2.4.8). Thus, it becomes clear that the user is dependent for specific dimensioning on the current data from the manufacturer catalogues or information obtained from the internet.

4.3 Linear Positioning Drives

A reliable estimate of positioning capability is very often of the utmost importance in the use of linear belt drives. The accuracy strongly depends on the type and size of belt and on the specifications of the electronic control system. The most popular applications are those requiring point-to-point motion and this can be accomplished using a teaching process to memorise waypoints with a known repeatability which can be achieved during positioning. Off-line programming however requires repeatability in absolute coordinates. The variances that occur here are referred to as positional variances. For applications in arc welding or laser cutting, the main criterion is the path accuracy. Application specific permissible variances are given in Table 4.1.

Table 4.1 Application areas

Painting	Path accuracy	<5 mm
Palletising	Repeatability	<2 mm
Spot Welding	Positional accuracy	<1 mm
Joining	Path accuracy	<0.2 mm
Printed circuit board assembly	Positional accuracy	<0.1 mm

Calculating the Positional Accuracy

With the calculations below, it is possible to estimate positional accuracy at any desired point of the drive travel. Since the action mechanisms are independent of the chosen timing belt profile, the equations can be applied to any belt type. The sample calculation presented in Chapter 4.5 refers to the AT-profile, the industry's most widely-used belt. First, the terms associated with the positional capability and the effects of variances on the positional accuracy are defined. Assessments of linear movements use the expected position variance P_a (also called the positioning accuracy), the positioning repeatability P_s, the hysteresis U and the positional uncertainty P_u. These values are used for machine tools in [21] and defined in [119] as well as [121] and with the same meaning but with different names for industrial robots and can also be applied correspondingly to linear drives. While the positioning variance P_a is determined after a single start-up, the repeatability P_s is determined by repeatedly approaching a target position under the same conditions of load, speed and such like, as in Fig. 4.7.

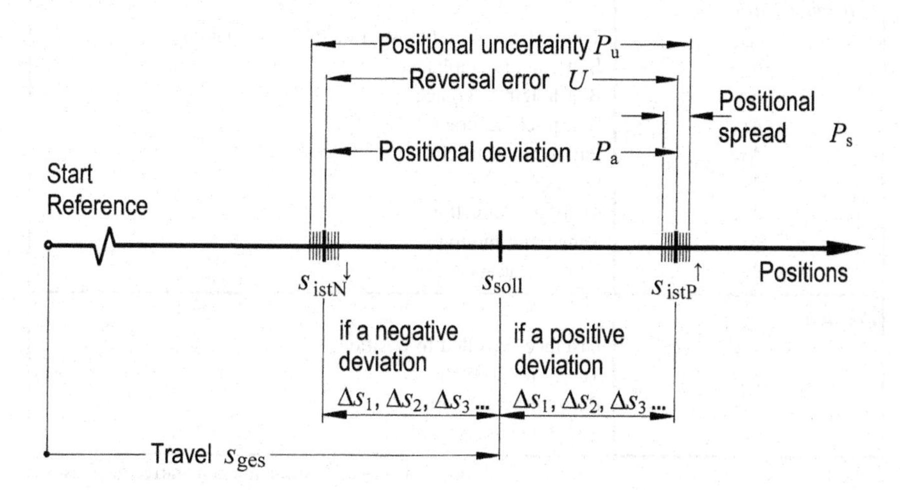

Fig. 4.7 Positioning parameters

The random variances between the actual rest positions give the value for the repeatability P_s. The hysteresis value U is the systematic variance between the average value and the positional uncertainty P_u, the highest measured difference of the actual positions in opposite starting and force directions in the systematic individual differences contribute to the difference s_{soll} from the target position under consideration, and are given the designations Δs_1, Δs_2, Δs_3 .... Δs_n (Table 4.2).

The positional variance is made up of both the positive and negative actual variances:

$$P_a = s_{istP} + s_{istN}. \tag{4.20}$$

The positive and negative variances in each case are the sum of the individual variances:

$$S_{istP} = \Delta s_1 + \Delta s_2 + \Delta s_3 + \ldots \Delta s_n \tag{4.21a}$$

$$S_{istN} = \Delta s_1 + \Delta s_2 + \Delta s_3 + \ldots \Delta s_n. \tag{4.21b}$$

Table 4.2 Positioning behaviour parameters

Positional variance (positional accuracy) P_a	Sum of all systematic variances as maximum difference between all average actual rest positions
Repeatability P_s	Size of the range around an average actual position lying within the identified random variances
Hysteresis U	Systematic variance of two actual positions subject to different force and starting directions
Positional uncertainty P_u	Maximum distance between the actual positions that may occur when moving to a target position with varying force and direction
Individual variances	
Δs_1	Tooth stiffness and tension member elongation
Δs_2	Belt clamp compliance
Δs_3	Belt length tolerance
Δs_4	Belt pitch accuracy
Δs_5	Pulley concentricity
Δs_6	Polygonality
Δs_7	Thermal elongation
Δs_8	Tangential motion
Δs_n	Further variances
Position	
s_{ges}	Distance travelled to position
s_{soll}	Target position
s_{istP}	Positive actual variance[a]
s_{istN}	Negative actual variance[a]

[a] A positive variance increases the actual position to the target position and a negative decreases it

When determining the sum of each element Δs, it is necessary to consider whether it is a positive and / or a negative actual variation. This data is used in Eqs. 4.21a and 4.21b. Such hints will be explained further into the chapter.

Tooth Stiffness and Tension Member Elongation Δs_1

The tooth stiffness and the tension member elongation impact on the target position s_{soll} as force dependent variables. They will be treated here together, so that the values are viewed in the context of Chapter 2.13, where transmission accuracy is described for rotary/rotational timing belt drives.

A single proportional variance Δs_1 to the target position can be practically demonstrated in any real linear drive with guide rails. On moving the linear unit, the frictional forces act opposite to the direction of F_R. In addition, process forces, holding forces for assembly tasks and gravity for vertical motion also take effect. $\sum F$ is designated as the sum of all forces on the linear slide.

On the basis of the tooth stiffness and tension member elongation the individual variance is calculated from

$$\Delta s_1 = \frac{\sum F}{c_{ges}}, \tag{4.22}$$

whereby the overall rigidity c_{ges} comprises of the rigidity of the belt teeth c_P in the arc of contact of the drive pulley and the belt tension member stiffness c_B at the neutral line in the belt and is calculated from

$$\frac{1}{c_{ges}} = \frac{1}{c_P} + \frac{1}{c_B}. \tag{4.23}$$

The tooth stiffness of the belt teeth in the arc of contact on the pulley is determined from

$$c_P = c_{Pspez} \cdot z_e. \tag{4.24}$$

In Eq. 4.24 the specific stiffness values c_{Pspez} for the ATL timing belt are listed in Table 7.2, Chapter 7.3. The maximum number of teeth in mesh, to be considered for this calculation, is $z_e = 24$.

The tension member stiffness c_B is subject to special dynamic conditions. The movable linear mass is coupled between two belts spans which are part of the overall belt length l_B as part belt lengths l_1 and l_2, see Fig. 4.1. Depending on the direction of rotation of the driving pulley, a loaded and unloaded belt length is formed. With each new position of the linear slide the ratio between these lengths changes. The linear slide can be considered to be virtually clamped between two belts and the stiffness behaviour is given by

$$c_B = \frac{l_1 + l_2}{l_1 \cdot l_2} \cdot c_{Bspez} = \frac{l_B}{l_1 \cdot l_2} \cdot c_{Bspez}. \tag{4.25}$$

which demonstrates the variable stiffness of linear drives. Any position of the linear slide will have its own spring rate. The stiffness c_{Bmin} is a minimum where l_1 and l_2 have equal lengths. In this case the relationship is

$$c_{Bspez} = \frac{4 \cdot c_{Bspez}}{l_B}. \tag{4.26}$$

The specific stiffness value c_{Bspez} for AT timing belts is listed in Table 7.2, Chapter 7.3. To determine the individual variance Δs_1 you need to analyse in which direction (plus or minus) $\sum F$ is used in the calculation. The friction force F_R will vary depending on its directional sign of travel. Functional and braking forces must also be considered depending on the direction of action with "$\pm$" or with only "+" or "$-$". The force direction of gravity is obvious. This means that two or more results for Δs_1 are conceivable, which will be used in Eqs. 4.21a and 4.21b, with the appropriate relevance to the target argument.

Belt Clamp Compliance Δs_2

When under load, the belt teeth experience the greatest deformation in a clamp plate and less when a fixed timing pulley is used as an end clamping fixture, see Fig. 4.5. In such clamping fixtures the tension member also exhibits small elongations. The sum of the forces $\sum F$ affects the linear slide position and the displacement of the belt teeth in the clamping fixture and thus their presence is position dependent.

Since elastomers are incompressible when clamped on all sides they exhibit only small, force dependent compliances (see Chapter 2.13). It is to be assumed that all teeth in the clamping fixture are involved in the transmission of power. This is achieved by the very large and relatively static clamping force which results from pressing the belt teeth into the tooth gaps of a clamping fixture. This leads to jamming of the belt teeth, resulting in a very high tangential stiffness. It is recommended that clamping plates of this type have a minimum clamping length of six belt teeth. Because of the small impact on overall compliance in the linear drive Δs_2 can be given the value of Zero, thus $\Delta s_2 = 0$. As a substitute measure for the calculated figure, the belt lengths l_B and l_1 and l_2 can be extended by three teeth for each clamping fixture.

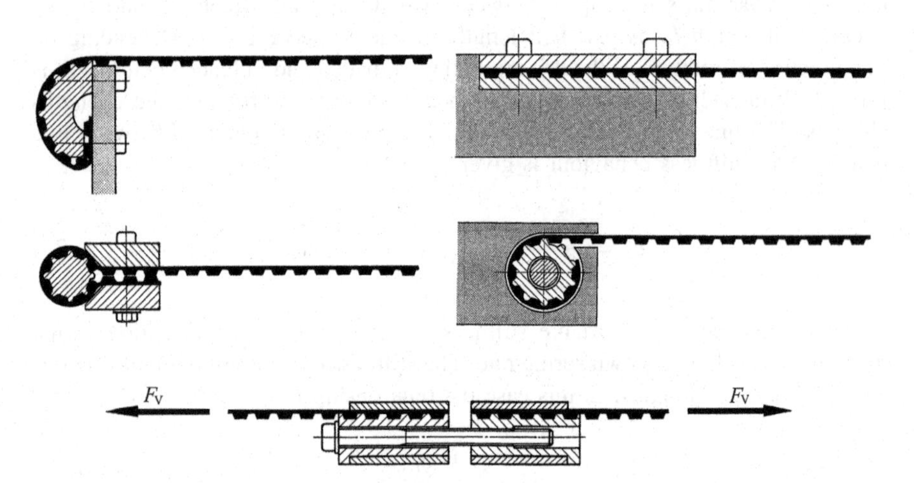

Fig. 4.8 Belt clamping examples

Belt Length Tolerance Δs_3

Measuring the pitch length is dealt with in Chapter 2.17. Accordingly, the belt length tolerance, in an unloaded state (or under measurement pre-tension) is a manufacturing process variation which is spread evenly over the entire length of the belt in all pitches. In the installed condition, the adjusted pre-tension force affects the current length tolerance. Timing belts used in the linear drives are therefore, usually, provided with a negative pitch, i.e. the pitch in the relaxed state is smaller than the nominal pitch. The optimal installation situation is reached when the applied pre-tension leads to an actual pitch which coincides with the nominal pitch.

This pitch equalising force can be found from the manufacturer's literature. The desired installation conditions can be achieved by the application of the pre-tension force while at the same time as the pitch is checked. Only in rare cases does the optimal pre-tension force result in an accurate nominal pitch. In field use it is usual to add a larger initial pre-tension. If the initial pre-tension is too small then a belt of another tolerance class should be selected. Possible differences in the nominal pitch of pre-tensioned belts can be calculated as a single deviation Δs3 and can be determined by

$$\Delta s_3 = \frac{\Delta L_{1m}}{1{,}000} \cdot s_{ges}, \tag{4.27}$$

where $\Delta L_{(1m)}$ is the actual tolerance in the pre-tensioned state per 1,000 mm belt length in the calculation. s_{ges} stands for either linear distance travelled or for the distance to the point of reference of which the single deviation Δs_3 refers to. If a tolerance is known in the installed state then its actual deviation will now have a concrete value. This should then be inserted for use into Eqs. 4.21a or 4.21b. With unknown tolerance position (±) both equations should be considered.

Belt Pitch Deviations Δs_4

Slight fluctuations in the tensioned belt pitch affect the tolerance of adjacent pitches. These are non-accumulative. The mating interference between the belt and the drive pulley over several teeth in causes the deviations to average themselves and reduce in field use. The maximum possible deviation is

$$\Delta s_4 = 0.05\ \text{mm}. \tag{4.28}$$

The single deviation Δs_4 is entered into Eqs. 4.21a and 4.21b using the same value.

Pulley Concentricity Deviations Δs_5

The toothed pulleys as well as return and idler rollers all take part in the transmission of motion and their radial deviations f_R lead to output positional variations. This particularly depends on the concentricity between the pulley bore and the outer surface of the pulley which usually has a deviation <0.05 mm. This causes periodically changing radii effects at the point of belt in-feed into the drive pulley and additional span elongations due to variations in the centre distance. This leads to sinusoidal fluctuations of the linear travel which could be affected by the unfavourable orientation of some or all the pulleys (n_z = number of pulleys) and could lead to cumulative variations:

$$\Delta s = n_z \cdot \frac{f_R}{2}. \tag{4.29}$$

The single deviation Δs_5 is entered into Eqs. 4.21a and 4.21b using the same value.

Polygonality Δs_6

The investigations in [30] treat in detail the influence of concentricity fluctuations and the polygon effect on the transmission variations in timing belt drives. It provides options for calculating the angular deviation between the input and output. The mathematical equations give a result for the polygonality but they are such small values that according to [123] the linear deviations have no practical relevance and so can be ignored. Thus $\Delta s_6 = 0$

Thermal Elongation Δs_7

The ambient temperature and its changes ΔT affect both the belt and the associated drive structure equally. Temperature increases will trigger a material-related elongation of the centre distance determined from the design elements. Thus, the distances in the linear system change proportionately and the corresponding deviations are calculated from

$$\Delta s_7 = \alpha_A \cdot \Delta T \cdot s_{ges}. \tag{4.30}$$

Length–temperature coefficient of steel	$\alpha_A = 12 \cdot 10^{-6}$ in 1/K
Length–temperature coefficient of aluminium	$\alpha_A = 24 \cdot 10^{-6}$ in 1/K

A drive structure in steel when compared with steel-reinforced belts will give the same value for the length/temperature coefficient α_A. While the distance change and thus the positioning behaviour of the linear drive is exclusively based on the expansion behaviour of the structure, the pre-tension force remains essentially constant. As temperature changes in the individual components take time to adjust to the new environmental conditions according to their mass, only temporary tensioning deviations result in the system.

With a support structure in aluminium and a timing belt with steel tension members, an increase in the pre-tension can be expected due to an increase in temperature.

A temperature increase or decrease equally leads to deviations Δs_7 in the Eqs. 4.21a and 4.21b with the same values used. If only a temperature increase is to be assumed then only the corresponding elongation value from Eq. 4.21a is used.

In Eq. 4.30, where appropriate, add (or subtract) elongation to the travel s_{ges} that results from the distance measurement between the starting or reference point and the fixed support, see Fig. 4.7.

Tangential Movement Δs_8

The tangential backlash c_{m1} is illustrated in Chapter 2.15, Fig. 2.28 and it is formed from the shortest distance between the pulley tooth gap flank and unloaded belt tooth flank, when opposite loaded flank rests on the working edge. The tangential movement Δs_8 refers to the centre position of the belt tooth in its pulley tooth gap and whose size is determined by half the value of c_{m1} (see Table 4.3). The absolute values for tangential backlash are rarely published by belt manufacturers. So far the only disclosure is found in the MULCO document "Linear Technology" [85] for the AT-belt profile. The tangential backlash c_{m1} affects the meshing between the belt and pulley, located approximately in the neutral line of action. Thus, one can, without making the result invalid, transfer the individual variance Δs_8 to the plane of the tension member. This is

$$\Delta s_8 = \frac{c_{m1}}{2}. \tag{4.31}$$

The single deviation Δs_8 is entered into Eqs. 4.21a and 4.21b using the same value.

Table 4.3 Tangential backlash c_{m1} in mm

Pitch	Normal-tooth gap	SE-tooth gap	Zero-tooth gap
AT 5	0.2	0.1[a]	0[a]
AT 10	0.4	0.2[a]	0[a]
AT 20	0.8	0.4[a]	0[a]

[a] The manufacturer's documentation should be consulted when using the narrower "SE" tooth gap or the Zero tooth gap. Due to ongoing developments in individual product programs, changes to the technical data are more than possible. It is therefore recommended that the latest documentation is requested from the chosen manufacturer

To optimize a backlash-free drive when using the AT-profile it is recommended to use the Zero pulley tooth profile up to and including at maximum 24 teeth, the SE tooth profile from 25 to 48 teeth and the normal profile from 49 teeth and upwards. Due to the applied pre-tension, the supporting forces of the bottom lands of the pulley's tooth gaps cause a tooth profile widening in the belt teeth. Also, as a result of the pre-tension, this creates changes in the position of the belt teeth in their respective pulley tooth gaps, located close to the pulley in-feed or out-feed areas, so that a backlash-free situation results, caused by "bracing" (see Fig. 2.24 in Chapter 2.13). If the motor pulley tooth gap is optimized according to these recommendations and the drive assembly is given, where appropriate, an additional pre-tension increase (Eq. 2.37), a backlash-free operation can be expected and the individual deviation from Eq. 4.31 is changed to

$$\Delta s_8 = 0. \tag{4.32}$$

Further Deviations Δs_n

In real-world linear applications further system deviations are possible. They are often negligible (very small) in their effects when compared to the total variance or they occur relatively rarely. Therefore they should be mentioned here but not treated in detail.

(a) The compliance of the drive support structure can have a negative impact on the positioning error. However, in general, the stiffness of the structure's centre distance determining components, when compared with the belt, is higher by one or more orders of magnitude. Thus their respective deviations are negligible.
(b) The location of the pitch line spacing in the belt is subject to manufacturing variations, which occur over a periodic length of the belt that is approximately the size of the belt profile [123]. These deviations are very small and are not taken into account.
(c) A linear transmission with a "creep speed" can have adverse effects on the positioning behaviour due to stick-slip effects. See Chapter 2.11 for more on this effect and recommended action.

Summary

Conclusions and measures to improve the accuracy of timing belt drives can be derived by evaluating the results of Eqs. 4.20, 4.21a and 4.21b. Particularly instructive are the intermediate results from the various calculation steps because they show the respective components of deviation on the slide position. For economic reasons the areas especially targeted for major improvements are those where change can be expected at relatively low cost. Good practical knowledge of the mechanisms will aid functional dimensioning.

The statement on the positioning accuracy quality should be taken more as an estimate. In particular, all points in the calculation process with property-related variables, such as tooth and tension member stiffness, have significant variations in tolerance. The current pre-tension and the belt length tolerance also have an effect on accuracy. There are also product brand differences and possible effects due to the belt types "Standard" and "with polyamide tooth facing". These factors were not taken into account in these calculations. The result is therefore a benchmark, with a possible variation of about 30%.

Influence of the Control System

The positioning behaviour of the entire drive system is determined by the type of motor, the servo technology and the associated electronic control system. Technical standards may include, for example, rotary encoders on the drive or motor shaft. The smallest step (or half-step) gives the minimum uncertainty of the travel that a linear slide can repeatedly approach and the motor steps that are seen by the encoder correspond to the resolution. Also software-controlled stepper motors can be used where sequences of absolute point-to-point control with prescribed accelerations and velocities are programmed. Additionally, pause cycles with a restart can be set externally. Rotational angle control systems of this type which operate in conjunction with the drive motor can be regarded as being robust in design and inexpensive to procure. It is clear, however, that with this type of control there is no direct access to the actual carriage position and as a result the impact of the previously described geometry and load-dependent individual deviations still apply.

Much higher accuracy can be achieved by control systems which detect the actual travel position directly at the linear slide. Such versions are relatively complex and sensitive as the measurement sensors have to be placed in the immediate area of movement in the mechanical guidance systems. This results in higher costs.

4.4 Dynamics and Vibration Behaviour

Timing belt linear drives are capable of highly dynamic behaviour such as changing direction in very short cycle times. Chapters 2.1 and 4.1 deal with the favourable characteristics of timing belt drives under extreme acceleration and braking conditions. These characteristics depend on the nature of the application as well as the size of the masses and the distribution of inertia in the drive system. Furthermore, the drive torque and the motor's own inertia also influence the overall dynamics of the system. Timing belt linear drives can handle accelerations well above 20 g.

Experience with many linear drive applications points to a significantly vibration-insensitive behaviour for all types of designs and sizes. Due to rapid alternating of the belt slack side to the loaded side, the stiffness and stress conditions are those of the belt spans as well as the natural frequencies. There is therefore no time for the build-up of dangerous vibration.

Undesirable vibration in the form of transverse belt fluctuations can appear in drives with large span lengths, such as those found in portal drives and storage and retrieval systems. The starting point is usually an obvious sag in a belt section which can be caused by dynamic forces such as the motor starting torque, to which the belt responds with a corresponding elongation. Since a constantly acting stimulus is missing, there is no build-up to a steady-state vibration. Rather, the elongations are characterized by a brisk damping behaviour after a few periods. The force fluctuations in the running drive are usually temporary and act randomly on the belt section. Only an unfavourable combination of an already elongated drive belt and additional vibrational build-up can cause interference with adjacent parts. Furthermore, elongations of this kind in the tension member, with partial fluctuations in belt tension, can affect the linear accuracy of the predetermined drive movement. The first recommended reduction measure is to increase the pre-tension, which reduces the vibration amplitude but will not eliminate it altogether. A particularly effective reduction measure is to physically support the sagging belt section (see Chapter 4.6 Linear positioning systems). In a closed linear drive, the belt span is constrained and supported within the aluminium extrusion and the belt lies under its own weight over the entire length of the tooth or back side within the aluminium profile. Any vibration is already suppressed by the fact that the belt is always stretched along the support. The effect of friction or frictional wear behaviour is usually without significance due to the low specific weight of the belt.

When halting at a stop point an undesirable vibrational fluctuation may result. This is essentially based on the sizes of the spring-mass system, the braking force and the friction conditions. The fluctuation exists during the necessary time for the vibration to subside. Stopping at the target position cannot be predicted, whether to a standstill, a single overrun, or a repeated fluctuation of position. When started from a defined direction, a hysteresis derived positional deviation will result due to the residual friction but in an indefinite direction from the actual position. This described resonant behaviour increases significantly with larger belt lengths and linear masses. The effective direction of vibration is longitudinal, i.e. it affects the tension member with repeated additional forces. In part, these linear transmissions are inclined to similar behaviour during starting. Especially with soft start and at creep speed, a smooth motion is not always possible (see in Chapter 2.11 describing “stick-slip effect”). A successful remedial measure is to change the stiffness of the belt, for example, by using a larger width.

Other possibilities consist of improvements to the motor control programme characteristics including both the start and braking ramps. Where appropriate “soft” power management through intelligent servo controllers is beneficial instead of “hard” starts and positioning with sinusoidal acceleration profiles.

4.5 Positioning Accuracy Calculation Example

The positioning ability of an automated tool transfer station in a work centre is to be proven. The specifications are (see Fig. 4.9):

Mass of the linear slide	m	$= 250$ kg
Belt length	l_B	$= 20{,}000$ mm
Design linear travel	s_{ges}	$= 8{,}000$ mm
Coefficient of friction	μ	$= 0.1$
Frictional load	F_R	$= 245$ N
Acceleration	a	$= 12$ m/s^2
Speed	v	$= 5$ m/s

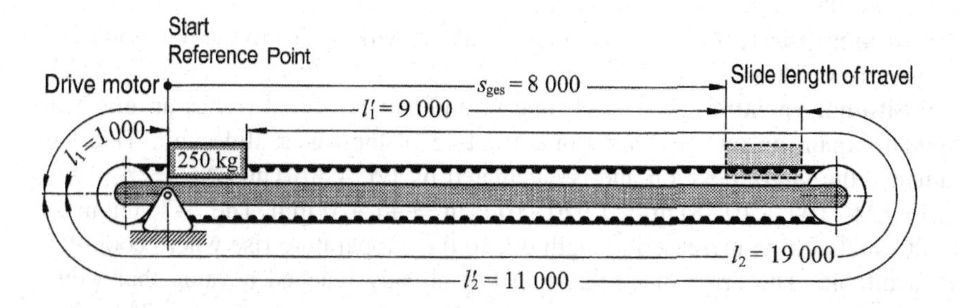

Fig. 4.9 Drive geometry for calculation example

A timing belt 50 ATL10/20,000 M is used with a drive pulley $z = 32$ of SE-toothform. The support structure of the linear slide unit is constructed of aluminium profile. The maximum allowable error is reached at the stop-point of the tool transfer station (after a travel of $s_{ges} = 8{,}000$ mm) with a positional error of $P_a \leq 2$ mm (or ± 1 mm). During a working shift the machine temperature can change from the ambient temperature of 20–30°C. The control system uses an angular encoder in the drive unit. Movement to the target position is under frictional load. With the following calculation example, the positional variations that can be expected are to be determined with a point to point control in the stationary condition and the positional error P_a is to be proven.

Positional variation Δs_1: While under way and travelling to the stop-point, errors are to be expected due to the load-dependent elongation behaviour of the belt caused by frictional forces. Thus the compliance of the belt drive has to be determined. The tooth stiffness in the arc of contact of the drive pulley is calculated by Eq. 4.24. Thus $c_P = c_{Pspez} \cdot z_e = (600 \cdot 10^3 \cdot 16)$ N/m $= 9.6\ 10^6$ Nm^{-1}.

The cable stiffness of the tension member at the position of the tool transfer station is calculated from Eq. 4.25. Thus $c_B = l_B \cdot c_{Bspez} / l_1 \cdot l_2 = (20 \cdot 2{,}800 \cdot 10^3 / 9 \cdot 11)$ N/m $= 566 \cdot 10^3$ N/m. (The values for c_{Bspez} and for c_{Pspez} are each taken from Table 7.2).

The total stiffness of the linear drive is calculated from Eq. 4.23. Thus $1 / c_{ges} = 1 / c_P + 1 / c_B = (1 / 9.6 \cdot 10^6 + 1 / 566 \cdot 10^3)$ N/m; $c_{ges} = 534 \cdot 10^3$ N/m.

This yields a Positional variation due to tooth in mesh deformation and tension member elongation from Eq. 4.22. Thus $\Delta s_1 = \sum F/c_{ges} = (245 / 534 \cdot 10^3)$ m $= 0.445 \cdot 10^{-3}$ m $= \pm 0.445$ mm.

Positional variation Δs_2: The stiffness of the belt clamps are not taken into account due to the low values. Thus $\Delta s_2 = 0$.

Positional variation Δs_3: A belt with a minus pitch tolerance is used which will be tensioned to the nominal pitch. Thus $\Delta s_3 = 0$.

Positional variation Δs_4: The influence of the tolerances of adjacent tooth pitches are according to Eq. 4.28 with $\Delta s_4 = \pm 0.05$ mm.

Positional variation Δs_5: The linear drive uses two pulleys and because of radial eccentricities there can be errors according to Eq. 4.29. Thus $\Delta s_5 = n_z \cdot f_R / 2 = (2 \cdot 0.05 / 2)$ mm $= \pm 0.05$ mm.

Positional variation Δs_6: The running radii variations due to the polygon effect are very low. Thus $\Delta s_6 = 0$

Positional variation Δs_7: A change in ambient temperature has an effect on thermal expansion on the aluminium structure of the linear slide unit. This will influence the target position and is calculated by Eq. 4.30. Thus $\Delta s_7 = \alpha_A \cdot \Delta T \cdot s_{ges} = (24 \cdot 10^{-6} \cdot 10 \cdot 8)$ m $= 1{,}920 \cdot 10^{-6}$ m $= +1.920$ mm. The "+" sign in the result stands for an increase in length due to the temperature rise where cooling is not available. The error of +1.920 mm has already reached a value that would nearly take up the maximum allowed positional error of ≤ 2 mm. Therefore, firstly, a change in the support structure material to steel is recommended, and secondly, to move the bearing point of the linear unit to shorten the transfer station (see Fig. 4.10) Δs_7: Due to the above changes the linear thermal expansion with the new conditions is calculated from Eq. 4.30 again as $\Delta s_7 = \alpha_A \cdot \Delta T \cdot s_{ges} = [12 \cdot 10^{-6} \cdot 10 \cdot (8 - 4)]$ m $= +480 \cdot 10^{-6}$ m $= +0.48$ mm (Figs. 4.8, 4.9, and 4.10).

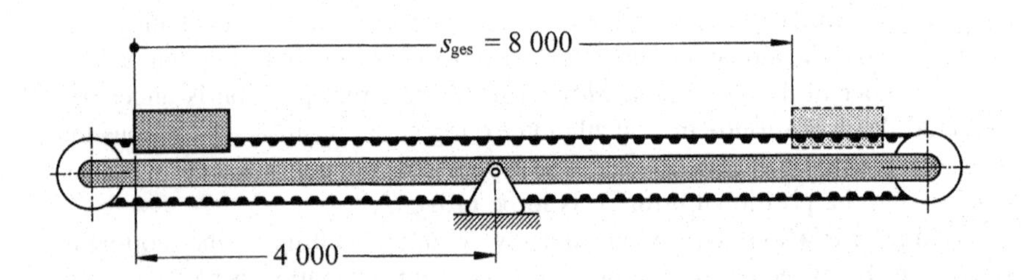

Fig. 4.10 Drive geometry for calculation with the changed design details

With the bearing point change, a clear improvement is achieved in the positional variation at the stop-point of the slide. With the constructional change, however, a position shift accompanies the start and/or reference point, i.e. the length increase acts from the fixed bearing point in both directions and in each case with half the value.

Positional variation Δs_8: Due to the use of the narrow "SE-toothform" on a drive pulley of $z = 32$ then a backlash-free tooth meshing assumed. Thus from Eq. 4.31 $\Delta s_8 = 0$.

The sum of all positive variations formed from Eq. 4.21a:

$$\begin{aligned} s_{\text{istP}} &= \Delta s_1 + \Delta s_2 + \Delta s_3 + \Delta s_4 + \Delta s_5 + \Delta s_6 + \Delta s_7 + \Delta s_8 \\ &= (0.478 + 0 + 0 + 0.05 + 0.05 + 0 + 0.48 + 0)\,\text{mm} \\ &= 1.058\ \text{mm}. \end{aligned}$$

The sum of all negative variations formed from Eq. 4.21b:

$$\begin{aligned} s_{\text{istN}} &= \Delta s_1 + \Delta s_2 + \Delta s_3 + \Delta s_4 + \Delta s_5 + \Delta s_6 + \Delta s_7 + \Delta s_8 \\ &= (0.478 + 0 + 0 + 0.05 + 0.05 + 0 + 0 + 0)\,\text{mm} \\ &= 0.578\ \text{mm}. \end{aligned}$$

The positive and negative actual deviations are added together in Eq. 4.20 to give the positional variation:

$$\begin{aligned} P_{\text{a}} &= s_{\text{istP}} + s_{\text{istN}} \\ &= (1.058 + 0.578)\,\text{mm} \\ &= 1.635\ \text{mm}. \end{aligned}$$

Summary: According to the calculations, a positional deviation of $P_{\text{a}} \leq 2$ mm ($\leq \pm 1$ mm) is possible. Thus the feasibility of the design is validated.

4.6 Linear Units

Open Construction Linear Units

Open design linear units consist of a moving carriage with guide rails and mountable end attachments, as in Fig. 4.11. Often the structure becomes a sub-structure at the point of application. Ball-return, roller-bearing linear units in lightweight, compact or heavy-duty series are available from a range of manufacturers. Mating guides can be solid shafts, hollow shafts or supported shaft and rail units. The timing pulleys are fitted in the end attachments and the driving shaft is generally *smooth* and without a keyway. The preferred choice of shaft-hub connections to the motor are conical locking bushes due to the fluctuating torques. The drive and idler pulleys do not have to have flanges in shorter linear units since the belt clamping points on the linear carriage hold the belt to the desired track.

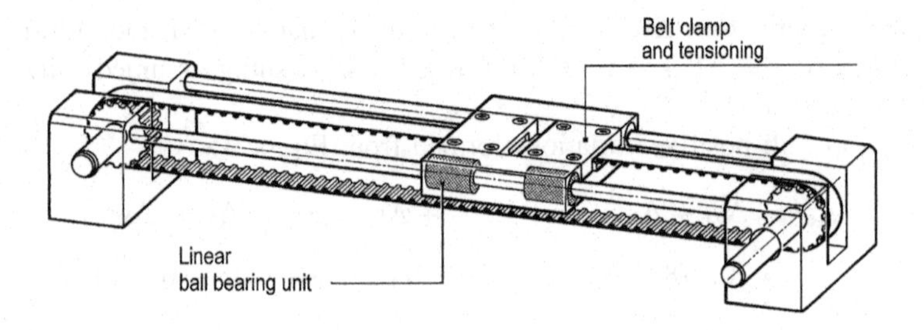

Fig. 4.11 Open construction linear positioning unit

Only at centre distances greater than 30 times the belt width are flanges / guides recommended (see also Chapter 2.10). All makes of timing belts are suitable for linear units of this type of construction.

Closed Construction Linear Units

Closed construction linear units, as in Fig. 4.12, are self-driven motion subsystems which are appropriate for driving, guiding and positioning within a larger system. The lateral forces due to weight and payload are taken up by a continuous aluminium profile in conjunction with an integrated linear guide. These guides, depending on the construction, are single or multiple shafts of rolled, hardened and ground steel. Also bolted-in, profiled guide rails can be used as guides. The common characteristic of the driven axes, such as Fig. 4.12, is that the belts can only be installed in a closed system profile by threading it through the profile. Thus, for this type of design, only a single belt can be used.

Linear units of this type are used in a number of specialized industries in the materials handling sector. With similar external characteristics, as shown in Fig. 4.12, there are differences in the detailed design and in the classification, depending on the masses to be moved. There are both heavy duty models and the so-called mini units available. The optional features differ in the quality of the linear guide, which can either be equipped with rollers, ball bearings or cylindrical roller bearings. The closed design protects the in-built components and at greater lengths the timing belt is supported on the inner wall surfaces. Belt sag is therefore not possible as the belt is always flat and span vibrations cannot develop. The friction caused by the weight of the belt is negligible.

For units of increasing cantilever length, the section moduli of the aluminium profiles for resisting torsion or bending, need to be known. In most cases, all brands are available in variable lengths. Standard lengths to about 4 m and in some cases up to 8 m are available with multiple units assembled into greater lengths. The units can be used horizontally, vertically or in any orientation.

Complete packages include the associated electronic controllers with flanged servo and stepper motors and options for entering speeds and distances. The user

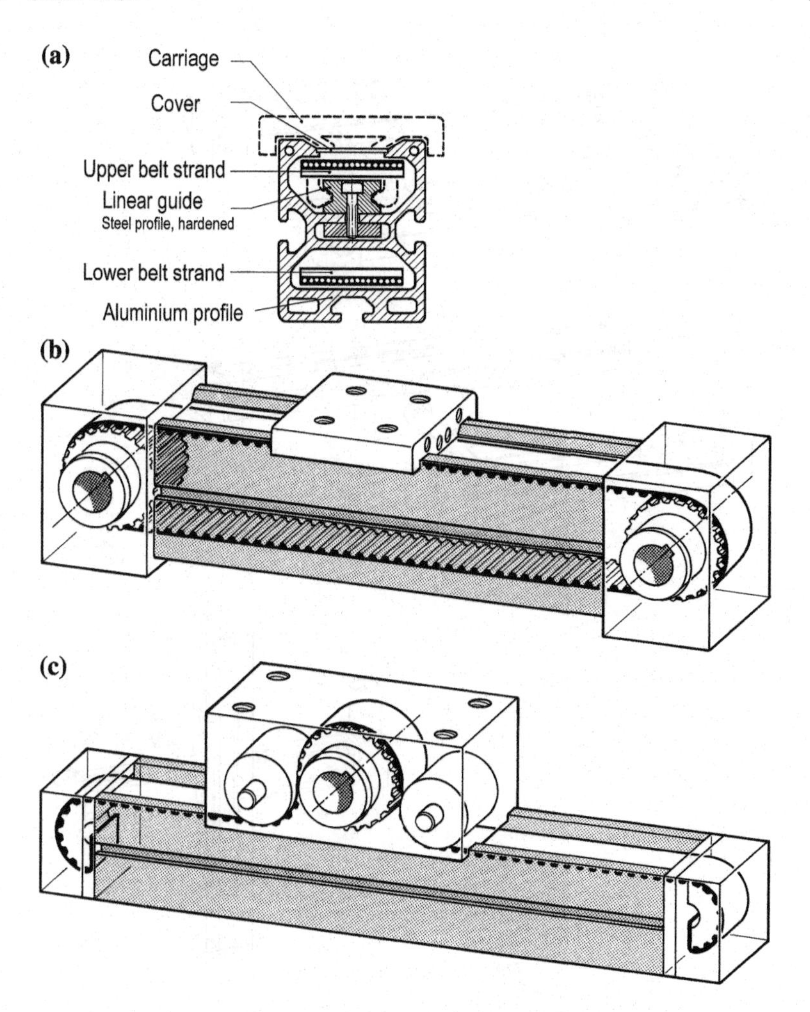

Fig. 4.12 Closed construction linear positioning system: **a** Sectional view, **b** carriage type, **c** bridge type

can move the carriage manually to the target position, for example, and then use the teach-programming functions.

No matter which toothed belt type the linear units are equipped with, their operation is always principally the same. Estimates of the positioning behaviour of all types are possible by referring to Chapter 4.2–4.5 The user needs to solve automation tasks optimally and safely through sound design. This applies, in addition, to product-specific features such as installation conditions, mass to be moved, speed, cycle time, accuracy and lifetime as well as the acquisition and operating costs, in order to select the appropriate linear unit specifically from an overall point of view [41].

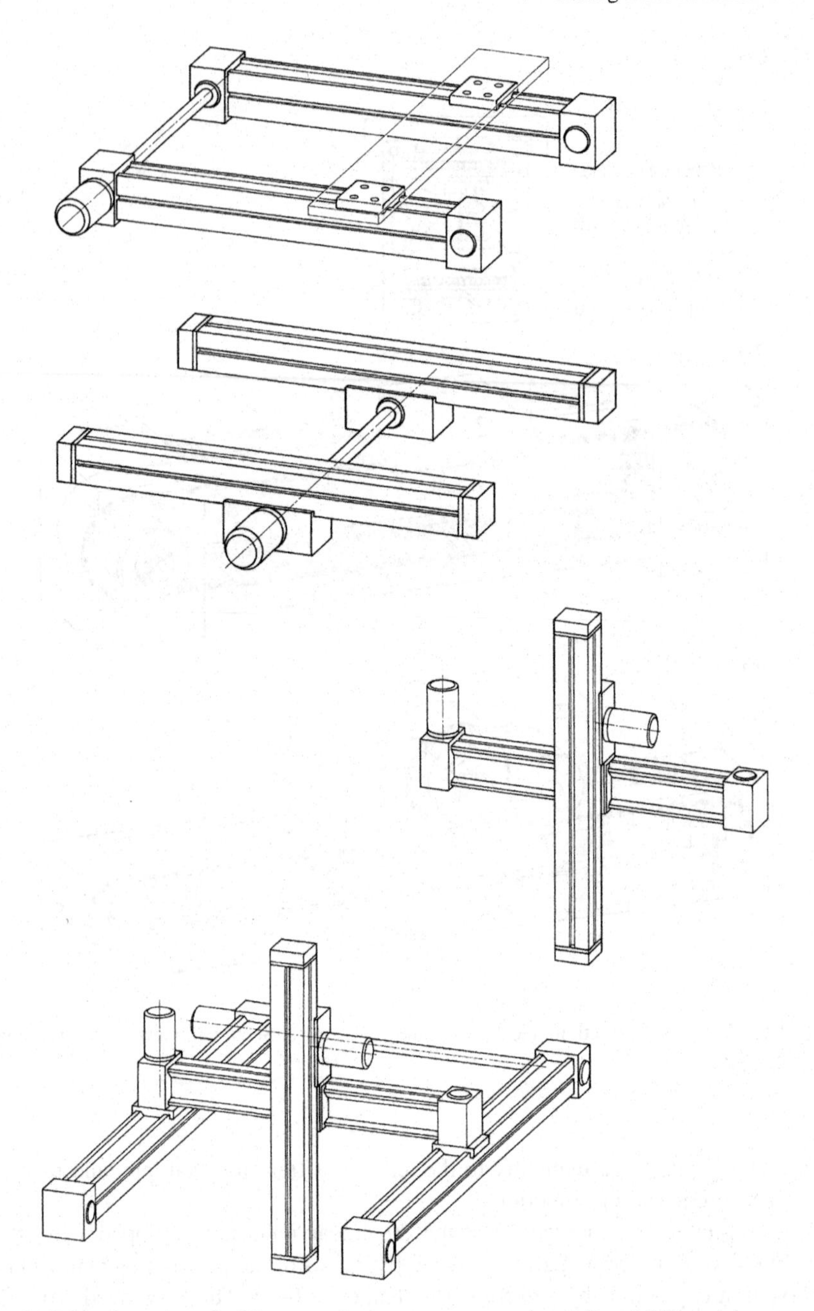

Fig. 4.13 Configurations for Linear motion systems, arranged as 2-D and 3-D portals

Closed construction linear units, for example, can be used in pairs in automated assembly lines as portals with a stationary guide profile and a moving carriage or as a stationary carriage and moving body profile. By combining two or more units, 2-D or 3-D portals can be created. T-slots on three sides of the extruded aluminium profiles facilitate this sort of system design. Double-ended hollow shafts enable the motor to drive from either side. The preferred application area is automation operations such as gripping, transferring and storing parts and feeding into and removal from manufacturing cells. The following examples illustrate the diversity of use, see Fig. 4.13.

Examples of Implemented Linear Units

The analysis of build space, loads, stroke length, cycle time and positioning requirements affects the choice of a linear system. If this includes delivery and pricing the project engineer will prefer tested components, see Figs. 4.14 to 4.30. In general the use of certified components increases the reliability and reduces design effort and time. The arrangement of two or more linear units in X–Y and/or in X–Y–Z configurations means surface and/or 3-D portals can be accomplished,

Fig. 4.14 Linear unit from Bosch Rexroth Group [7]. Depending on the application different variants are available: Assembled length; Custom servo, three-phase or stepping motor; Choice of planetary gearbox integrated in the timing pulley; Position detection; Control system

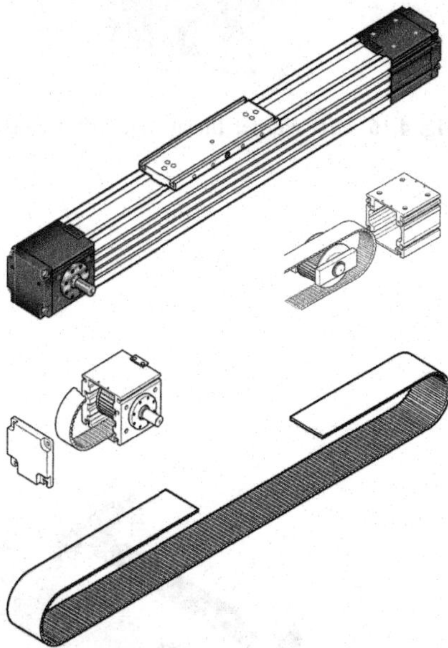

Fig. 4.15 Construction of a linear unit from Bosch Rexroth Group [7] consists of: Aluminium-profile system including guides; Drive end assembly with shaft; Idler end assembly; Timing belt

see also Chapter 4.12. Selected construction methods are shown below. Manufacturer web pages are given in the bibliography.

The optional selections can also apply to other examples. Through selectable gear ratios in the planetary gearbox, the customer can adjust the drive torque to the requirements of the application. This makes it possible to fine-tune the external masses and the motor inertia in an optimal manner. This is especially important for the optimization of the drive controller and facilitates highly dynamic drive systems.

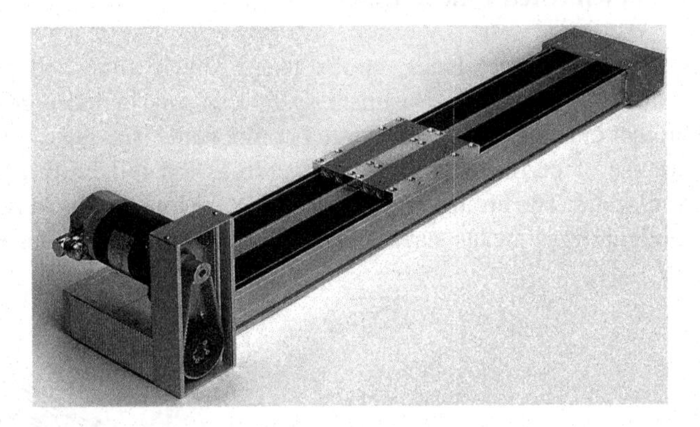

Fig. 4.16 IEF-Werner linear unit with two parallel belts to increase the stiffness [58]

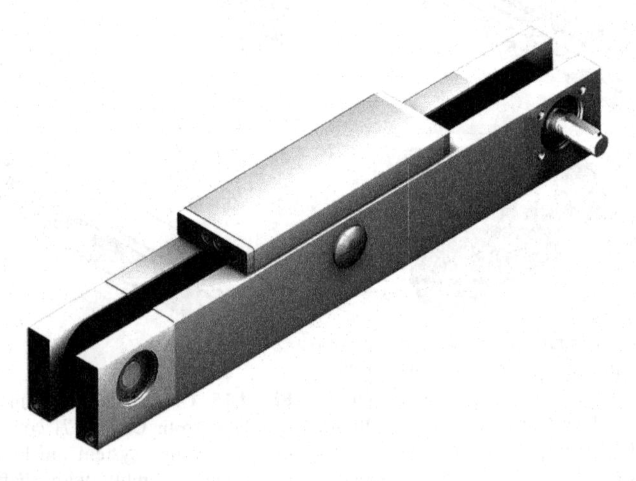

Fig. 4.17 Bahr linear unit [1]

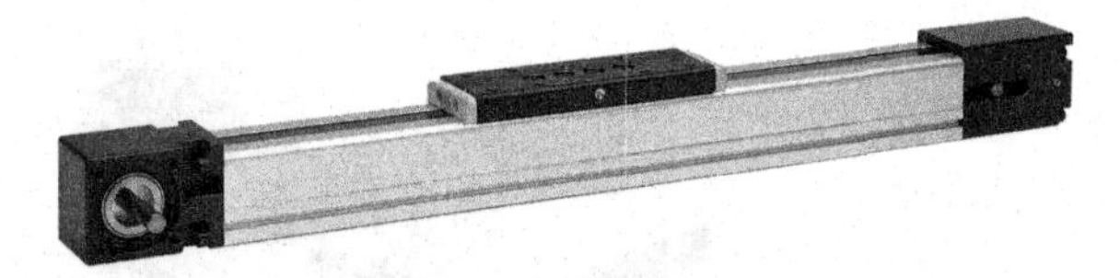

Fig. 4.18 Dreckshage linear unit [23]

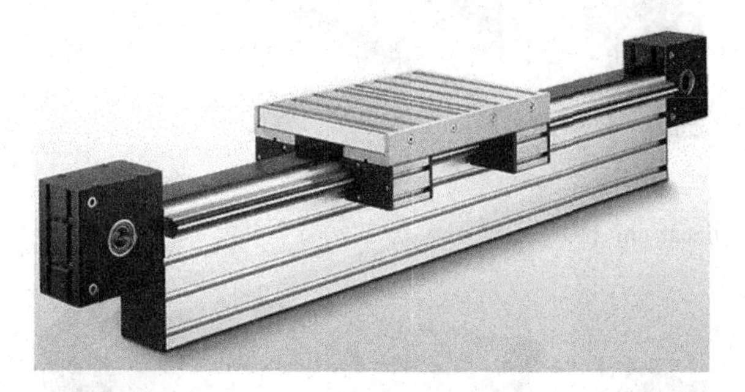

Fig. 4.19 Item linear unit [56]

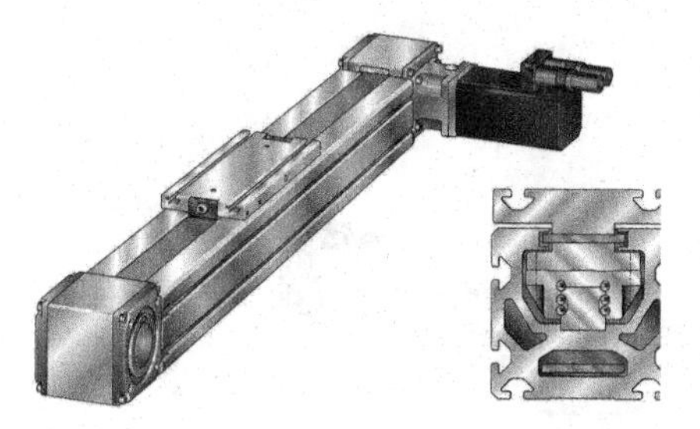

Fig. 4.20 INA linear unit with planetary gearbox and servo motor. Connections for position measurement and motor controller [52]

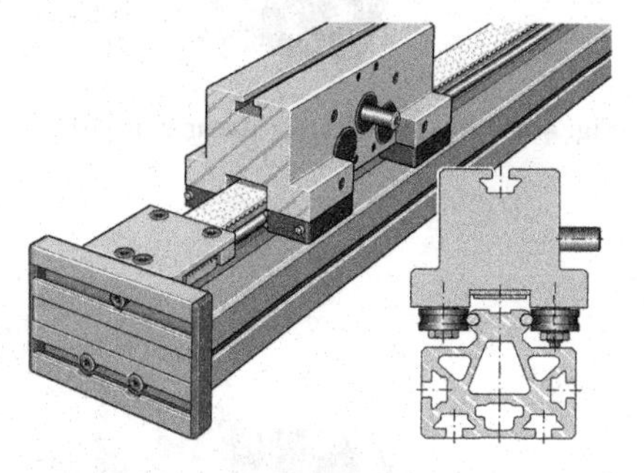

Fig. 4.21 INA bridge linear unit. Omega drive installation in the carriage. Four rolled-guide shafts with roller bearings [52]

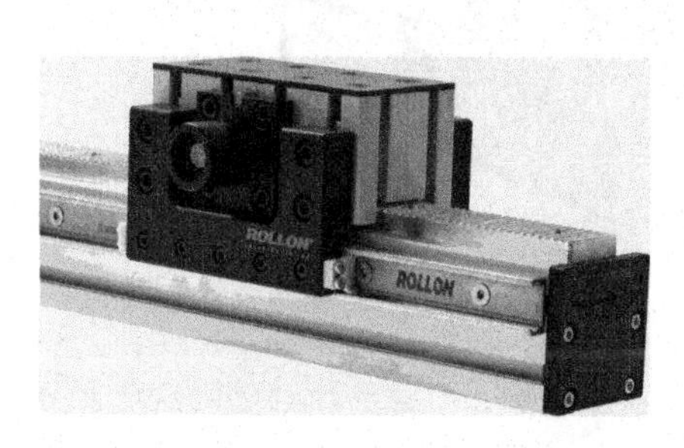

Fig. 4.22 Rollon bridge linear unit [103]

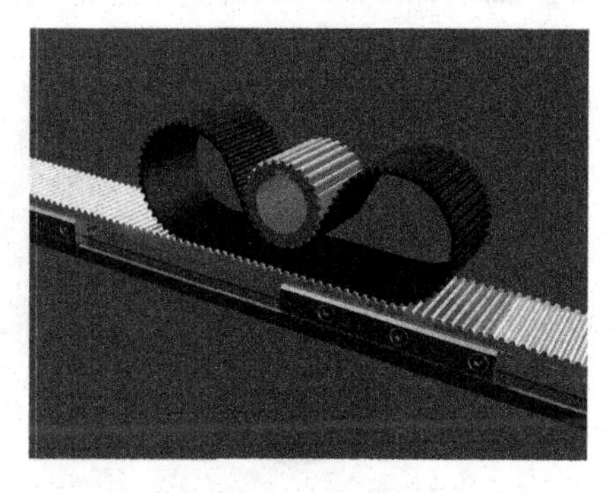

Fig. 4.23 Rollon omega drive system with intermeshed linear rack [103]

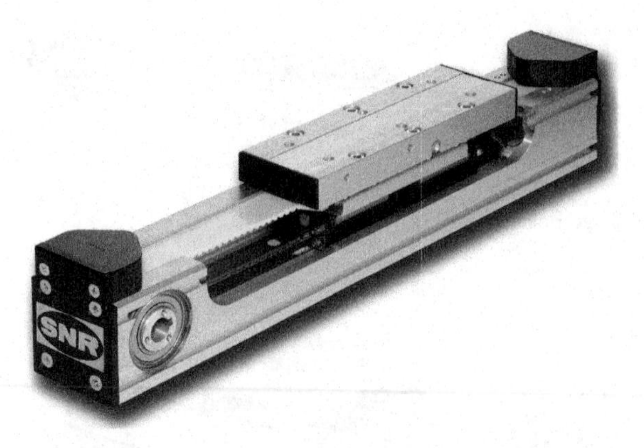

Fig. 4.24 SNR linear unit [106]

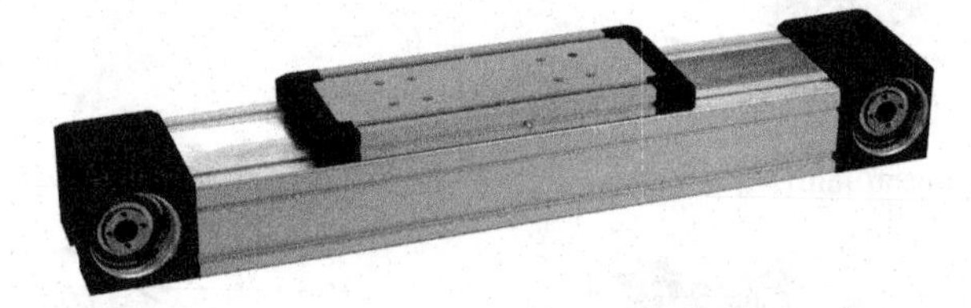

Fig. 4.25 Rose + Krieger linear unit [104]

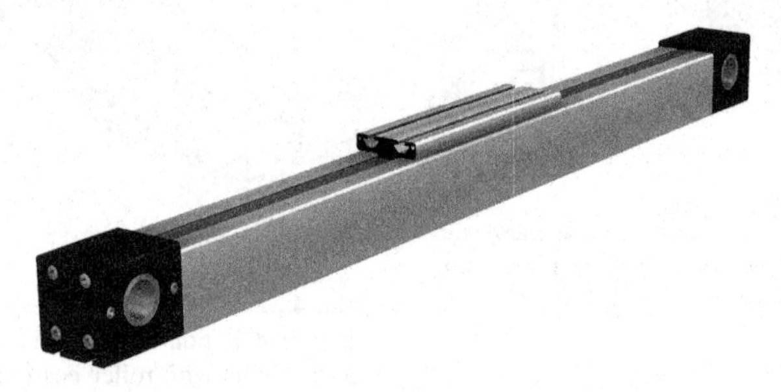

Fig. 4.26 Paletti linear unit [95]

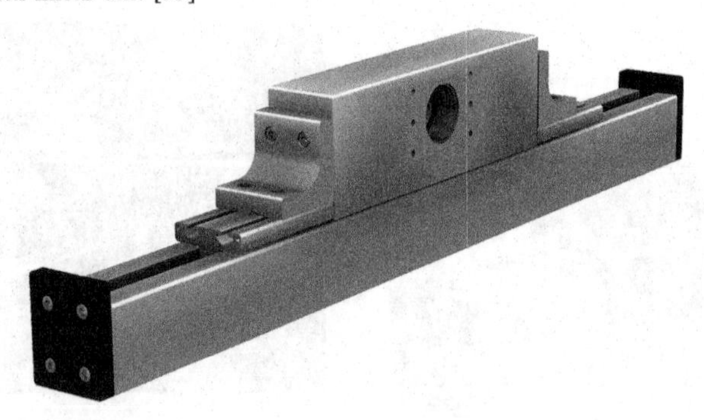

Fig. 4.27 Paletti bridge linear unit [95]

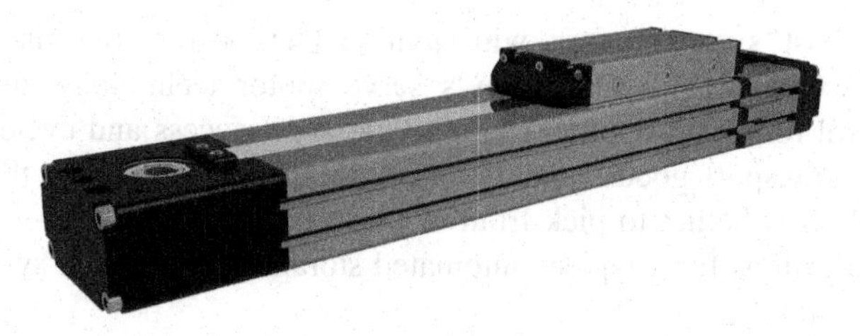

Fig. 4.28 Motus linear unit [82]

Fig. 4.29 Isotec long distance unit with *two* driving carriages [51]

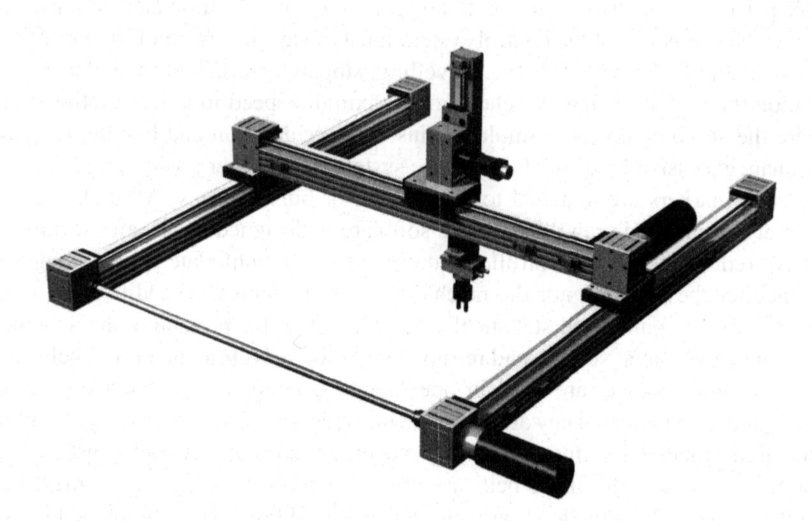

Fig. 4.30 Hepcomotion X–Y–Z pick and place motion system. Constructed from Hepco Motion linear and drive components [47]

4.7 Automated Storage and Retrieval Systems

Automated storage and retrieval systems using stacker cranes are basically unmanned, computer-controlled warehousing facilities. The most common constructions involve rail-mounted devices utilising single columns of welded box sections. Only one travelling beam is usually used to reduce maintenance. Traction and control signals are transmitted via contact lines or towed energy chains. The best systems work with non-contact power transmission via an inductive power supply. Using today's servo motor technology and advanced process control results in optimal routing with short access and cycle times. The aim being to transport goods quickly, accurately, safely and with the minimum effort from A–B, whether to pick from or place into storage.

The specifications for a typical automated storage and retrieval system using a stacker crane are:

- Stacker length : up to 150 m
- Stacker height : up to 40 m
- Vertical speed : up to 3 m/s
- Horizontal speed : up to 10 m/s
- Work rate : up to 500 stock movements per hour

The given data are maximum values because, due to the huge variety of different parameters possible, an individually designed system would not be representative. Factors to be considered include whether the system is for pallets, containers or small components or whether intermediate products are also planned for inclusion.

A particular security concern relating to unmanned automated storage and retrieval systems is that the control system must always be aware of the location of the crane and its current state (i.e. travelling, stopped, etc.). To arrive at a specific location the system is first switched from maximum speed to a deceleration phase where the servo controllers simulate sinusoidal acceleration and braking ramps to eliminate excessive loads on the dynamic system. For further positioning accuracy, reference markers are attached to the racking or bin locations. All picking takes place at a controlled crawl speed. The software is designed to be safe, secure and to have redundancy. The controller is designed to accommodate power outages so that neither the program nor the mechanical systems crash. In addition, all faults are recorded and analyzed statistically to derive data for predictive maintenance.

In such automated storage and retrieval systems, polyurethane timing belts with steel tension members are used in open lengths, either as a driving or control/positioning component. They are available in any length. The steel tension members offer high values of stiffness with high accuracy/repeatability and constant low-maintenance operation. If the belts are used as a control/positioning element then lifting and other dynamic loads are undertaken by cables and the timing belt is used only for confirmation of current speed, or to establish the position. The dynamic belt loads are therefore low and any deviations due to differing elongations between belt and cable are negligible. Technical data on the timing belts used to date:

- Longest system length to date : over 200 m
- Positioning belt types : 16T10, H8M-20
- Load bearing belt types : 150AT20, H14M-100
- Maximum allowable tensile load : over 30,000 N
- Maximum break load : over 100,000 N

The designer will be interested in potential belt arrangements for such special equipment. The following examples in Figs. 4.31, 4.32, 4.33, 4.34 and 4.35 show

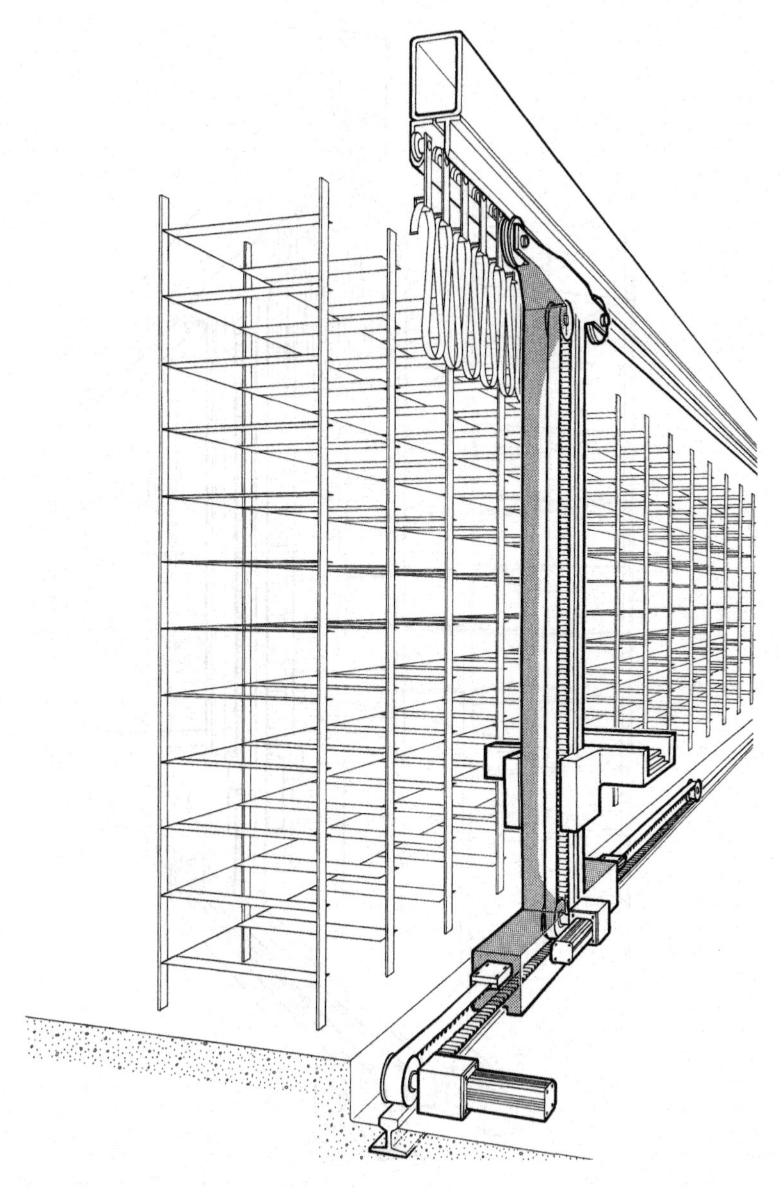

Fig. 4.31 Stacker crane with *two* independent timing belts. The horizontal drive motor is stationary (in the foreground). The hoist drive motor moves along the guide beam below the crane pillar

five options for different horizontal-vertical concepts each with its own characteristics. The different belt drive geometries also require differing control strategies. What is suitable for one particular application does not necessarily suit another and it is recommended to decide the list of requirements in each case.

The five designs give the user diverse options for the drive geometry. The different solutions presented cover the same X–Y movements as also found in

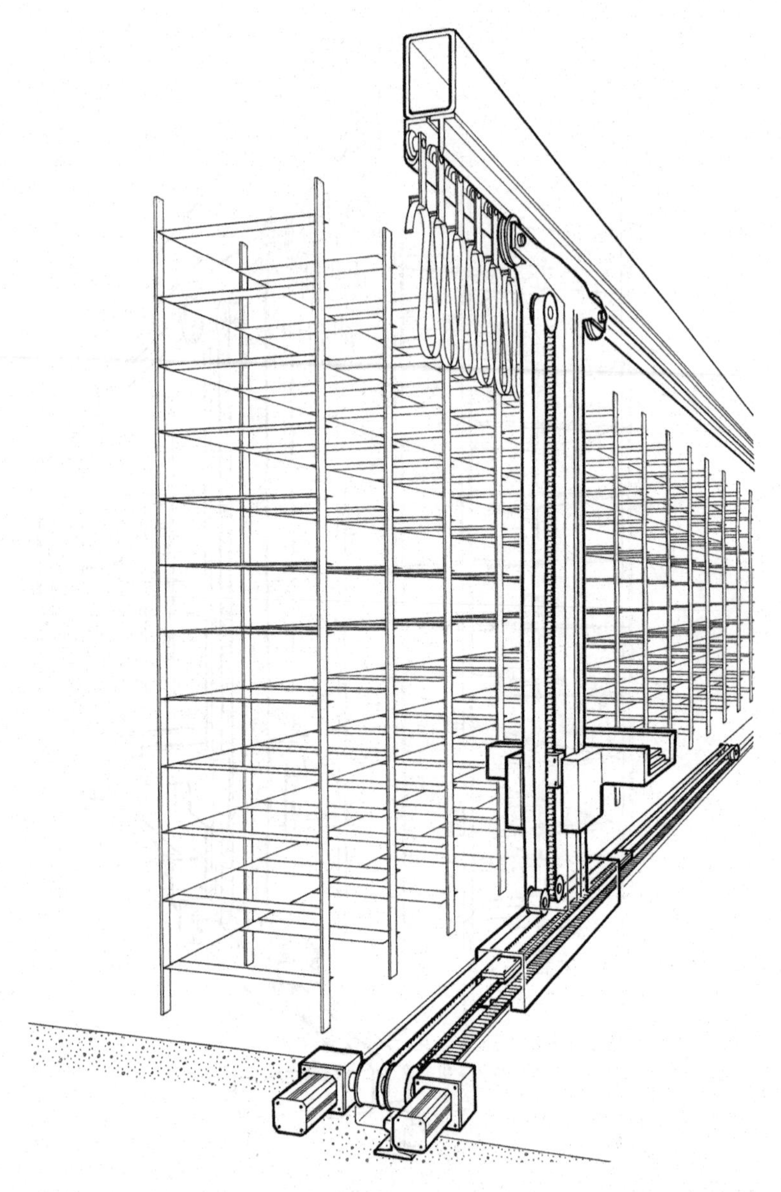

Fig. 4.32 Stacker crane with *two* independent timing belts. Both motor drives are stationary with the *left* motor controlling the *vertical* motion and the *right* controlling the *horizontal* motion of the crane. To move the crane horizontally both motors must rotate synchronously, however to raise the vertical axis only the left motor needs to be used. Diagonal movement is achieved by a combination of motion in both axes

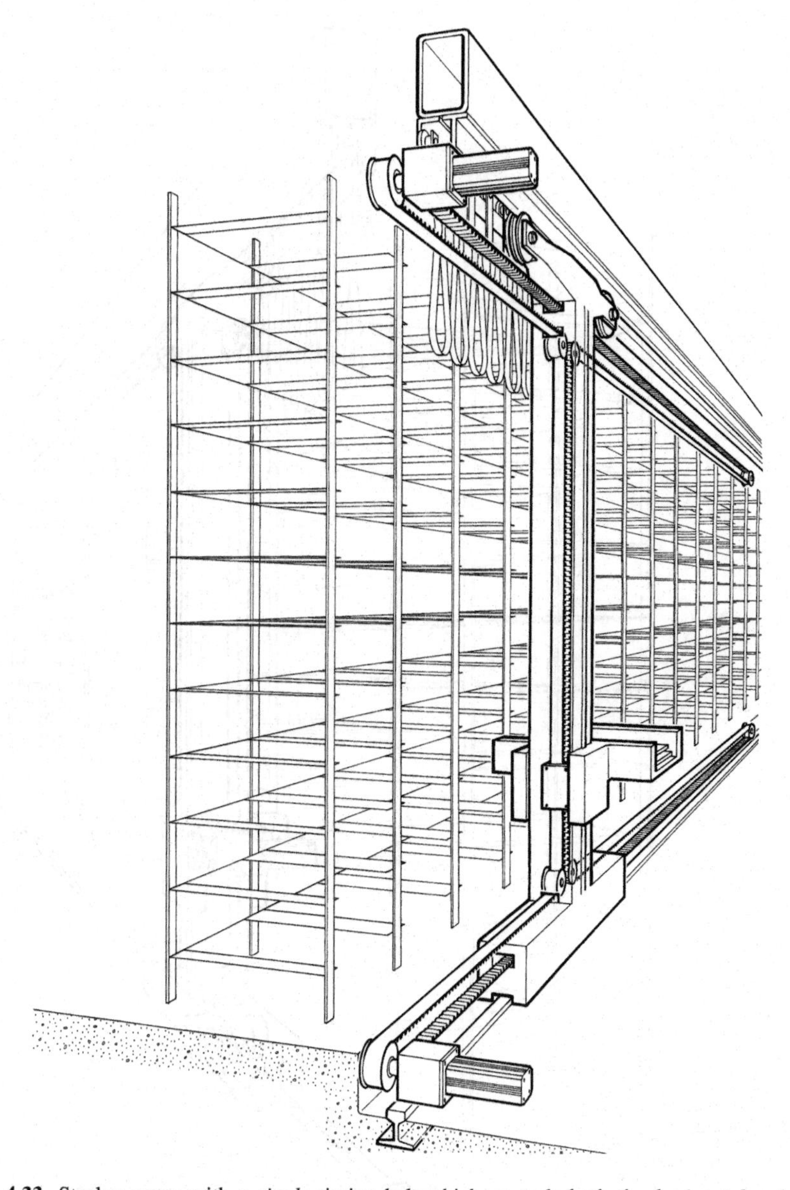

Fig. 4.33 Stacker crane with a *single* timing belt which controls both the *horizontal* and the *vertical* motion. The drive consists of an open-ended length of belt with each end attached to the crane truck. The drive layout is similar to an "H" pattern with a drive motor "above" and "below" at the ends of the horizontal axes. When the motors are actuated in opposite directions, at the same speed, the crane moves horizontally and then vertical movement is achieved with the motors rotating in the same direction. Any other combination of different speeds and/or directions of rotation move the crane diagonally

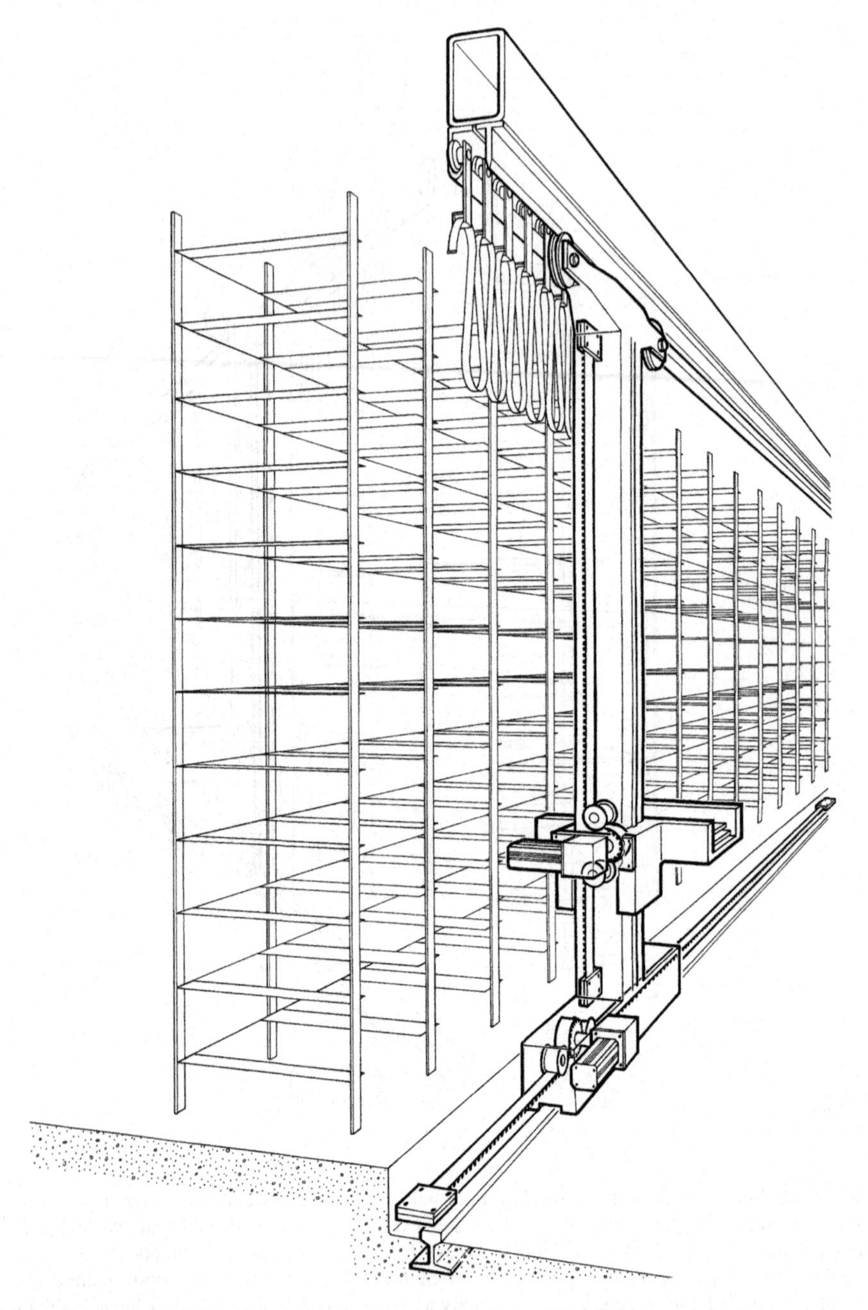

Fig. 4.34 Stacker crane with *two* independent timing belts. The geometry of the belts meshing with their respective drive pulleys is similar to the shape of the Greek letter *Omega*. Both motors travel along their axis of control. The ends of each belt are clamped rigidly to their axis. The horizontal timing belt lays parallel to the ground

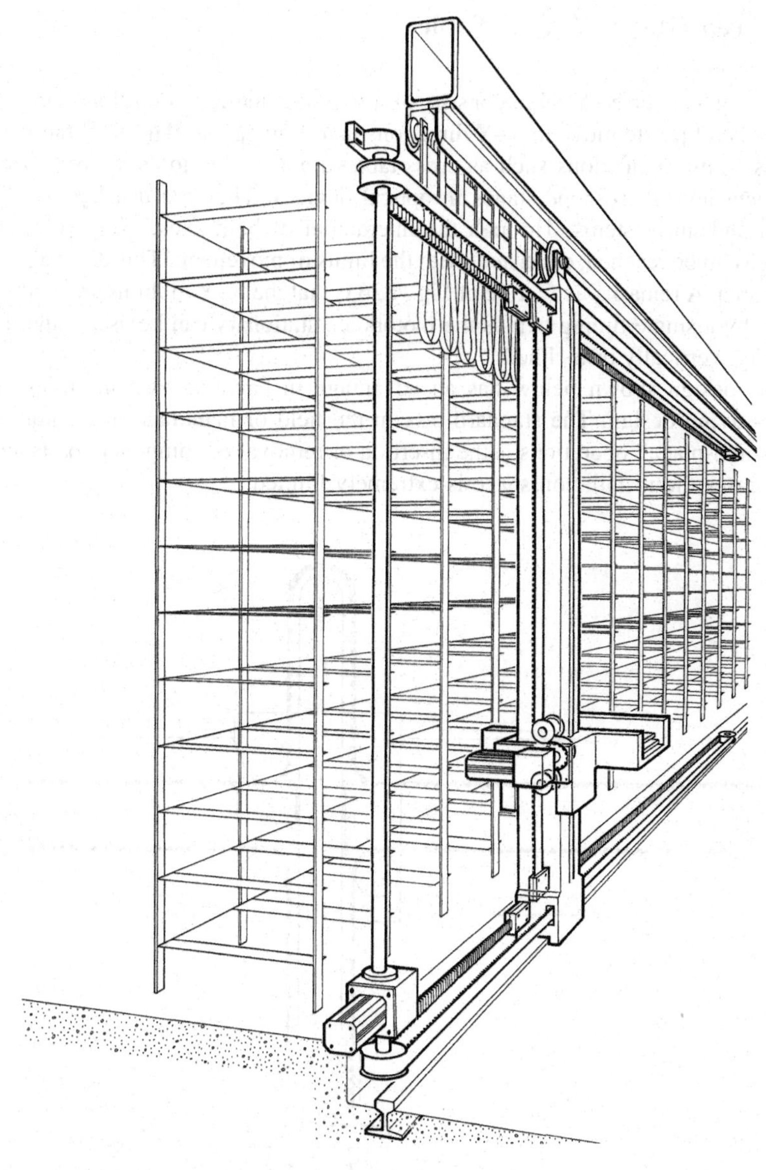

Fig. 4.35 Stacker crane with *three* separate belts. This design does not require a travelling top beam because it is driven at both the *top* and *bottom* by *one* shaft. The horizontal movement of column is achieved through the parallel pairing of the top and bottom belts

co-ordinate tables or portal robots. All these drive concepts have the common feature that they always are of great length with very long belts. The inevitable belt sag is countered by the fact that the belt is simply stored on a suitable supporting structure. Belt compliance, positional accuracy and stiffness are discussed in Chapter 4.2–4.5.

4.8 Area Gantry / X–Y Table

An area gantry or X–Y table consists of a two-axis motion control system which allows an object to move in X–Y directions in a single plane. The X–Y table name comes from applications such as cross-tables for machine tools as well as measurement and microscope stages in optical devices. They are usually moved by worm and nut systems. Through the interaction of both axes, every position in surface can be reached, consistent with the limits of movement. This determines the work area. A remarkable feature of Fig. 4.36 is that the X–Y motions are controlled solely by a single timing belt as the motors are stationary can be used either horizontally, vertically or inclined.

The design shown below has an advantage in handling systems through its space-saving design. The standard task in the field of manufacturing automation "pick-and-place" operations is the insertion or removal of component parts and in most cases the installation space is extremely limited.

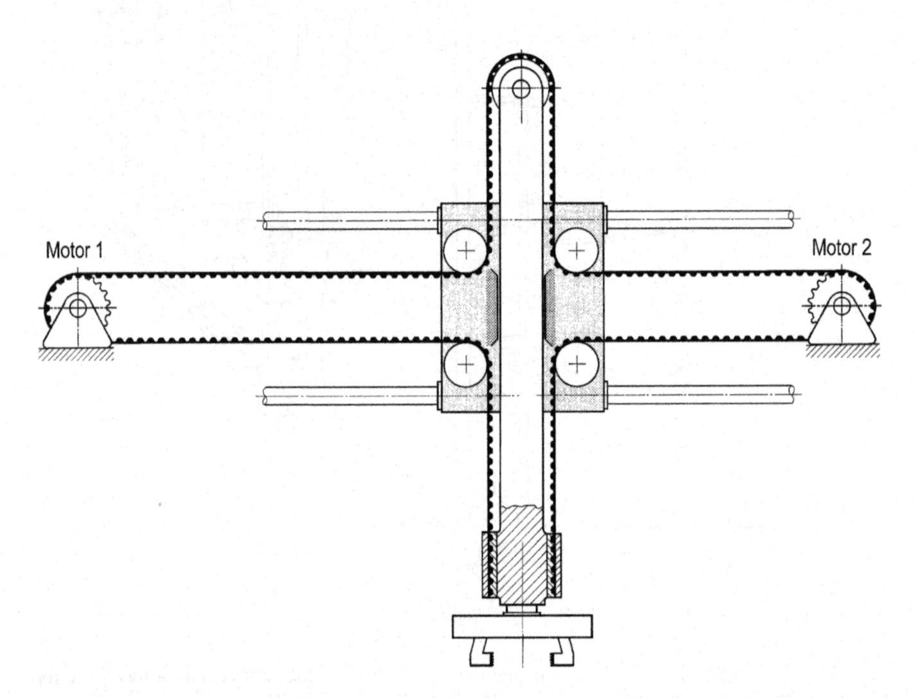

Fig. 4.36 X–Y handling system. Positioning: the same motor *rotational* direction and speed moves the gripper *horizontally*, *vertical* motion occurs with the motors rotating in *opposite* directions at the same speed. Any other combination of *different* speeds and directions of rotation will make the gripper move *diagonally*

The design of Fig. 4.37 is properly described as an X–Y handling device or area gantry. A belt drive geometry in character of "H"-pattern. Systems of this kind can also be constructed from single-axis devices, see Chapter 4.6.

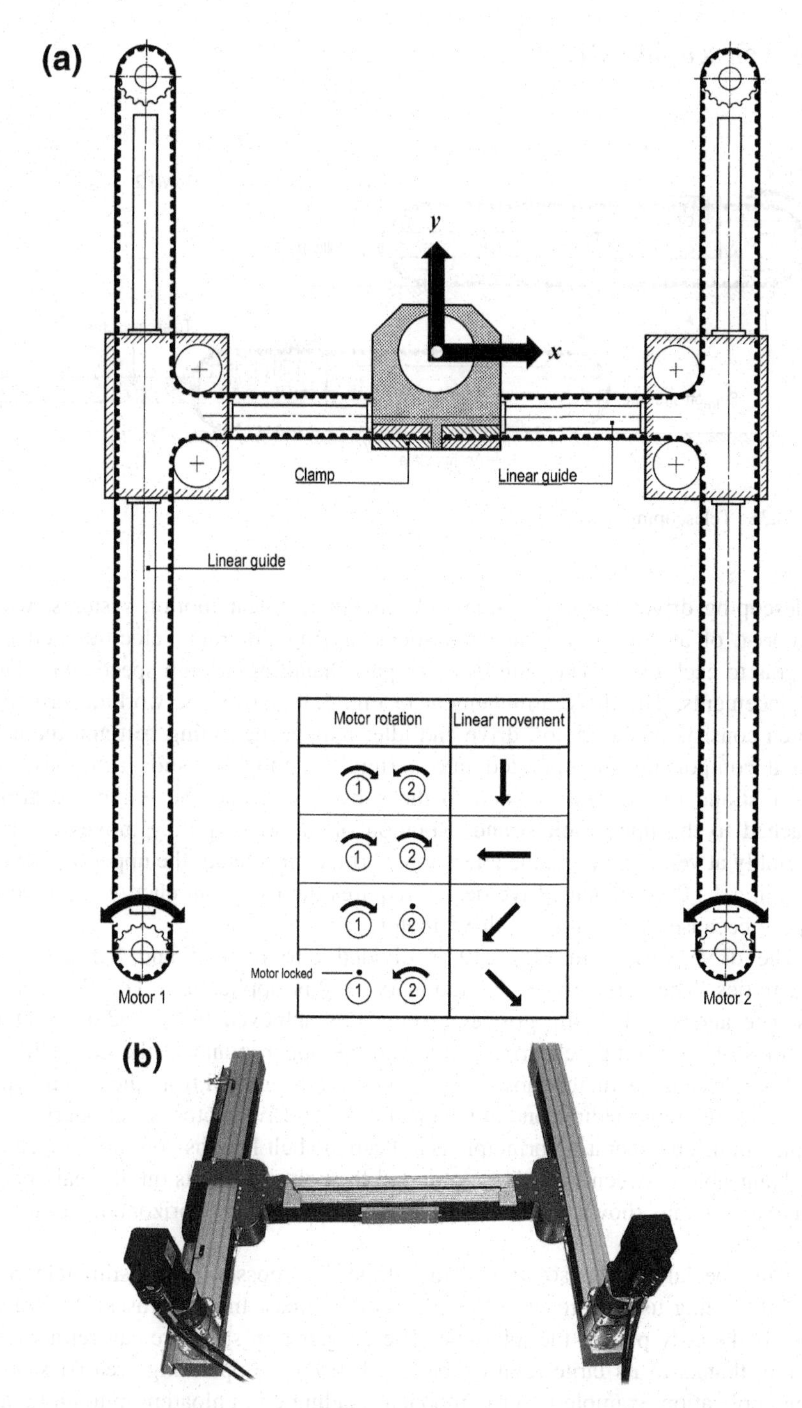

Fig. 4.37 **a** Area gantry. *Top view*. Compare this with Fig. 4.33 "Storage and retrieval system" **b** Area gantry. X–Y linear drive in character "H"-pattern from Isotec [51]

4.9 Telescoping Drives

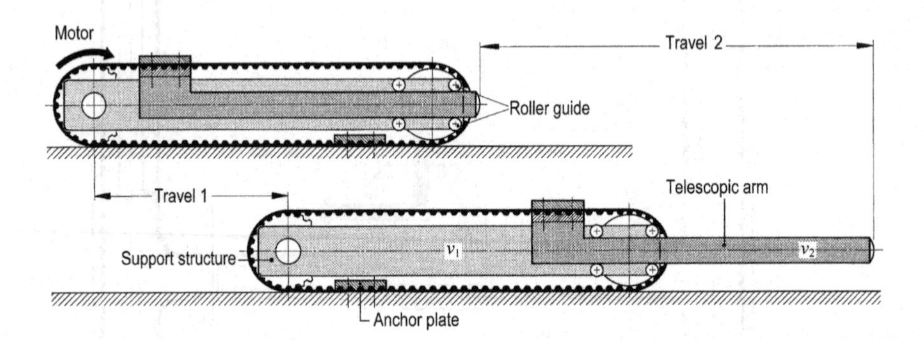

Fig. 4.38 Telescoping drive

Telescoping drives are extendable and retractable linear motion systems, which comprise of at least two connected slides, having different velocities and path lengths to each other. They are used for parts handling at high speeds with large displacements. The drive arrangement in Fig. 4.38 shows the working principle, which consists of the motor, drive and idler pulley, the timing belt and the associated components incorporated into a rigid but movable slide assembly. The lower strand of the belt is fixed to the substructure and the telescopic arm is attached to the upper belt strand. Rotation of the drive pulley moves the base assembly at velocity v_1 and the telescoping arm, connected to the upper belt strand, at v_2 in a 1 : 2 ratio. This drive design requires the use of an energy chain power supply because of the integral drive motor.

The drive variant in Fig. 4.39 is divided into toothed and flat belt drive assemblies. The drive motor is stationary in this application. The doubling of distance and speed of the gripper assembly is achieved by the addition of the motions of the timing belt driven slide and the upper strand of the flat belt.

The telescopic actuator shown in Fig. 4.40 is constructed from the two movable elements, the intermediate and the main slide. The drive motor is stationary in this application. The operating principle is to keep the belt lengths both unchanged and unchangeable between the drive motor and the belt end points on the main boom. All the systems shown can be used in any orientation (horizontal, vertical or inclined).

This special motion solution demonstrates the possibility of slim telescopic modules being used to reach difficult assembly areas. In most cases, the “reaching” is the only part of the job done. The gripper arm structure can retract completely thanks to its large return travel back from the operating area (risk area). One application example is for component loading and unloading into and out of

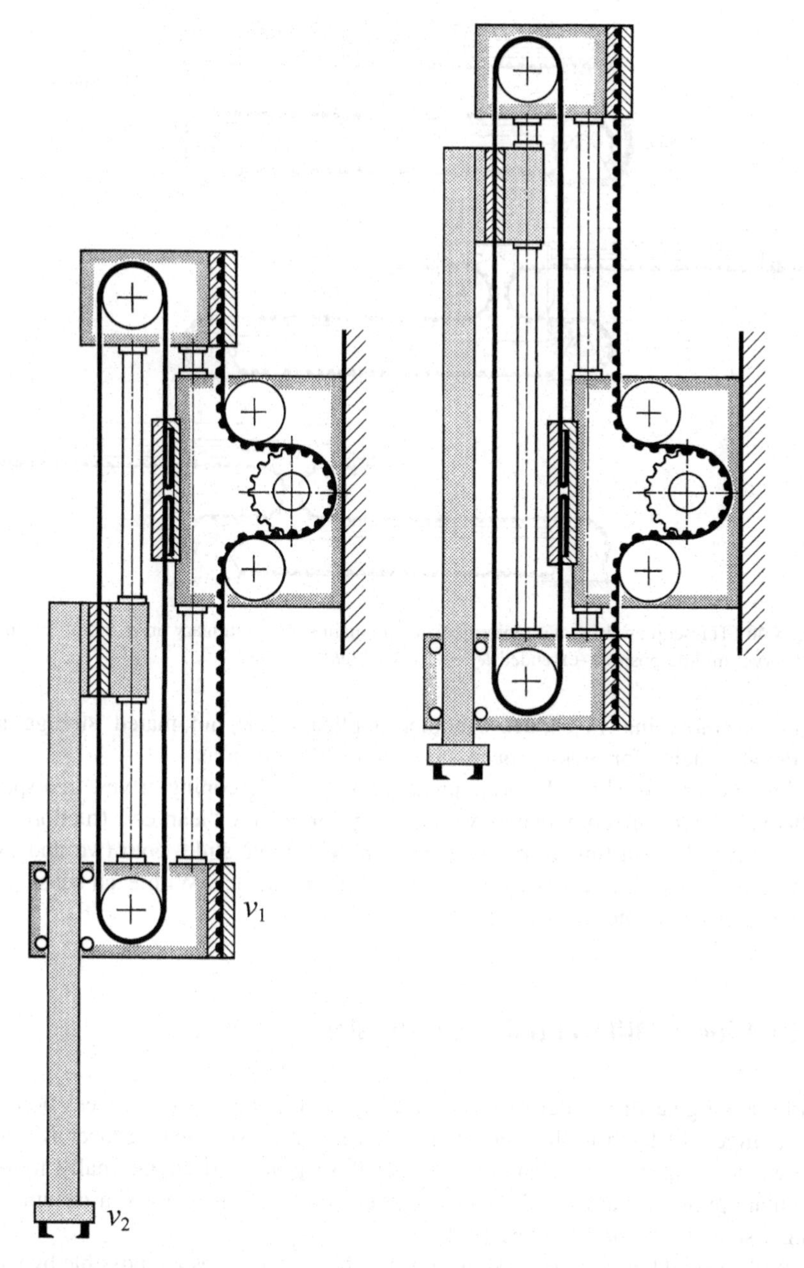

Fig. 4.39 Telescoping drive using a coupled toothed and flat belt variant. The drive motor is stationary in this application

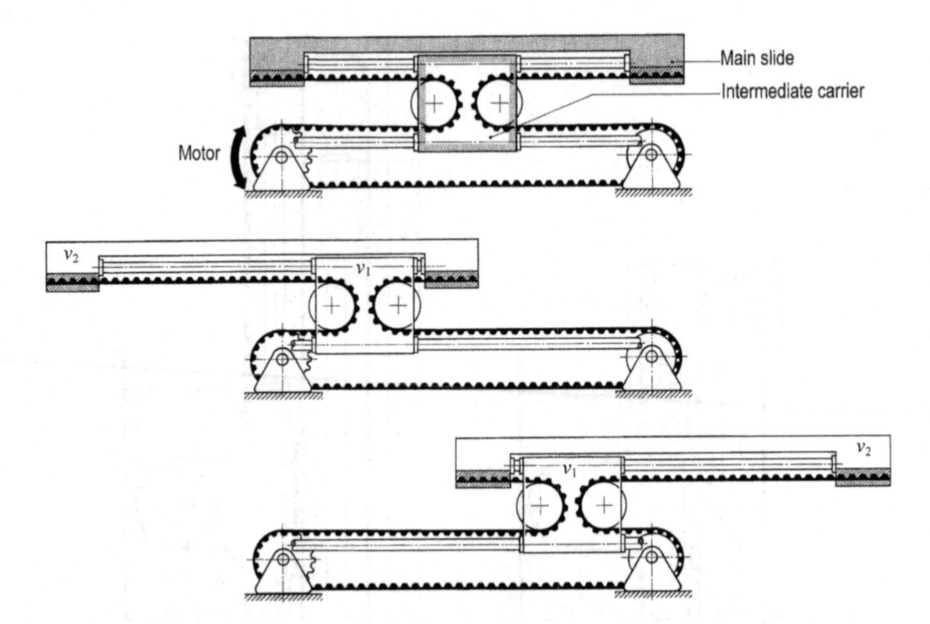

Fig. 4.40 Telescoping drive with intermediate and main slide. Stationary drive motor. From *top* to *bottom*: middle position–extended *left*–extended *right*

injection moulding machines. Another application is automated storage and retrieval systems for picking or re-stocking of bin locations.

These examples show the principal solutions for telescoping drives. In a space-efficient design, the components would be optimized for "lean construction".

Additionally, starting with a single-acting telescopic slide, both two and multiple-stage telescopic drives can be achieved in one device. With each subsequent stage both speed and distance travelled doubles.

4.10 Linear Differential Transmission

Differential gear drives derive their ratios by subtracting the differences between the numbers of teeth in their gears. The Harmonic Drive® gear reducer is a well known application which has an external flex spline and an internally toothed circular spline in diametrical contact with each other. It is a co-axial design with single-stage ratios of 30–320:1 [28].

With special belt / pulley designs, linear differential drives are possible by using timing belt drives. The configurations in Figs. 4.41 and 4.42 show layouts which each contain two-stage differential ratios. Ratios of 2–2,000:1 are possible using this type of drive geometry, depending on the size and arrangement of the selected pulleys. The drive motor is preferably fixed at z_A. The pulley sets z_1/z_3 and z_2/z_4 are rigidly interconnected and form their own related rotational units.

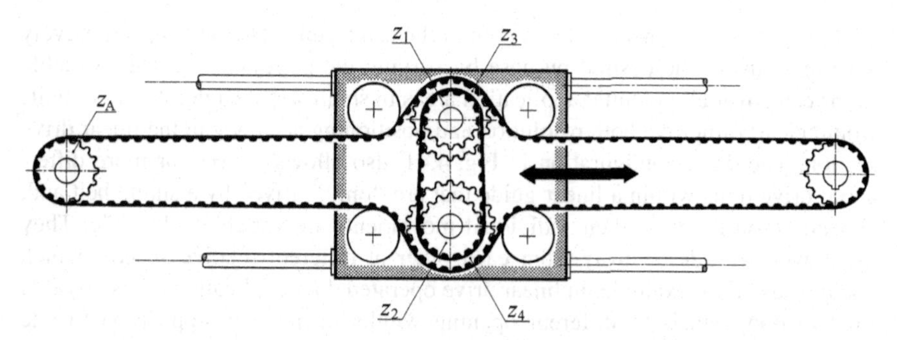

Fig. 4.41 Differential drive with the pulley reduction inside the linear unit (*travelling*). Cybertron [18] was the originator of this interesting motion system

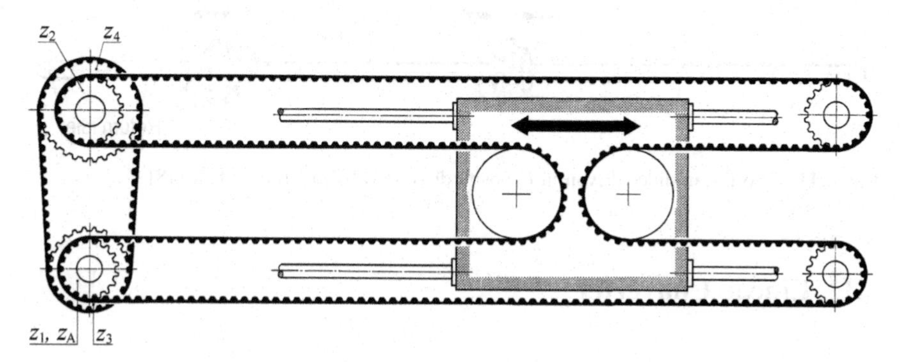

Fig. 4.42 Differential drive with pulley reduction outside of the linear unit (*fixed*)

The ratio i for the above linear drive can be derived from the number of teeth in z_1–z_4 by

$$i = \frac{1 + \frac{z_1}{z_2} \cdot \frac{z_4}{z_3}}{1 - \frac{z_1}{z_2} \cdot \frac{z_4}{z_3}}. \tag{4.33}$$

The distance travelled s_{ges} by the linear slide is determined from the rotation angle δ of the motor drive pulley with z_A teeth and the belt pitch p in conjunction with the ratio from by Eq. 4.33:

$$s_{\text{ges}} = \frac{z_A \cdot p}{i} \cdot \frac{\delta}{360}. \tag{4.34}$$

With the angular velocity ω, the velocity v of the linear slide can be calculated

$$v = \frac{z_A \cdot p \cdot \omega}{i}. \tag{4.35}$$

The speed of the linear slide, in both of the drive geometries above, is relatively easy to adjust to the desired purpose by varying the pulleys z_1–z_4. The available ratios can cover all potential application areas in small steps. High ratio drive units would clearly improve both resolution and positioning accuracy in the linear drive section. The drive configuration in Fig. 4.41 also allows for two or more differential drive units within a linear guide that are "only" driven by a single belt and, depending on the task, their individual movements are variably selectable. They can travel towards or away from each other at different relative speeds, which could be used, for example, in linear drive operated door applications. It is possible for two door panels of different opening widths to move in opposition to one another and each at their own speed (see Fig. 4.43).

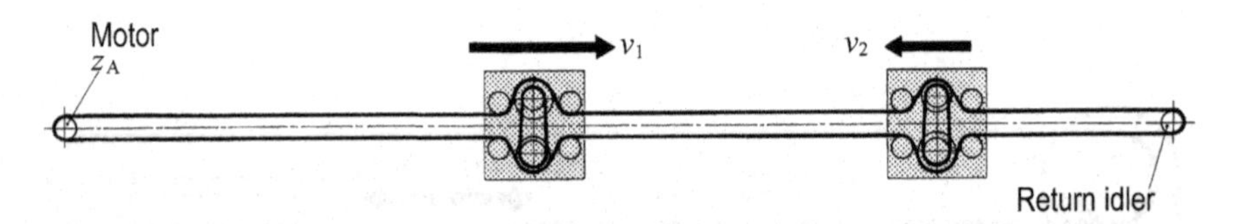

Fig. 4.43 Two linear slides driven in opposite directions by *one* timing belt [18]

4.11 Linear Converter

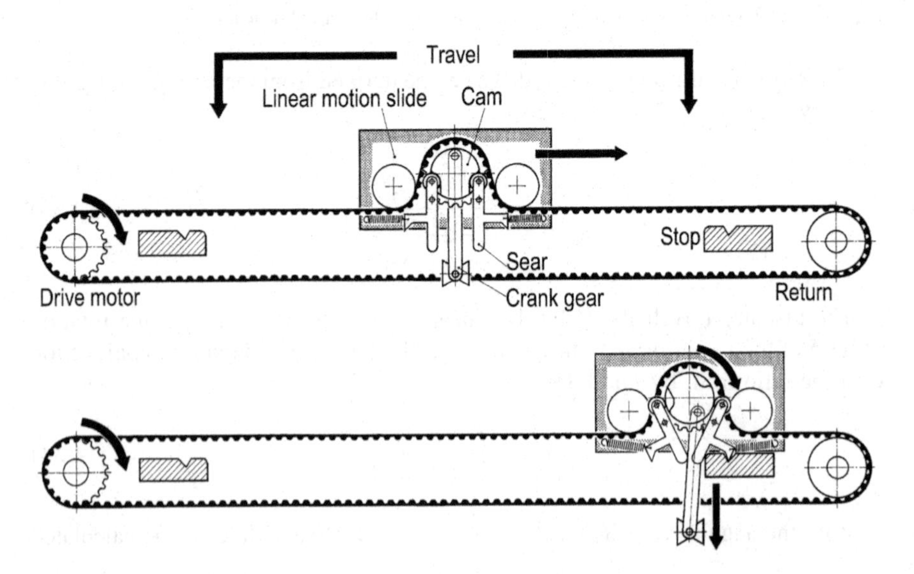

Fig. 4.44 Linear converter with lifting, travelling and lowering motions. The length of the horizontal travel is adjustable by moving the stops

A linear converter is a device with a main horizontal motion which also gives a vertical stroke at each end of its working length. Composite lifting movements are always in demand in transport technology to automate the process of handling sequences for

- Access and lifting
- Horizontal processing
- Lowering and dropping

These processes are possible by using the linear converter in Fig. 4.44. It can be used to help, for example, if pallets are to be moved on to a conveyor belt as lateral motion is not always sufficient. Furthermore, handling operations often have the requirement for moving or de-stacking products, within or at the end of repetitive manufacturing processes.

The illustrated linear converter consists of a complete sequence of mechanical actions. The first illustration shows the linear slide in the middle position with the drive motor rotating. The linear slide mounted cam pulley is locked in position by the latches and is in the rest position. On reaching of the right limit stop (with the motor still turning) the horizontal movement is terminated and the actuated latch releases the cam. On its axis, the final rotation of the toothed pulley results in a motion of the connecting rod which triggers the vertical stroke. With the passage of the vertical limit switch actuator to its lowest point the motor rotation is reversed. With the latch-pulley cam-follower principle, horizontal movement to the left is only allowed if the lifting stroke has returned to its upper starting position.

The special feature of this particular motion system is that only one motor is necessary for all functions.

4.12 Portal Drives

Portal robots are automated motion devices which are constructed of both fixed and movable structures. Their main composite linear movements serve a rectangular work area. The portal-like structure, called the bridge module, offers flexible automation options through the possibility of feeding both machines and production processes from above. Area portals are used for covering large workspaces and linear portals, for example, where the work space of an articulated arm robot needs to be expanded (Fig. 4.48). Robot axis portal types are usually used in large systems and can have lengths of 2–20 m and often significantly longer. The positional accuracy of these designs, with no position detection systems, is about 0.1 mm per metre of travel.

The operating parameters should be optimized in relation to the belt stiffness behaviour by careful design (see Fig. 4.46).

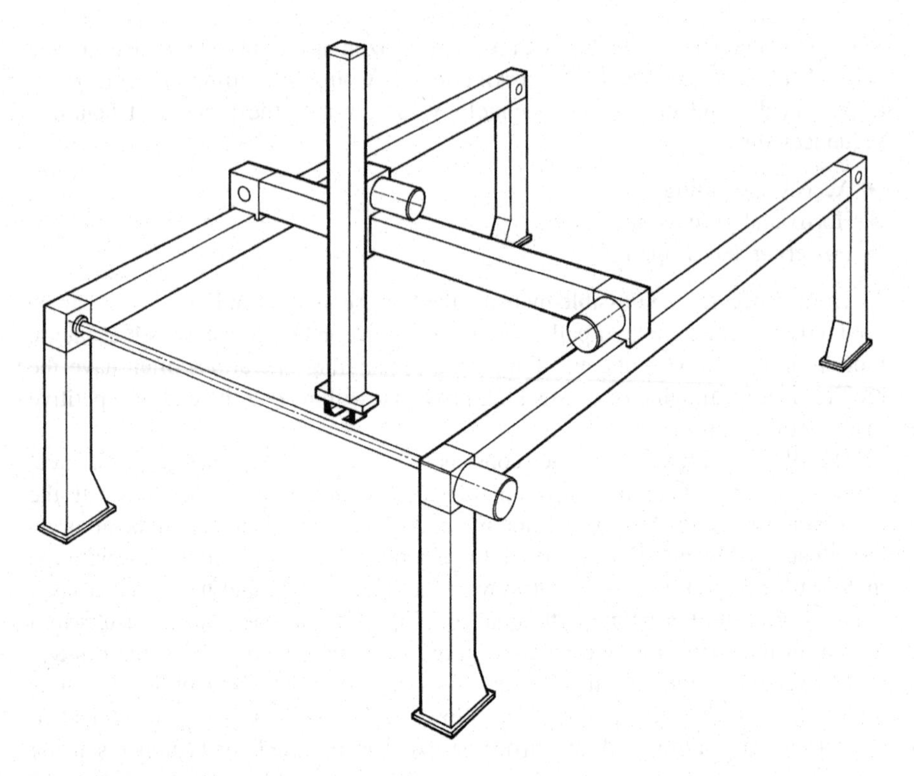

Fig. 4.45 Portal robot

A commonly used belt arrangement is shown in Fig. 4.46a where the stationary motor drives the linear slide through the top and bottom belt strand. In Fig. 4.46b the belt length is halved with a travelling motor and the use of a Ω-drive system. Variant Fig. 4.46c focuses on T and AT timing belt profiles, which mate together in a linear fashion so that there only very short belt sections under unsupported tension. The AT-belt is attached to the machine frame or is bolted to using an ATN-profile (see Chapter 2.3.16). The idler pulleys of the drive unit enable contraflexure for the Ω-wrap and as pinch rollers to ensure the mesh of the belt in the drive pulley. In variant Fig. 4.46d the mating linear toothed component is a machined rack. Figure 4.46e shows a linear actuator drive with a double-sided belt running on a rack. The guide rail ensures a uniform bearing area on the meshing teeth. In the system shown in Fig. 4.46f the belt teeth are engaged by pinch rollers into a rack. The design of Fig. 4.46g uses a pulley instead of a rack. Pinch rollers are not required as the pressure of the drive pulley ensures meshing through the arc of contact. The drive in Fig. 4.46f is described by Tilkorn, M [116] and refers to timing belt based trolley drives on bridge cranes.

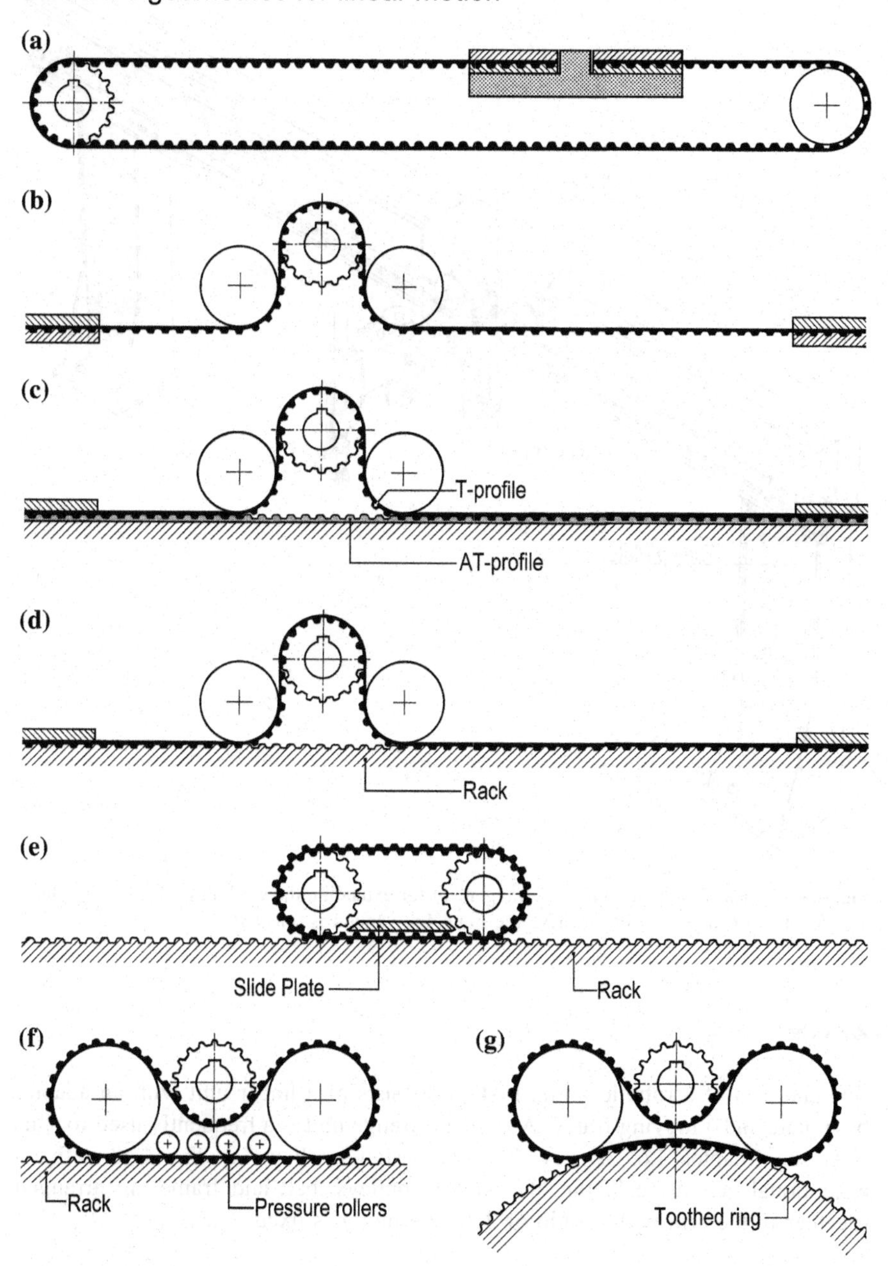

Fig. 4.46 Portal drive variants **a–g**

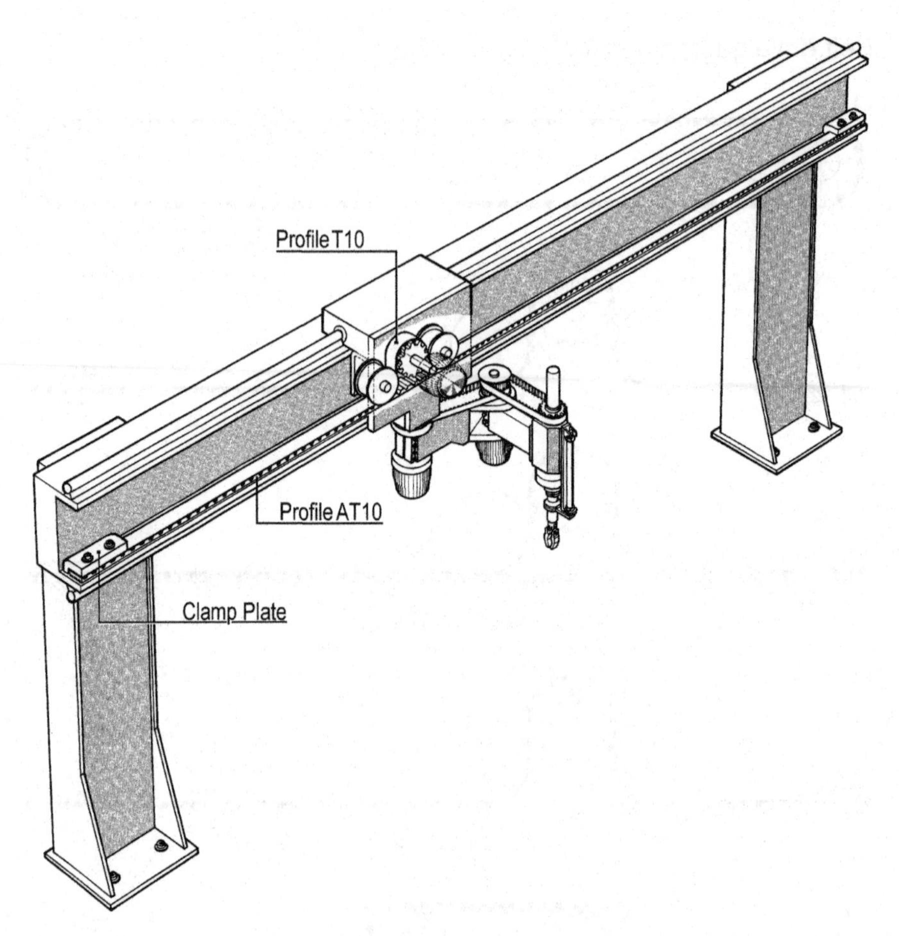

Fig. 4.47 Linear Portal. A common arrangement for extending the work envelope of articulated arm robots. For the portal linear drive the Fig.4.46c version is selected

Z–Axis

The end arm of a robotic portal system consists of a linear unit with an adapter base plate and a moving frame. A standard bridge unit is a frequently used for this task, see Chapter 4.6. A very different design is shown in Fig. 4.48 with an exceptional belt drive layout. The driven pulleys, belt and frame are arranged together in the arm with a ratio of 2:1. The motor is fixed.

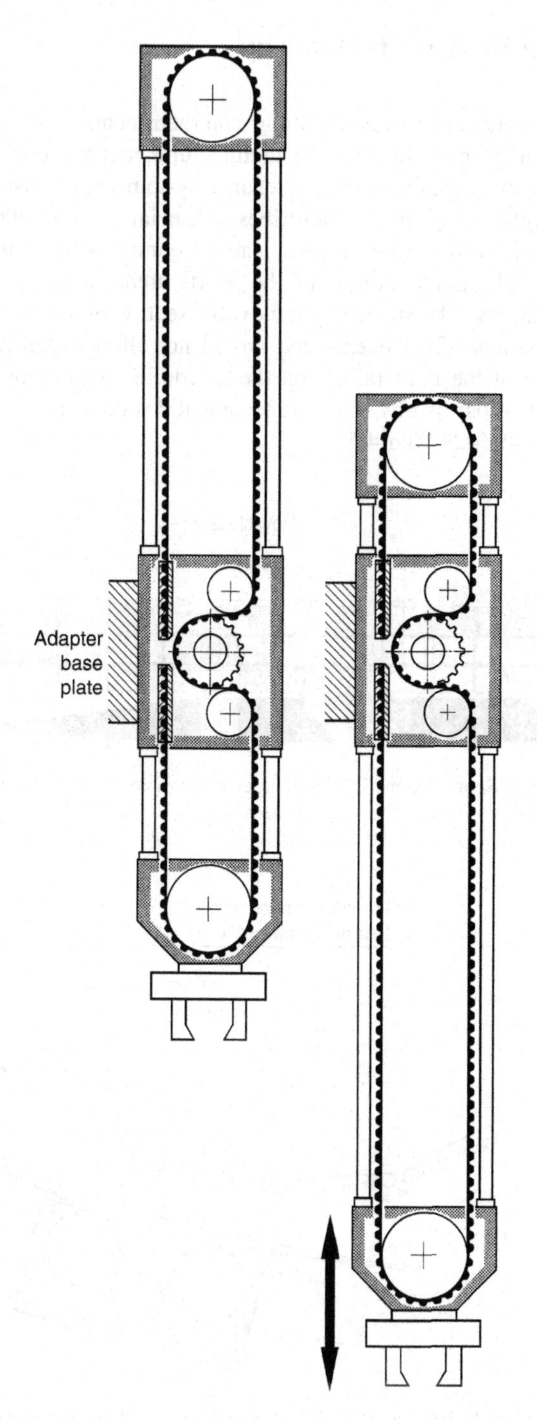

Fig. 4.48 Belt drive layout for Z-axis with a rate 2:1

4.13 Building Systems Technology

The use of photovoltaic panels on buildings and their technical installation have to be compatible and compliant with the architectural requirements. This includes appropriate fasteners, cleaning and maintenance systems and measures to optimize the incidental light. This chapter describes a special solar tracking technology project developed by one of Europe's leading companies in building systems technology [13]. The glass rotunda of the sports arena in the south area of the Oldenburg building will be shaded by a movable section of the translucent facade. Direct sunlight would affect events and would not allow video recordings. The simultaneous use of the light falling on the facade in conjunction with the solar photovoltaic technology is obvious. The technical design of the moving facade is shown in Figs. 4.49, 4.50 and 4.51.

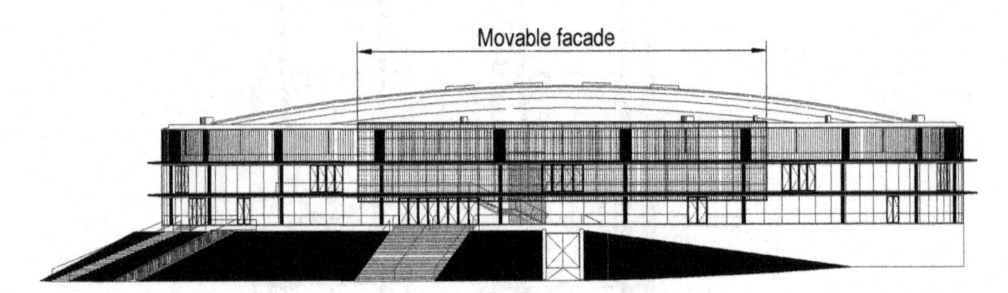

Fig. 4.49 Oldenburg Sports Arena, 72 m diameter movable facade, 13,500 kg in weight, mounted on 36 rollers to support the vertical loads

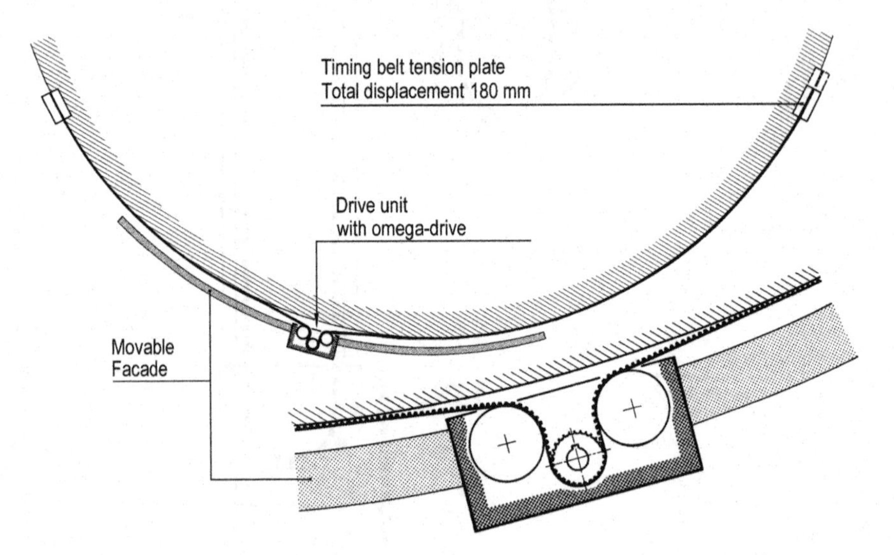

Fig. 4.50 Updated facade, belt type100 ATL20, length 118 m, elongation preload 1.5%, preload distance 180 mm

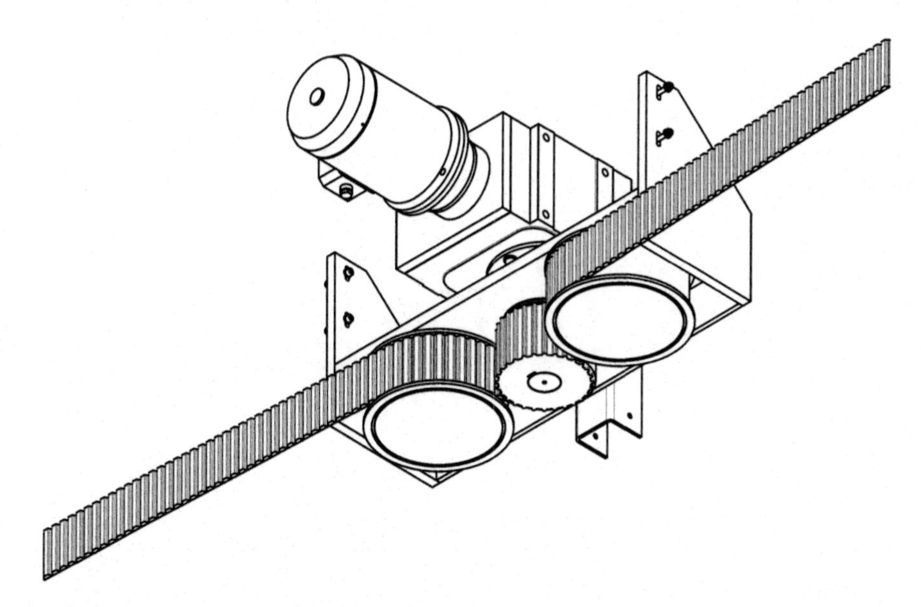

Fig. 4.51 Colt facade Ω-drive [13], installed torque 1100 N·m, 12,300 N drive force, speed 5.5 m/min. 100 ATL20 toothed belt with $F_{zul} = 31{,}500$ N, drive pulley $z = 28$ with $d_0 = 178.25$ mm, back tension pulleys $d = 250$ mm. The highest forces arise under external wind loads

Chapter 5
Timing Belts in Product Transportation

Abstract The path to the fully automated factory contains many logistical challenges, mainly in the areas of product transportation, handling and feeding. In this chapter, timing belts are considered in both their range of applications and variety, and described in detail for specific transport tasks. The explanations offered are detailed regarding the different friction conditions on both the transport and running side of the belts.

5.1 State of Art

The timing belt offers many opportunities for solving transport problems because of the technical advantages of its slip-free operation and has gained wide acceptance in automated factory systems. In most cases, the timing belt's capabilities such as profiles, separating, handling and positioning are used to supply components and sub-assemblies already positionally orientated to production processes. This versatility finds applications almost all industry sectors.

It is possible for the back of the belt to have special coatings or transport profiles. Moreover, the timing belt can be prepared for other tasks by mechanical reworking on the toothed side and/or the back of the belt.

Almost all transport belt applications use a design with a slide bed (see Fig. 5.1). The group of polyurethane belts with trapezoidal profiles such as T, AT and inch pitches are preferred to use for this type of application because their tooth geometry has a flat tip surface. To achieve the lowest coefficient of friction possible with the support rail, it is recommended that the belt has a polyamide tooth facing (PAZ).

Figure 5.1 shows the basic layout of the drive motor variations. The height of the hatched lines above the belt back indicates the associated tension member loads, with the assumption of a uniform line load in the loaded span. The front-drive variant of the three types is preferred because of the lower bearing loads.

R. Perneder and I. Osborne, *Handbook Timing Belts*,

DOI: 10.1007/978-3-642-17755-2_5,

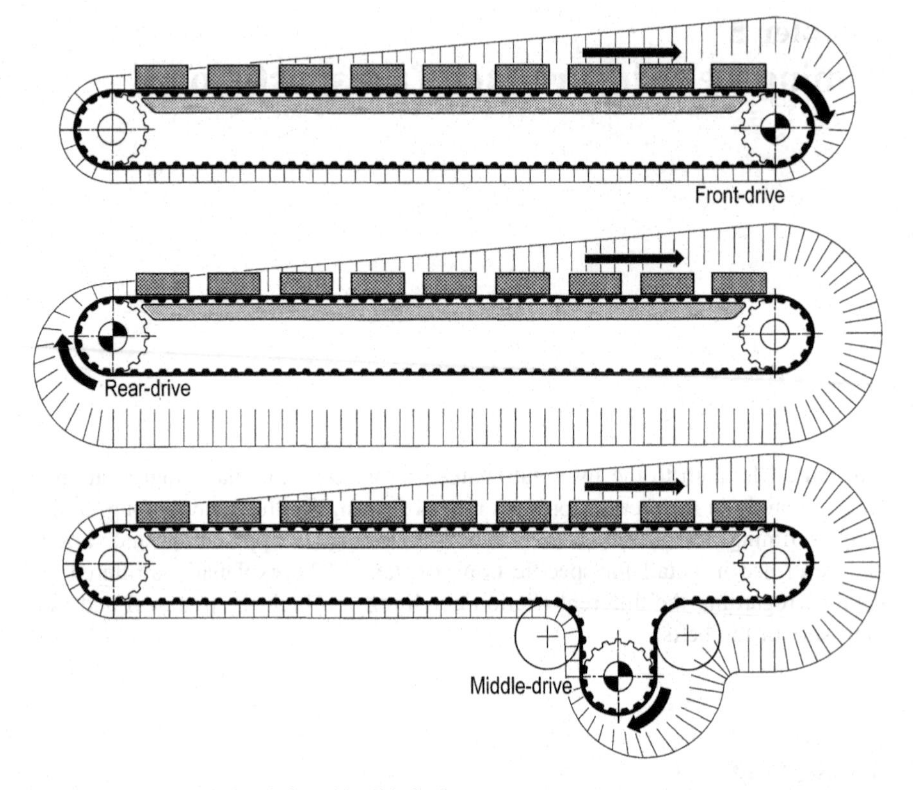

Fig. 5.1 Transport timing belts with slide beds

However, in a lot of applications the rear-drive design dominates because the front of the conveyor is often used as a discharge or transfer station and needs to be as light as possible. The final design layout can also be determined by other considerations and sometimes the motor and gearbox are then better placed elsewhere. The three designs presented here are to be regarded as equal options. The user requirements for a transport project are judged to be met depending on which variant best fulfils the most favourable conditions for efficiency and usability for "his/her" constructive design.

5.2 Transport Belt Design

To design transport drives, the sum of all motion resistance loads is used. The loads are composed of friction F_R, acceleration F_B and gradient forces F_H. The frictional resistance caused by the sliding motion of the belt on the support rail is especially important. The weight of the transported goods exerts a force on the back of the belt which can be either a point or line load.

The weight is experienced through the belt onto the support rail and the associated drag is reflected as a surface pressure on the belt's tooth tip surface. In this contact region, under normal operating conditions, the frictional forces act opposite to the drive direction. In the belt calculation, the tangential force F_t from the motor is determined and it must be greater than the sum of all the resistances. The same value is applicable for the belt design and it must be ensured that the belt is able to absorb the full motor tangential force without damage.

The necessary tangential force, both from the motor as well as applied to the belt is calculated from

$$F_t \geq F_R + F_B + F_H \tag{5.1a}$$

$$F_t \geq m \cdot g \cdot \mu + m \cdot a + m \cdot g \cdot \sin\varepsilon. \tag{5.1b}$$

In practice the calculation can be simplified for slow and horizontal transport applications where acceleration and gradient forces are not involved. The design is limited solely to frictional forces.

$$F_t \geq m \cdot g \cdot \mu. \tag{5.1c}$$

The corresponding torque and power can be calculated from the following equations:

$$M = \frac{F_t \cdot d_W}{2 \cdot 10^3} \tag{5.2}$$

$$P = \frac{F_t \cdot d_W \cdot n}{19.1 \cdot 10^6}, \tag{5.3}$$

where the units for F_t are in N, M in N·m, P in kW, d_W in mm, m in kg and n in min^{-1}.

The required tangential force determines the dimensions of the transport belt which is no different compared with a rotating drive belt where the same values of tooth shear strength, tension member stiffness and flexibility are considered. To calculate the minimum pre-tension the relative position of the drive unit is important (see Fig. 5.1):

$$\text{Front drive} \quad F_V \geq \frac{1}{2} \cdot F_t \tag{5.4}$$

$$\text{Rear drive} \quad F_V \geq 1 \cdot F_t \tag{5.5}$$

$$\text{Middle drive} \quad F_V \geq \frac{1}{2} \cdot F_t \text{ to } \geq 1 \cdot F_t. \tag{5.6}$$

During accumulation operations the additional frictional forces must be included when a relative motion between the transported items and the belt back takes place. Thus the maximum number of transported items or the total mass involved in the accumulation operation needs to be determined. All the frictional resistances are added together.

In order to calculate the tangential force F_t, the largest coefficients of friction should be used. These are usually formed when starting with static friction conditions and/or for overcoming breakaway forces. The larger the contact pressure, the stronger the micro fluctuations between the contact surfaces lock together. A lot of power needs to be applied in order to separate materials in contact for the first time.

Furthermore, the influences of friction and the resultant heat generated need to be considered under continuous operation (see next **chapter**).

5.3 Friction and Tribological Behaviour

The calculation of F_t in Eq. 5.1a is exact; however because its determination is unclear, the coefficient of friction μ is very vague. The real friction conditions [60] are primarily a function of the influence of wear over time and contact temperature and haven't been adequately investigated. With the belt running on a support rail made of steel or aluminium, a tribological material loss from the belt and a complementary polymer film formation on the sliding surface can be observed. Over a longer period both the polymer layer formation and significant levels of friction build up. This tendency for material transfer in continuous use leads ultimately to the complete removal of the polyamide tooth facing and thus to a new tribological pairing between belt and support rail. The increase in friction absorbs more drive power giving further increases in the contact temperature. Failure finally occurs by the belt melting. Recent studies suggest the limit for the contact temperature should be about 60°C and, in order to avoid thermal damage, this should not be exceeded.

A development institute of technology [60] has been working on the problem since 2006 with the aim of obtaining design safety recommendations for frictional heat and environmental heat acquired by conduction, convection and radiation.

Knowledge of limiting tribological factors could open up higher load applications for the users of timing belt technology.

To date, the friction conditions in general use have rarely led to problems. It is axiomatic that the items to be transported will decide the type of belt and the belt width. This results in high safety factors for the mechanical stress seen in the drive and that the friction and contact heat limits will not normally be reached. Exceptions to this are haul-off drives with high levels of friction ($p_{fl} \cdot v$) (see Chapter 5.11).

The best test results in relation to belt frictional wear behaviour, which are also confirmed by significant empirical results, are PAZ-coated belts on PE support rails. Although a polymeric film is not formed on the surface of the support rail, with continuous use, coefficient of friction increases from about

0.2 to 0.5 are still experienced with increasing surface pressure p_{fl} and haul-off speed v, along with the expected friction-generated temperature increases (Table 5.1).

Table 5.1 Coefficient of friction of polyamide tooth-faced PU timing belts (PAZ) in sliding on different materials [60]

Conditions	Support rail				
	Steel	Alu	PE	PA	PTFE
Short-term use	0.2…0.5	0.3…0.6	0.2…0.3	–	0.2…0.3
Continuous use	0.5…0.9	0.6…0.9	0.3…0.5	0.3…0.7	–

5.4 Conveying, Contact Surfaces and Backings

Since conveying conditions are very variable from task to task, both the properties and the optimization of the timing belt surface contact areas can be considered important.

Belt Teeth

Belt tooth profiles are subject to precise geometric specifications within tight tolerance limits and additional post-manufacture coatings should not be used because the correct belt/pulley meshing cannot accept any dimensional changes. Some sprayed coatings are available to reduce friction or give antistatic properties but as these are no more than a few microns thick they do not affect the drive functionality. For applications where the belt slides across a surface then polyamide tooth facings, incorporated during manufacture, are preferred since their coefficient of friction is low.

Belt Edges

Belt edges are in frictional contact with the pulley flanges and generate both tooth entry and exit resistances. With a self-tracking belt, the vee-guide is the belt contact surface which makes the belt track straight. The edges of the belt can also run between the sides of guide plates. Significant lateral forces during operating conditions can cause camber resistance[1] and deflection or direct lateral shifting of the goods transported. These will increase the frictional forces acting on the belt and increase the tendency to edge wear. Since in all applications there is product close to the conveying surface of the timing belt, abrasion and dirt formation should be avoided. Thus, the use of polyurethane as a belt material is preferable as its stability against friction and its wear behaviour is clearly superior to other types of elastomers.

[1] Camber Resistance: Using support rollers or idlers so that the belt always comes to rest on the same edge. The contact force x coefficient of friction is the camber resistance.

Belt Backings

The back of the belt is the active contact surface for conveyed products and can be optimised for the desired application by being covered with a special backing. Most timing belt suppliers fulfil differing customer requirements with a wide range of backings, which relate to the modification of base belt material values such as Shore hardness, friction levels and thickness. Suppliers are ready, in most cases, to provide the accompanying material resistance characteristics and changes in bending values for the overall belt design. In practice even a standard timing belt or one with the standard back coating of PAR (low friction polyamide for accumulator applications) can operate without a special backing (see Fig. 5.2) in many conveying applications.

Fig. 5.2 Slide bed parallel conveyor with product carriers. Timing belt AT10-PAZ. Belt back is standard PUR uncoated. PE support rail. Highly stable with no wear. An effective design by mk [81]

If the timing belt is to be used for timed indexing on a slope or if certain parts being transported must not touch each other, then the belt back can be fitted with profiles in any arbitrary number and sequence (see profiled belts in Chapter 5.6–5.10).

5.5 Slide Beds/Roller Beds

Conveying over a slide bed, as in Fig. 5.1, represents a proven solution that has become generally accepted due to simple implementation as well as cost-effective construction. If, however, during the conveying process, high frictional forces per surface area ($p_{fl} \cdot v$) are to be expected then a roller bed design should be considered. However, it should be noted that the tooth side of the belt is unsuitable for direct contact on the supporting rollers as it reacts with high levels of vibration as the tangential motion of each belt tooth and gap impacts on the surface of the roller. This changes the support level of the belt intermittently, which changes the tangent point of the roller support on the belt tooth face to the belt tooth gap.

The designs represented in Figs. 5.3 and 5.4 are therefore recommended for the construction of a rolling bed drive. The solution in Fig. 5.3 shows a companion belt in form of a flat band between the toothed belt and supporting roller. It is unpowered and it must have sufficient inherent longitudinal rigidity. The actively driven toothed belt, which should be regarded as a rider belt, carries the companion belt forward by friction. The support height level of the companion belt should therefore be a little higher than the tooth tip line of the toothed belt so that it is moved along securely during no-load operating conditions (without product). An alternative solution in Fig. 5.4 is proposed using an arc-toothed belt (see Chapter 2.3.13) or a double helix belt (see Chapter 2.3.14). The tooth position and/or tooth gap of both types of belts does not wholly coincide with axis of the support rollers. The tooth face surfaces always remain in contact with the support roller and with no intermittent contact of the belt teeth then no vibration can arise. These solutions are clearly more complex than a slide bed and should therefore be reserved only for conveying with high loads.

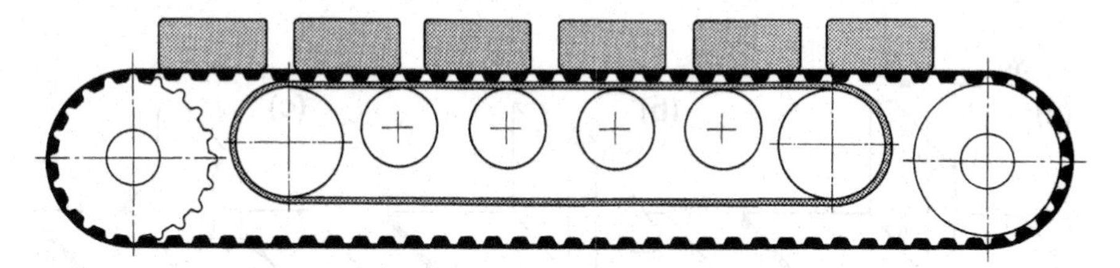

Fig. 5.3 Roller bed with support belt

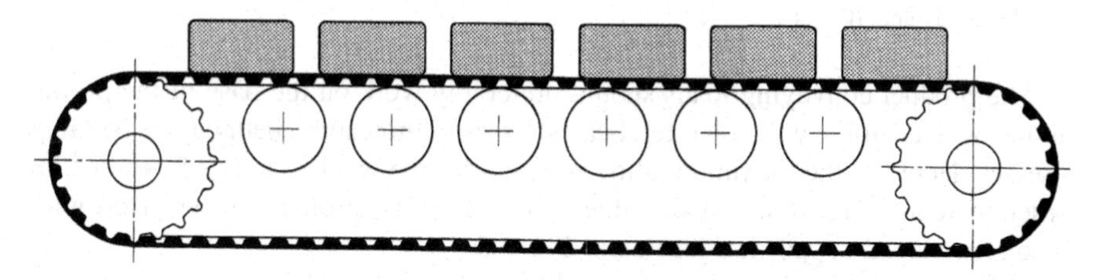

Fig. 5.4 Roller bed with arc- or double-helix toothed belt

5.6 Profiled Timing Belts

The back side of timing belts can be equipped with welded flights in any number and sequence. The nature and shape of the profile itself depends on the product and its purpose. Profiled belts are used to solve various feeding tasks in automated production equipment. These specially designed belts are also referred to as synchronous conveyor belts or indexing belts. They are only made possible because of the weldability of the polyurethane timing belt and the profiles, which are usually made from the same base material as the belt. In order to attach profiles to the belt back either contact or friction welding is used. To sustain the quality of the weld (full surface weld), the geometry of the profile base, its size and the forces acting on it should be considered. Other methods of profile attachment are bonding (rare) and mechanical attachment (see Chapter 5.8 and 5.9).

Any increase in the profile footprint in the longitudinal direction (profile thickness, see Fig. 5.5) is accompanied by a local decrease in belt flexibility. When the belt is meshed with a pulley, the pre-tensioning load forces both the belt and the profile weld area to accept the pulley curvature. The resulting surface stress should be taken into account in order to prevent the profile tearing off. This reduced flexibility was originally noted by Breco Antriebstechnik GmbH [10] in their catalogues. The relation of the profile thickness s_N to the number of teeth in the pulley (see Table 5.2) is limited to where the maximum edge stress in the weld zone equals 2.5 N/mm^2. This value and the minimum number of teeth are based on both experience and experimental observations and have been proven empirically in practice. Since the belt bending largely takes place in the area of the tooth gap, the preferred position for the welded profile is over a tooth.

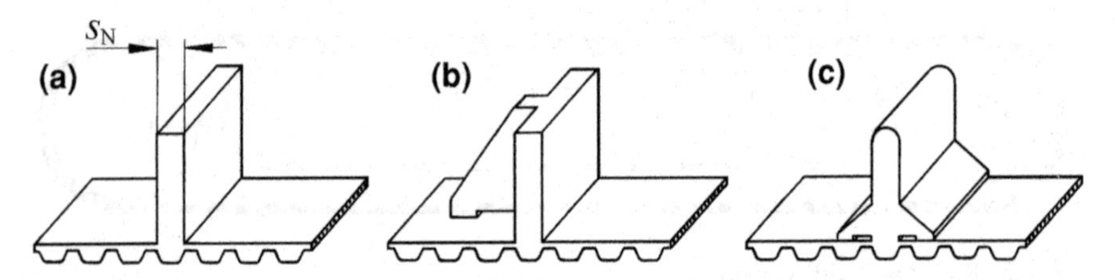

Fig. 5.5 Profile versions **a** standard right angle, **b** with single-sided support leg, and **c** with double sided support legs

The product conveying loads should preferably work on the base of the profile as the welded joint will tolerate shear stresses significantly better than bending stresses. Here too the maximum allowed stress for the weld zone is 2.5 N/mm^2. In order to reduce the base stresses due to bending, the profile can be made with specially shaped supporting legs as in Fig. 5.5.

The forces acting on the profile result from

- Inertial forces of the transported products in acceleration and braking, for example in indexing drives
- Friction during transport from either a support plate or guide rail
- Gravity from either angular or vertical conveying
- Centrifugal force due to the rotation of the pulley and associated parts
- Inertia of the profile in the transition between entry and exit of the arc of wrap of the pulley

Table 5.2 Profile thickness

Max. profile thickness in mm when welded over a tooth							
Belt designation	Minimum number of teeth in the pulley						
	20	25	30	40	50	60	100
XL	5	6	7	8	9	10	12
L	6	7	8	10	12	14	18
H	8	9	10	12	14	16	20
XH	13	14	15	18	20	23	30
T5, AT5	5	6	7	8	9	10	12
T10, AT10	8	9	10	12	14	16	20
T20, AT20	12	13	15	18	20	23	30
H5, R5, S5	5	6	7	8	9	10	12
H8, R8, S8	7	8	9	10	12	13	16
H14, R14, S14	9	10	12	14	16	18	20
Max. profile thickness in mm when welded over a tooth gap							
Belt designation	Minimum number of teeth in the pulley						
	20	25	30	40	50	60	100
XL	2	2	3	4	6	8	12
L	3	3	4	5	7	10	16
H	4	5	6	7	10	12	20
XH	5	5	6	8	12	20	30
T5, AT5	2	2	3	4	6	8	10
T10, AT10	3	4	4	6	9	12	18
T20, AT20	5	5	6	6	12	20	30
H5, R5, S5	2	2	3	4	6	8	10
H8, R8, S8	2	3	3	5	7	10	15
H14, R14, S14	3	4	4	5	8	10	20

The forces involved in each individual application should be fully investigated as to whether they act alone or in concert. Centrifugal force is rarely involved and then only at a very low level.

The inertia of the profile is often underestimated and regularly leads to an additional force on the welded joint in the transition area between the linear and circular motion of the belt. While belt speed in the tensile plane is constant the profile has a consistent speed only while in the linear range of the belt span, but when entering or exiting the pulley curvature it is subject to a considerable acceleration or a corresponding deceleration. In Fig. 5.6, the distance between the outer line and the belt surface represents the velocity curve of the profile centre of gravity. With increasing distance from the profile centre of gravity, the tensile forces grow with acceleration and act as bending stresses on the welded profile base. If the profile is equipped with add-on components or if the conveyed product remains in thrust contact with the profile when entering the pulley curvature then the additional forces of inertia must be taken into account. Thus, the designer needs to take special attention of the "velocity jump" in the design of the profiled timing belt and apply appropriate countermeasures in the design phase such as

- Optimise the profile centre of gravity
- Reduce profile mass
- Reduce mass of components
- Increase the pulley size
- Reduce profile thickness s_N (increase flexibility)
- Ensure shear forces act in the direction of the profile base weld

The term "velocity jump" is central to this concept since the transition takes place in a very short distance between the linear and circular motion of the belt. Due to the inherent elasticity of the profile and the belt back, the "jump" is phased in over a period of time. To calculate these forces of inertia refer to [11] in the bibliography, which deals with the centre of gravity or the level of force from the belt back and mechanically attached profiles of the ATN timing belt system (see Chapter 2.3.16).

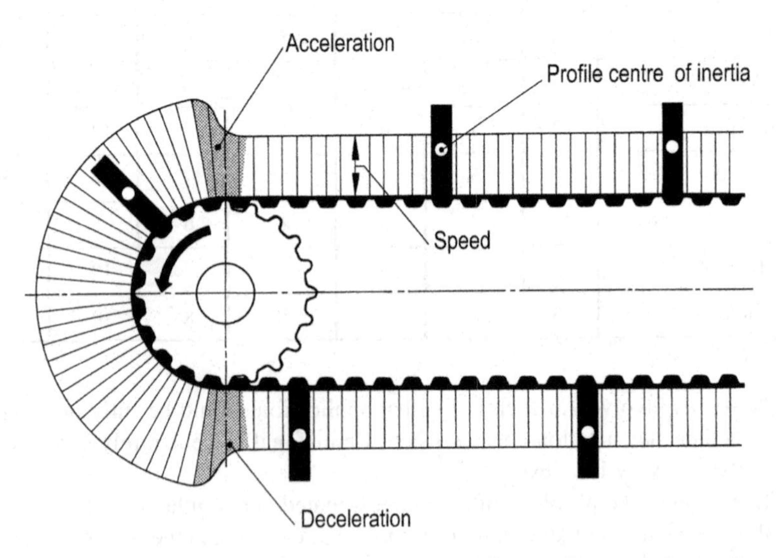

Fig. 5.6 Velocity curve of the profile centre of gravity while travelling around the pulley

Summary

A safe and well-designed welded profile is dependent on the profile shape and its expected duration of load in service and rigidity, which is influenced mainly by the material properties and geometry. The next chapters deal with application examples for profiled timing belts containing numerous illustrations depicting proven solutions.

5.7 Example Applications of Profiled Timing Belts

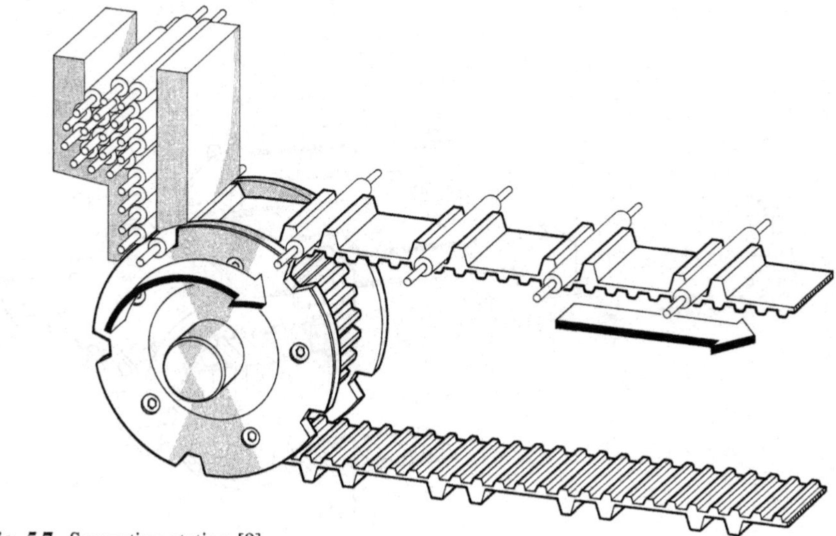

Fig. 5.7 Separating station [9]

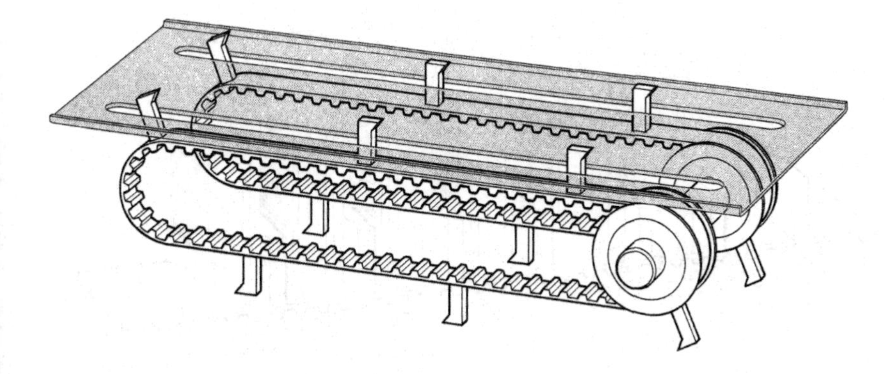

Fig. 5.8 Light goods conveyor with sheet metal support plate. Matched pair of belts [9]

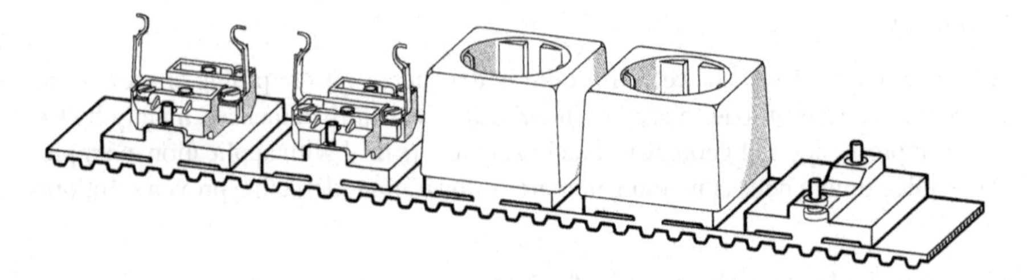

Fig. 5.9 Indexing belt in the electrical components industry [9]

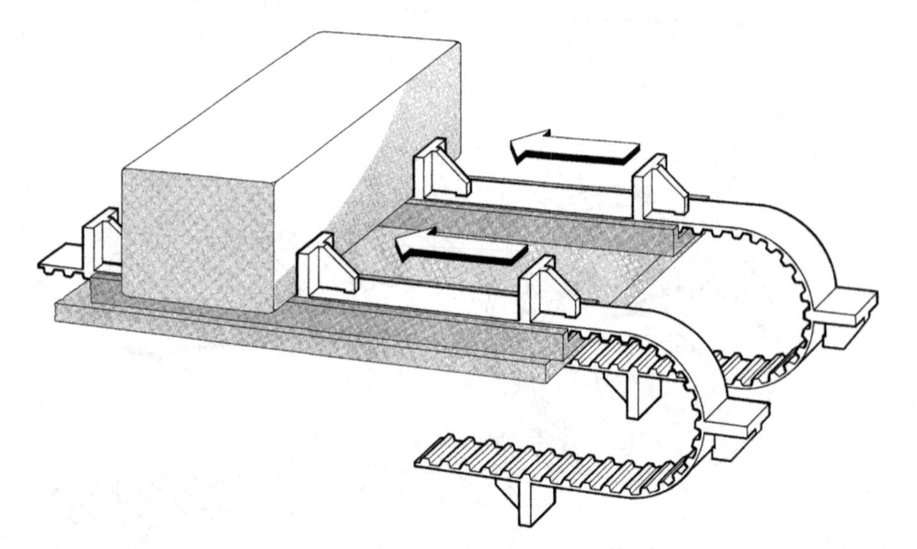

Fig. 5.10 Conveyor belt with a support rail and side guides. Matched pair of belts [9]

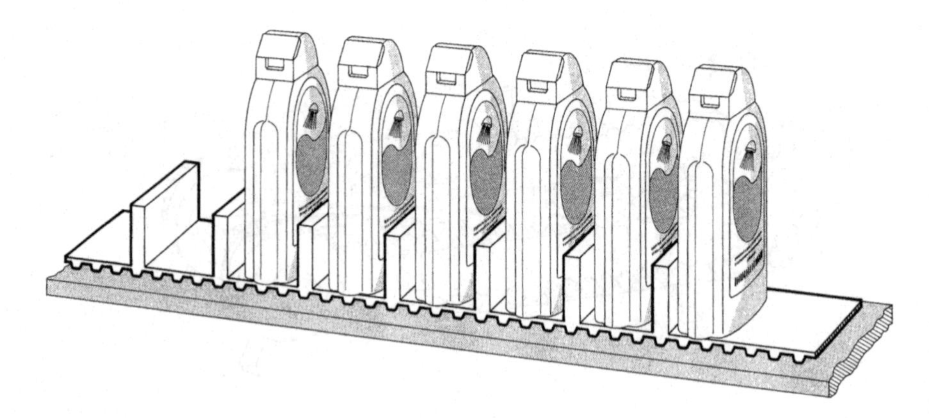

Fig. 5.11 Conveyor in the cosmetics industry [9]

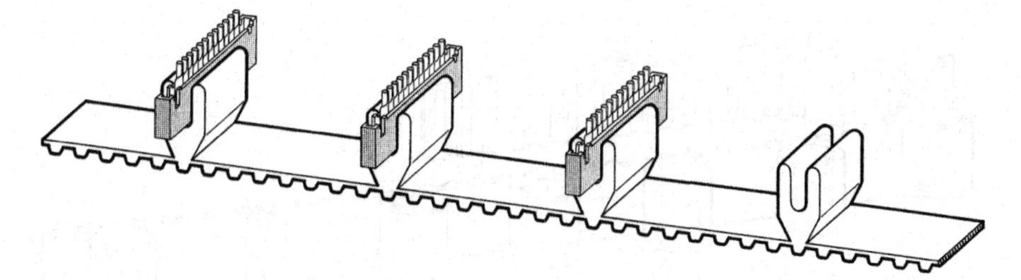

Fig. 5.12 Electrical connector assembly belt [9]

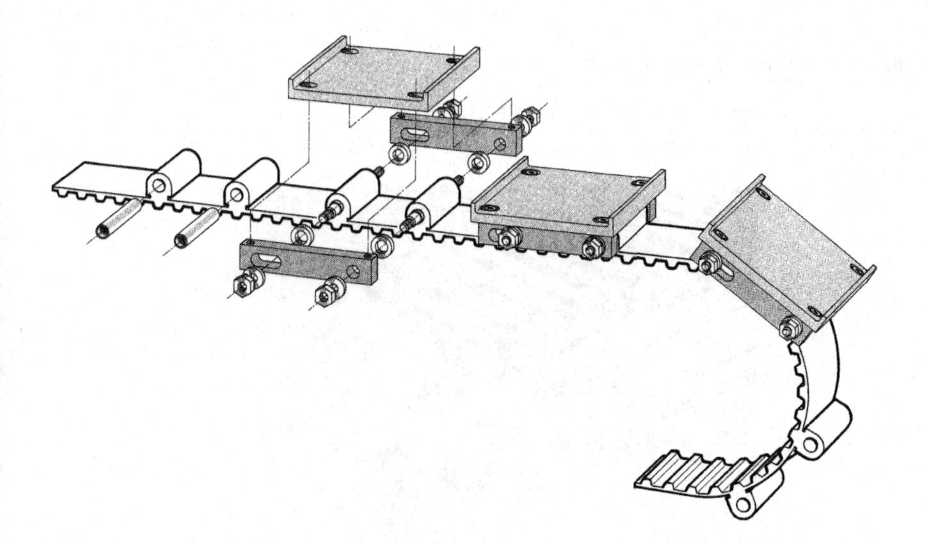

Fig. 5.13 Automation belt with product carriers [9]

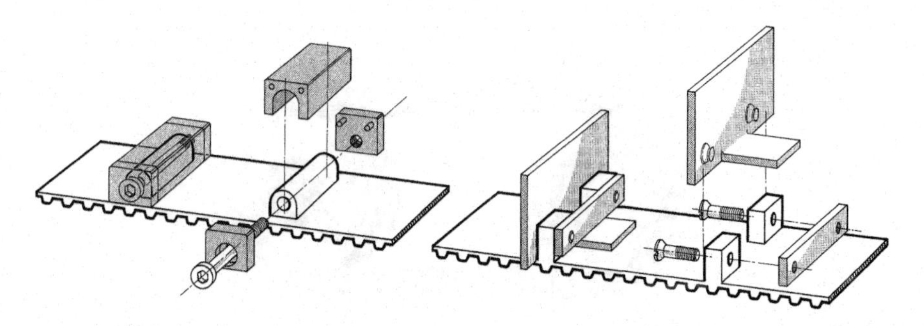

Fig. 5.14 Mechanical attachments [9]

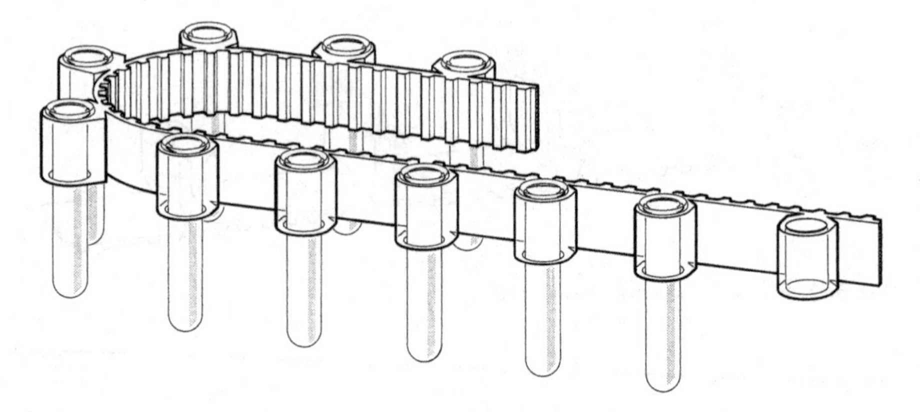

Fig. 5.15 Rotary magazine belt [9]

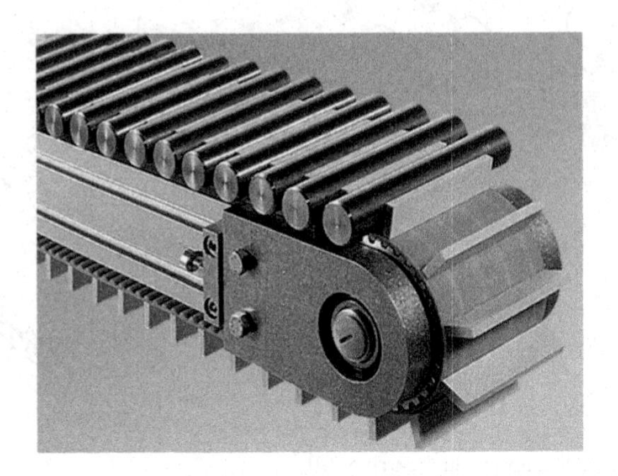

Fig. 5.16 Conveyor magazine belt [81]

Fig. 5.17 Brush belt with welded bristle profiles [74]

Figures 5.7–5.17 shows the use of profiled timing belts in different industries. The main applications are in production, automated transport and feeding for the consumer goods and mass production industries. The design specifications for such equipment is very varied as they depend upon the mass of the parts to be moved, the number of units per minute, conveyor length, continuous or pulsed feed, cycle time and positioning requirements. In indexing applications both product acceleration and deceleration loads act on the profiles whereas the items shown in the illustrations are rigidly carried within the product pockets.

Accuracy of Profile Welding

In the production of conveyor belts, the accuracy of each weld position is attained by alignment to the teeth in the belt. This is usually ±0.5 mm relative to the desired target position. Depending on the manufacturer, tolerances as low as ±0.2 mm can be achieved. If you look at the absolute deviation of a single profile to its profile neighbour, there can be deviations of the tooth pitch tolerances of the belt as well as the changes created by different preloads. When an indexing belt with profiles is to be installed and linked up, then the first requirement is for positioning accuracy and secondly further adjustment will take place at the desired manufacturing and/or transfer station. With long conveyor runs and multiple positioning then each station must be separately and independently adjustable.

Manufacture of Profiles, Further Processing of Profiled Belts

Small numbers of profiles with simple rectangular geometry should be considered for production from semi-finished sheet material; however, for larger quantities plastic injection moulded parts are common. In most cases, manufacturers offer an extensive range of different profiles from existing forms for repeat applications. The manufacture of the complete timing belt with welded profiles is usually dimensioned according to detailed drawings. The preferred welding position is in integral multiples of the tooth pitch with each profile "over a tooth". But any other spacing is also possible for any number and sequence of profiles (Fig. 5.18).

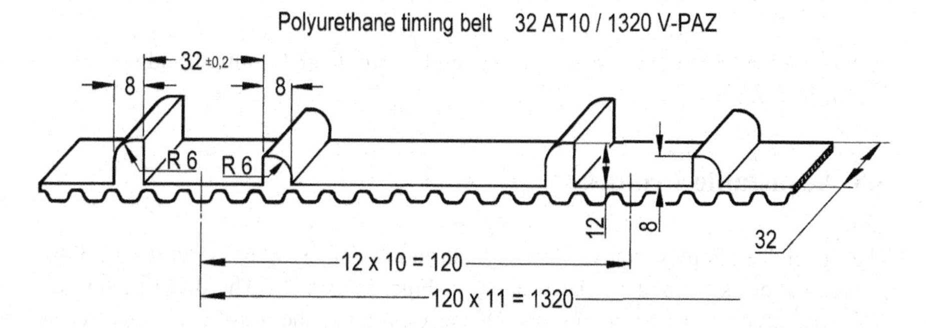

Fig. 5.18 Production drawing for a special belt with 11 profile pockets evenly distributed over the length of a polyurethane timing belt with the specification 32AT10/1320 V-PAZ

Profiled Timing Belt in a Production Application

A good example of profiled belts is shown in the packaging machine in Fig. 5.19 utilising two feeder belts coupled together. In the centre of the picture we see the transfer station with the conveyor working from left to right. The paired belt operation using narrow belts is equipped with simple right-angled profiles. They work overhead and hang in position in contact with the packets, which slide along the support rail made from straight and curved sections. At the transfer station, a wide belt whose back is separated into pockets, receives the packets.

The example shown is used in cigarette manufacturing. Behind the transfer belt are single and double-sided drive belts. In the upper field of view, details of the idler pulley arrangement can be seen. A characteristic of profiled belts working in parallel is that they are manufactured as matched length pairs. During conveying, the packets must fit accurately between the right and left profiles and be tilt-free whilst in thrust contact. Thus, the tooth positions of the pockets must be exactly in alignment with each other.

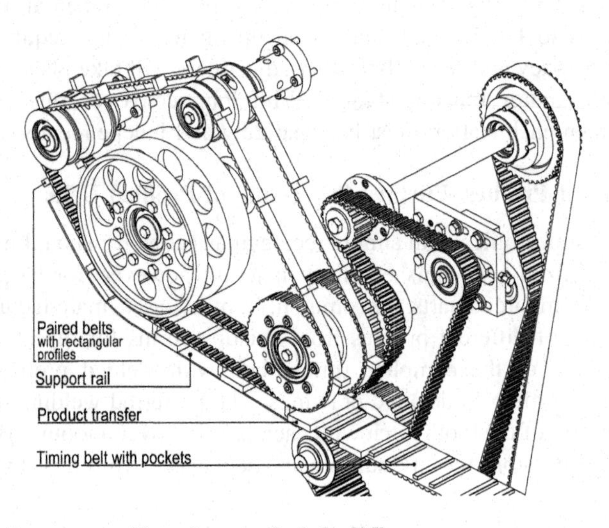

Fig. 5.19 Feeding system with packet transfer belts [35]

More information about conveyor examples, single and multi-strand systems is shown in Figs. 5.20–5.26.

5.8 Adjustable Profiles

The special examples below show synchronous conveyor belts with adjustable pockets or nests formed by the profiles in Figs. 5.20–5.22. The drive motor will typically have two or three pulleys. Of these pulleys, one is rigidly connected to the drive shaft, by a keyway for example, while the other(s) are rotationally adjustable on the same shaft.

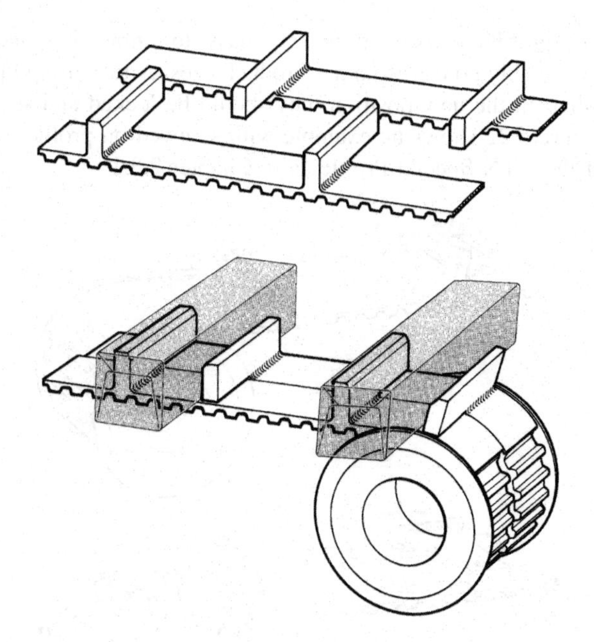

Fig. 5.20 Packaging machine with adjustable profile pockets

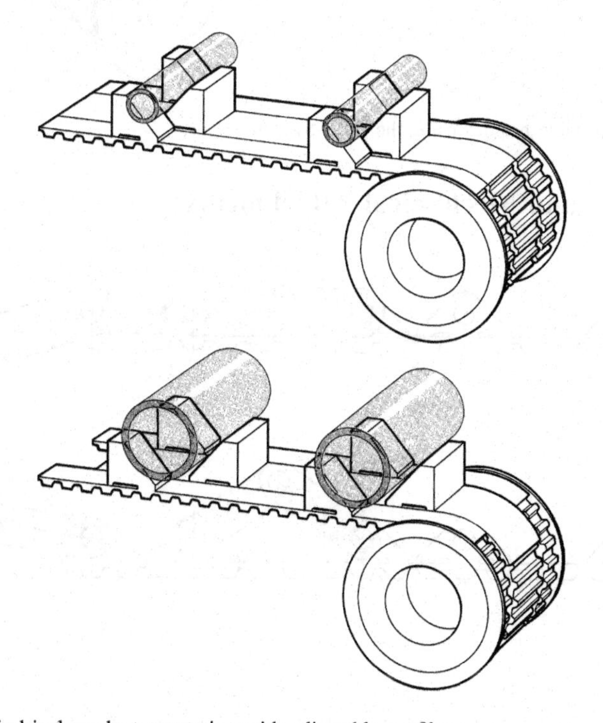

Fig. 5.21 Cylindrical product conveying with adjustable profiles

This sort of profile adjustment is very easy to implement and the whole installation can quickly convert to other product sizes. After setting up the pulleys on the drive shaft in the new position they should be locked and secured against movement. Figure 5.22 shows an example with screwed-on profiles which uses a timing belt of the ATN Special Profile as in Chapter 2.3.16.

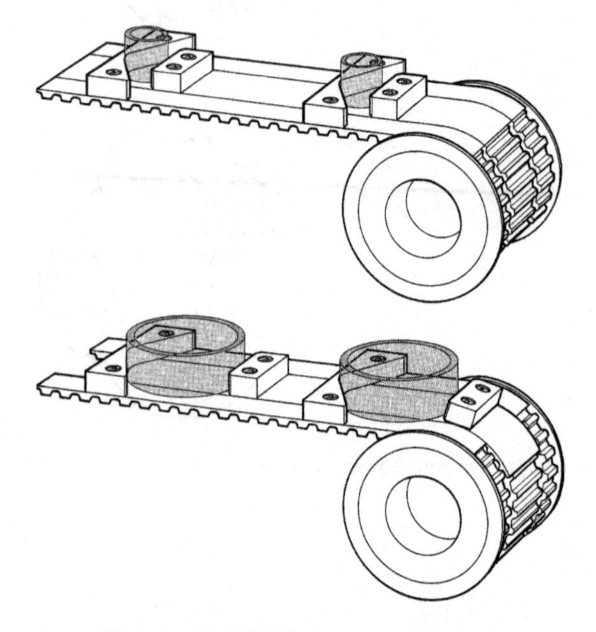

Fig. 5.22 Adjustable profiles using the ATN-System

5.9 Profile and Mechanical Attachments

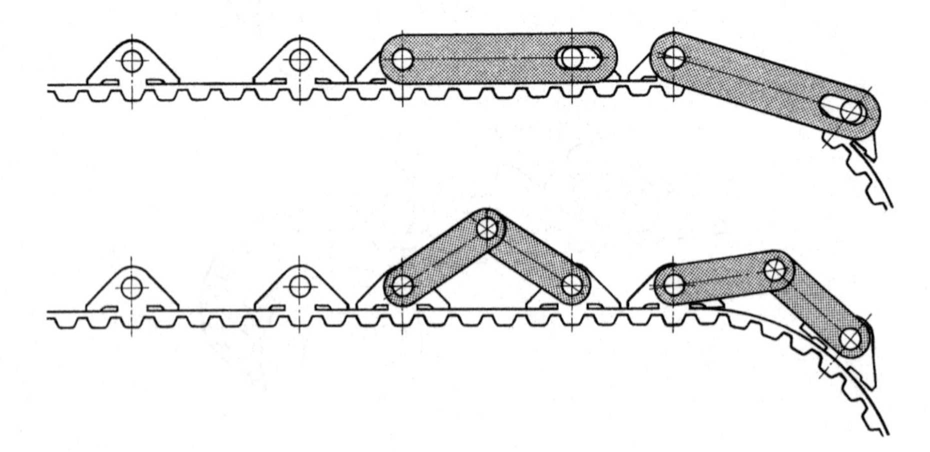

Fig. 5.23 Rotationally coupled attachments

A common application is in assemblies for transport and handling technology which are coupled to the timing belt and rotate with it (see the illustrations in Figs. 5.23 to 5.26)

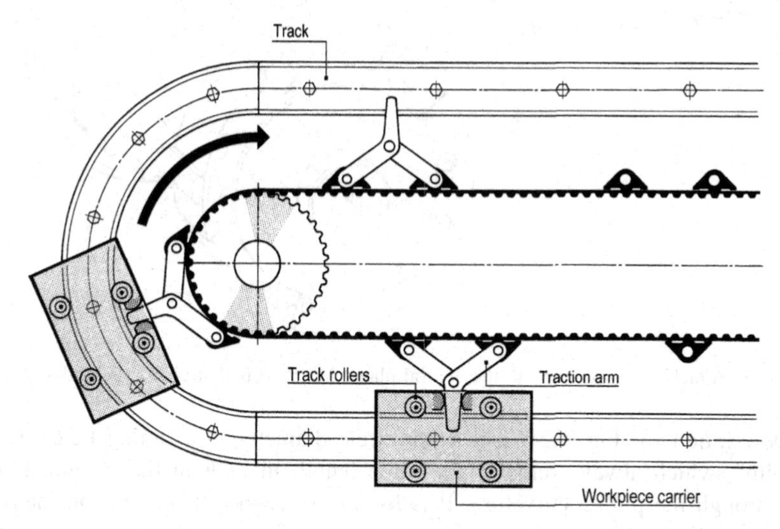

Fig. 5.24 Rotating hinged workpiece actuator coupled to the timing belt. A *top view*

The principle of attachment is no different whether *one* or *more* peripheral parts are arranged in a row. However the kinematic requirements of the rotational properties must be met and there are only a few possible solutions. Basically, the force on the profile must be at the smallest possible distance from the welded foot as the foot resists shear stresses much better than bending stresses. Additional support legs will improve the force transfer. Figure 5.23 shows two kinds of attachment with the necessary range of movement for both linear and circular motion. The link with the slotted hole is designed as an adapter for further attachments. In the case of the double link, the middle fulcrum allows rotational movement, see Fig. 5.24.

The attachments will experience speed differences for both of the above solutions between the transitions of linear and rotational movement. However, they do not experience a velocity jump as described in Chapter 5.6 (see Fig. 5.6). Rather, the acceleration takes place within the section of path that corresponds to the distance between the two profile attachments.

This chapter deals with types of attachments that give high quality solutions for large masses and forces. Of course, such designs are also sufficient for a multiplicity of fine-mechanical designs, where the force application is sufficient over a single profile.

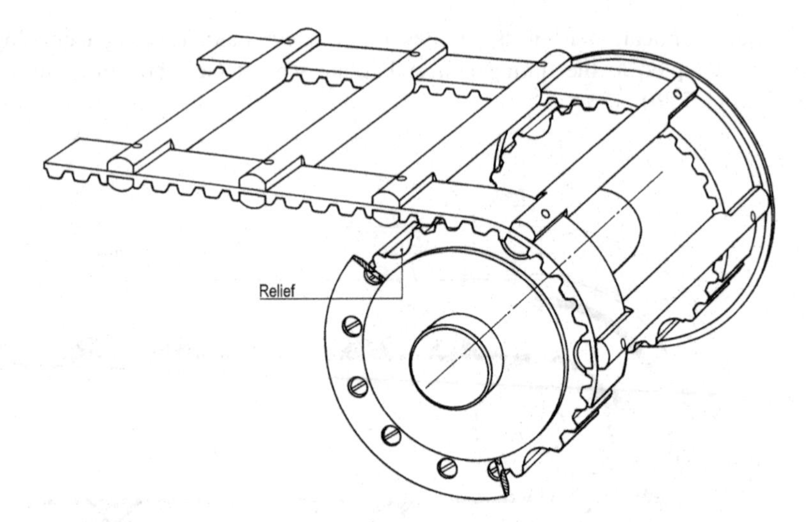

Fig. 5.25 Attachments mounted in the neutral plane. Matched belt structure. A study

Description for Fig. 5.25: The pitch circle diameter in a timing belt drive is that line, which always remains the same length in both in the extended span and through the pulley curvature. It is located in centre of the tension member. The pitch circle diameter viewed over the belt width can be considered, as in gear units, the pitch line or the so-called *neutral line*. In the rotating belt, the speed at any point is only the same in this plane. All other points with a radial distance to tension member centre show a speed jump in the transition between linear and circular motion, see the description in Chapter 5.6 in connection with Fig. 5.6.

It is evident that the neutral line concept is extremely attractive to solve many kinematic functions. For example, in the manufacturing process for wire, foil or paper, the need is always for constant product velocities in the linear and circular production segments.

The neutral line of the timing belt is only able to be put to use through indirect means. The centre of the bars exactly coincides with the centre of the tension member in Fig. 5.25. In order to solve tasks of this kind welded or mechanically attached profiles are used in connection with offset bars. When mechanically attached profiles are being attached on the toothed side of the belt then a matching undercut in the corresponding pulleys should be used.

The ATN toothed belt system [11] offers mountable and dismountable assembly options, see also Chapter 2.3.16. The toothed side of the belt is formed with pockets to receive the specially shaped plug nuts in any sequence. Up to four screwed connections per tooth can be used for the fixing of fittings on the transport side of the belt, depending on the width and the belt profile.

The application shown of a matched pair of belts in Fig. 5.26 refers to a rotating magazine for an automated feeder. The variety and possibilities of such belts in terms of transport length and width, and whether used horizontally or vertically,

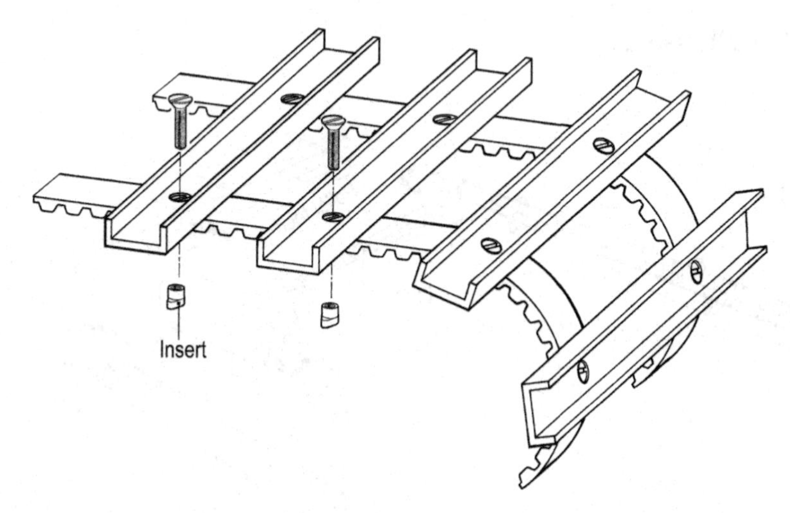

Fig. 5.26 Attachment assembly. Matched belt transport

are very diverse. The system can also be used to replace a two or multi-strand rotary magazine.

With projects using drives of two or more parallel belts, special requirements must be observed for the design and installation of the pulleys and idlers. Thus, all pulleys on the drive motor must have a matched tooth alignment. Since the backlash of the normal AT10 belt toothform is $c_{m1} = 0.4$ mm (see Table 4.3) teeth deviations pulley to pulley should not exceed that value at the neutral line. Any errors will cause tension on the profile's fixing elements which increases as the pulley diameters become smaller. Increasing the idler and pulley diameters decreases this effect and will also improve the smoothness of the belt significantly. This measure also leads to reductions in the forces of inertia of the associated component parts during the speed jump between linear and circular motion (see Fig. 5.4 in Chapter 5.6). Unloaded pulleys and idlers should preferably be smooth (without teeth) for simplicity but they can also be toothed pulleys. Examples of two strand conveyors can be found in Figs. 5.2, 5.8, 5.10, 5.19, 5.25, 5.26 and a multi-strand conveyor in Fig. 5.38.

To ensure the reliable function of two or more belts arranged in parallel within a machine, it is recommended to use only sets of belts with reduced length tolerances. The belt should be marked with a direction of run and with matched teeth so that they are assembled in the same position and direction as they were produced. Consequently, the entire set of belts should be changed if a drive change is necessary.

Requirements for the production of orders for two or more multi-strand belts:

Matched set of timing belts consisting of two (or more) equal length belts. Produced in pairs (sets). Tooth position and direction of motion marked.

5.10 Palletising

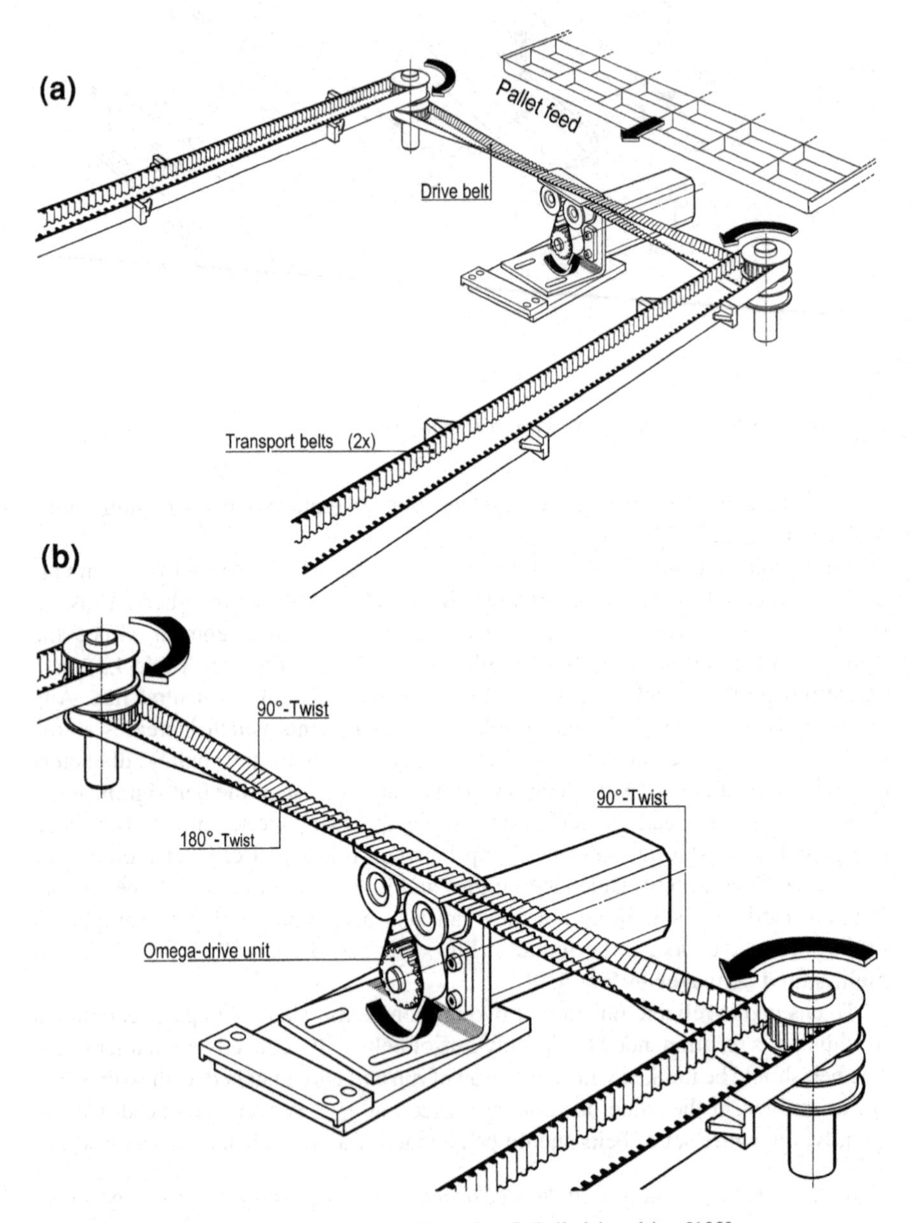

Fig. 5.27 above **a** Automatic palletiser [100], below **b** Palletising drive [100]

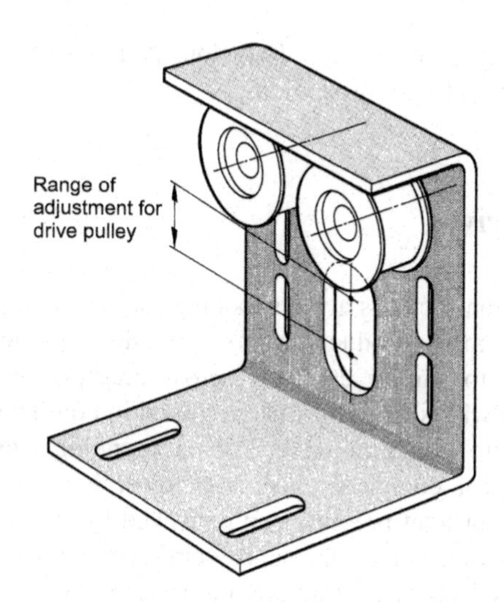

Fig. 5.28 Omega drive sub-assembly with idler pulleys for centre distance adjustment

A palletisation robot links manufacturing process production lines. Its application is in production lines where the individual processes have different cycle times. The pallet serves as a dynamic storage system and increases both availability and process stability. Once the process routine is established it can be maintained over the entire production run.

Automated palletising and de-palletising is effected via custom gripper systems where the parts are recognised, picked and placed from or to individual locations on the pallet and, when finished, the pallet is replaced with a new pallet within a specified time period.

Figure 5.27 shows the pallet being fed into the system which consists of two oppositely arranged conveying toothed belts with profiles. The belts, as well as pulley axes, are arranged in the vertical plane. The belts must work as matched pairs by positioning exactly with each other so that there is no possibility of the pallet twisting during the in-feed operation. The pallet is supported by and is pushed along support rails. The system programme takes care of positioning according to the demand-related cycle times.

A special solution was required to drive the two profiled belts around their pulleys on vertical axes in opposite directions. The designer found the "best" solution by utilising an unrestricted twisted belt drive layout (see Fig. 5.27). Only through this type of drive is a rotation reversal feasible. The timing belt has both a twist of 180° and two 90° bends in a right-angle drive. Compared with other systems this solution requires a significantly lower number of necessary components and therefore is more cost-effective. Furthermore, the central drive motor can be located well below the work area thus leaving the work surface undisturbed and allowing easy maintenance for the central drive belt.

The omega-drive system (sub-assembly with idler pulleys, see Fig. 5.28 must be accurately aligned to the belt running angle and thus it is easier to rotate the entire drive system to achieve this. The elongated positioning holes allow for plenty of adjustment and ease of servicing.

5.11 Haul-off Drives

The factory automation sector is always looking for new solutions for the efficient manufacturing of its many products. Many methods of production have become more prevalent due to similar demands across multiple sectors. For example, extrusion equipment with short haul-offs are found in continuous production lines which can incorporate cables or wires from a stranding facility. The design is essentially made up of complete upper and lower belts where the product is captured, put under constant pressure and conveyed by friction.

Equipment of this type is used for paper, semi-finished materials, foils, veneers and extruded profiles. The basic structure of upper and lower belts is also used for flat products such as glass or particle board, to transport them under contact pressure and allow work on the sides of the product (grinding, edging). Haul-off drives can be configured in different ways, see Figs. 5.29–5.31. The two separate drive belts are synchronised with each other.

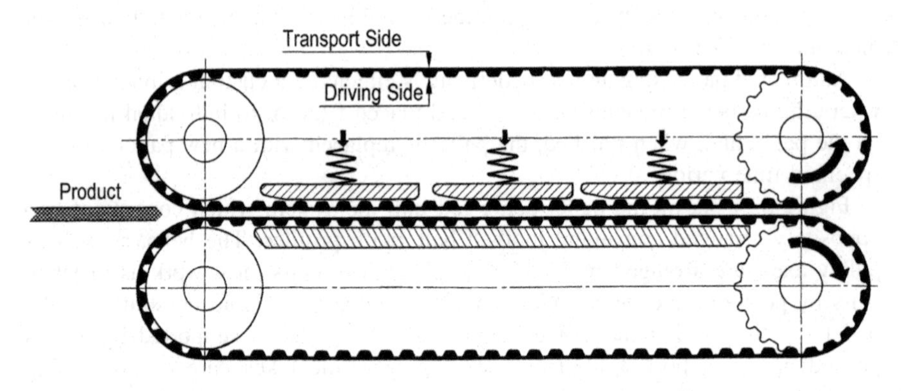

Fig. 5.29 Haul-off with slide support rails

The efficiency of a haul-off with a slide bed is so much higher, that the difference in the coefficients of friction between transport and the drive sides of the belt is cancelled out. With increased drive lengths, the necessary surface pressure required for safe frictional drag can be reduced. This measure is particularly applicable to the transport of pressure-sensitive products.

The belt teeth should have a polyamide tooth coating (PAZ) and the support guide rails or pressure plates should be made of PE or PTFE to give a low frictional value. For high friction on the back of the belt, natural elastomers and coatings of Linatex® or Pararubber® can be used. If profiled sections or cables are to be hauled-off then special cross-sectional grooves can machined in the back of the belt or additional longitudinal grooves incorporated. For reliable power transfer between belt and the guide rail the wide toothed contact face of the AT-profile offers obvious benefits.

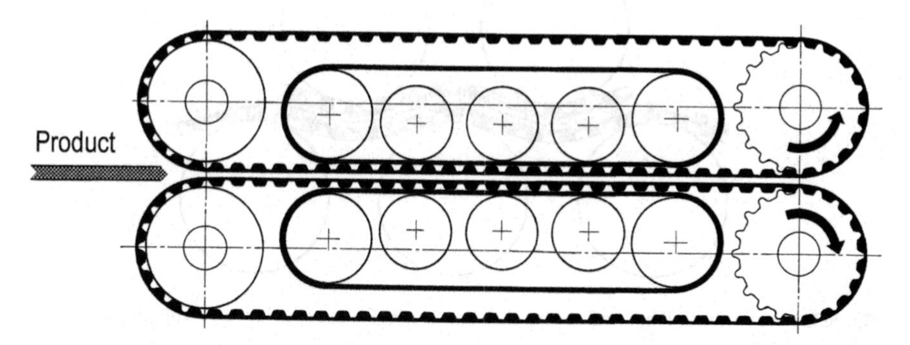

Fig. 5.30 Haul-off with rolling support belts

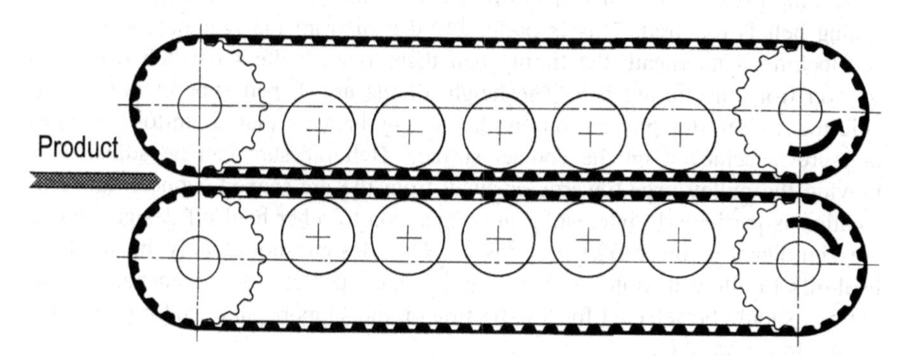

Fig. 5.31 Haul-off with rolling support pulleys. Only timing belt types such as "double-helical toothed" and "arc toothed" (BAT) should be used for this design

Haul-off drives usually work in continuous operation. With the use of sliding contact as per Fig. 5.29 the question arises over the mating friction between belt and support rail and their tribological stability (see also Chapter 5.3). With increasing contact forces and working speeds ($p_{fl} \cdot v$) the designs shown with rolling support pulleys are preferable. In Chapter 5.5 there is a detailed explanation, comparison and evaluation the two methods.

When designing a haul-off drive it is also important to ensure that the belt contact surfaces (the backs of the belts) are always working together in the same plane. With offset rollers the product would experience an alternating curved contact zone and with each bend of the belt would also see significant relative speed variations through the product thickness (see Fig. 5.32). For this reason support/pressure rollers should always be arranged exactly opposite each other.

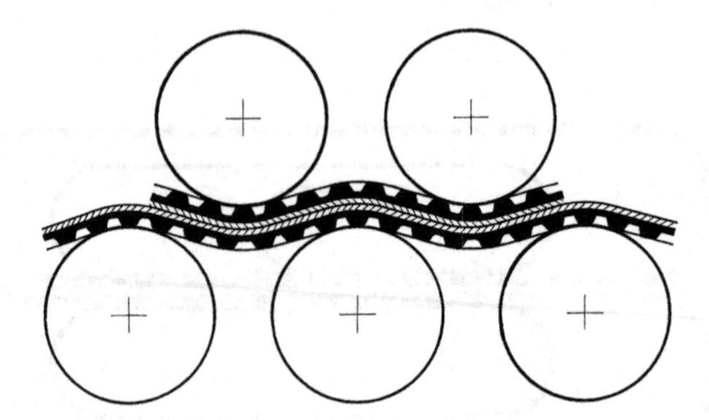

Fig. 5.32 Wave-shaped haul-off paths will not work. Product in contact with the belts will experience different speeds

A simplified version of a haul-off is shown in Fig. 5.33 where just a single timing belt is required. This is preferably driven from the rear pulley to avoid self-locking. Underneath the timing belt there is an endless flat belt driven by friction from the timing belt. The height of the arc-shaped support rail can be adjusted to set the pre-tension in the timing belt so that a uniform contact pressure is achieved on the contact surface. Belt pressure can be adjusted by moving the pulley axes towards or away from the arc-shaped support rail. This solution is particularly interesting in comparison to other haul-off designs due to the significantly lower frictional losses. This design should only be used for haul-offs to allow flexible products (films, tapes, cables) to be transported. The flat belt should be selected for low-friction on the support guide and high-friction on the transport side.

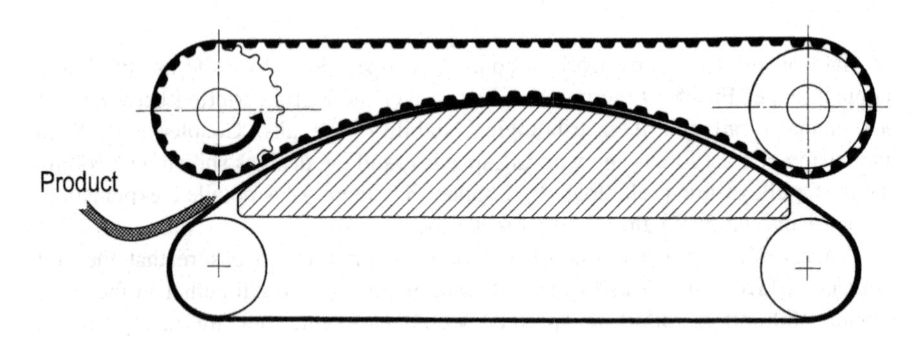

Fig. 5.33 Haul-off drive with a moving flat belt on an arc-shaped support guide

5.12 Vacuum Belts, Magnetic Belts

Vacuum belts offer a highly elegant conveying solution when light products such as roll-fed paper, foil or folded carton blanks are transported as individual items. The low mass of these parts mean they are not suitable for transport by friction alone and conveying must be achieved by other means. Using suction, a vacuum as low as 0.1 bar is often sufficient because the product generally has a large surface area to mass ratio. Conveying with vacuum is also suitable for sheet products of all kinds.

To achieve the desired result, the conveyor system design must allow air to be sucked through the belt. A good example of this can be seen with a machined vacuum groove on the toothed side of the belt and an associated guide rail containing suction channels (see Fig. 5.34). The longitudinal belt groove slides on the guide rail combining with the vacuum draw to provide an effective sealing contact area. The recommended material for the guide rail is a low-friction UHMWPe. The machined surface of the belt suction groove is smoothed by plasticizing and a low-friction nylon fabric facing is incorporated. When assembled, the active contact surfaces must be absolutely flat and parallel to each other. Even with careful design a completely loss-free system is rarely possible.

The physical properties of the belt back can be varied with a coating to achieve the desired co-efficient of friction and abrasion resistance while the use of slots

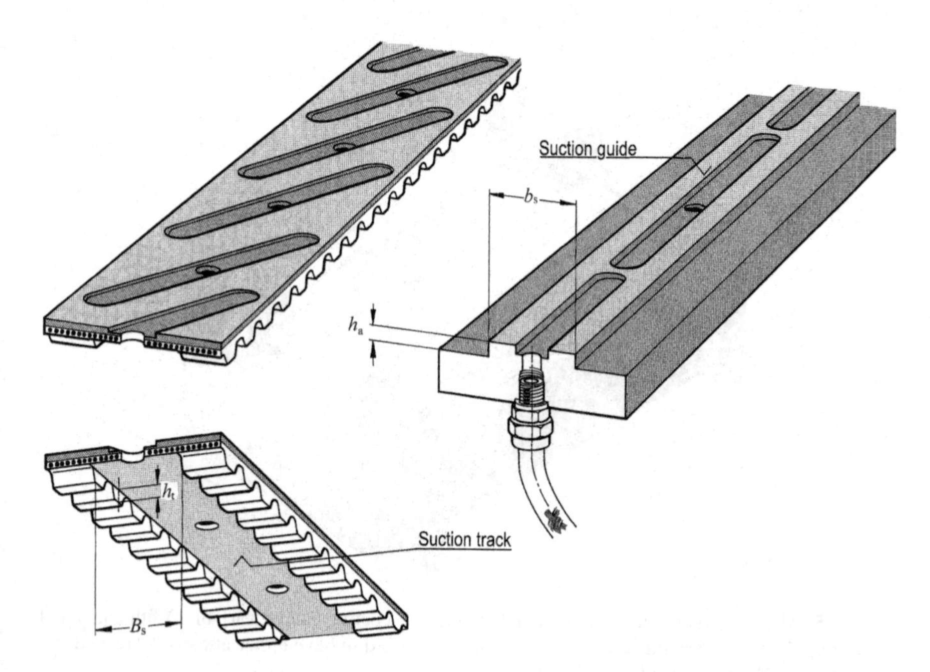

Fig. 5.34 Vacuum belts. Example for form-fill-seal machinery applications. Detailed design of vacuum groove and guide rail

instead of holes significantly increases the suction surface area. The design shown in Fig. 5.34 shows that, with the vacuum channels in an overlapping arrangement, a continuous suction over the entire length of the conveyor belt is possible. It should be noted that additional coating thicknesses will increase the minimum bending diameter of the belt.

The assembly of the belt vacuum groove and track vacuum guide results in a self-tracking arrangement. A minimum free play between the belt vacuum groove B_s and the track vacuum guide b_s is recommended

- Belt width <50 mm Minimum recommended free play +0.5 mm.
- Belt width >50 mm Minimum recommended free play +1.0 mm.

The vacuum provides both a suction draw for product as well as a contact force between belt and guide rail. The latter will increase the gradual wear of the sealing surface. The important dimensions of the vacuum conveyor are the tooth height h_t and the suction guide rail height h_a so that the head line of the belt does not run on the guide rail.

Another alternative for secure non-slip conveying is by the use of magnetic belts with a back coating of elastomers containing permanent magnetic materials. This method can be used to solve feeding or conveying tasks for film, paper, veneers and such like. Figure 5.35 shows that this type of belt conveyor can eliminate support rails. In operation, this sort of drive has correspondingly low frictional losses.

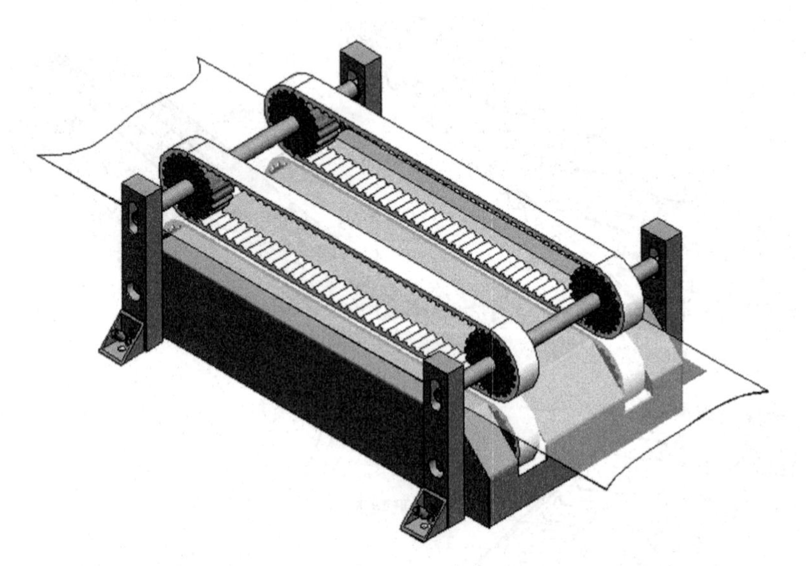

Fig. 5.35 Magnetic belts. The magnetic pull of the belt coating can be variable. With equal and opposite polarity of magnetization the belts can be arranged to have either attractive or repulsive properties. A Norditec development [92]

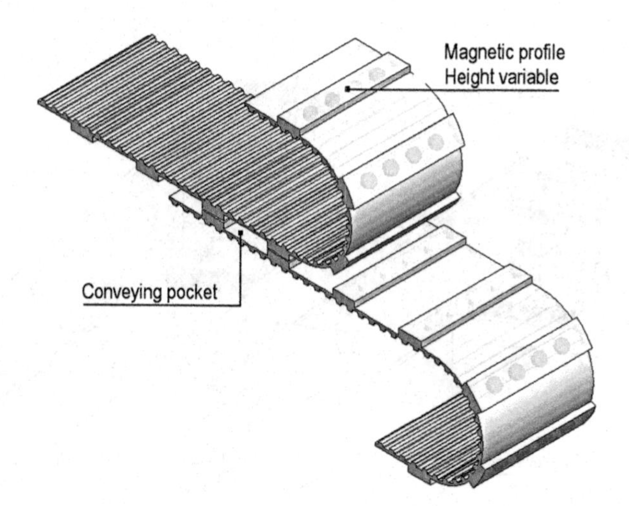

Fig. 5.36 Magnetic pocket belt. An example of use in collation or orientation in handling technology. A Norditec development [92]

Figure 5.36 shows a method for parts separation, while allowing the parts to be firmly gripped, using a magnetic pocket belt. By staggering the magnetic profile on each belt the size of the pocket can be of variable size.

Automated processes with extreme transport and feeding requirements need special solutions. Amongst such applications, where there are often very long distances to be bridged, is the feeding of armature plates in electric motor and transformer manufacture as well as the transport of sheet metal in automotive panel manufacture. Such an application of a magnetic transport system is described by Schuler Automation GmbH & Co. KG, Germany. The process of cutting sheet from coil happens at the beginning of the manufacturing process for automobile body panels. The production line processes steel sheet from 400 to 2000 mm width with a thickness of 0.5 mm.

The system comprises of an automated chain of such work stations until they are stacker unloaded via multiple parallel magnetic conveyors. They are available with up to six conveyor tracks and the entire system is width adjustable.

The single conveyor example in Figs. 5.37 and 5.38 comprises of the support structure, the operating elements and the belt and magnetic rails for sheet transportation. The magnet types can be either permanent or electromagnetic. Systems of this kind are usually specified with AT-profile polyurethane timing belts with steel tension members to achieve the necessary performance. The steel tension members assist the penetration of the magnetic field and resist sag in the return strand over long distances. Such a conveyor is also able to transport product on the return strand due to the use of magnets.

This type of conveyor is used for stacking, unstacking, for feeding press lines and the linking of individual presses.

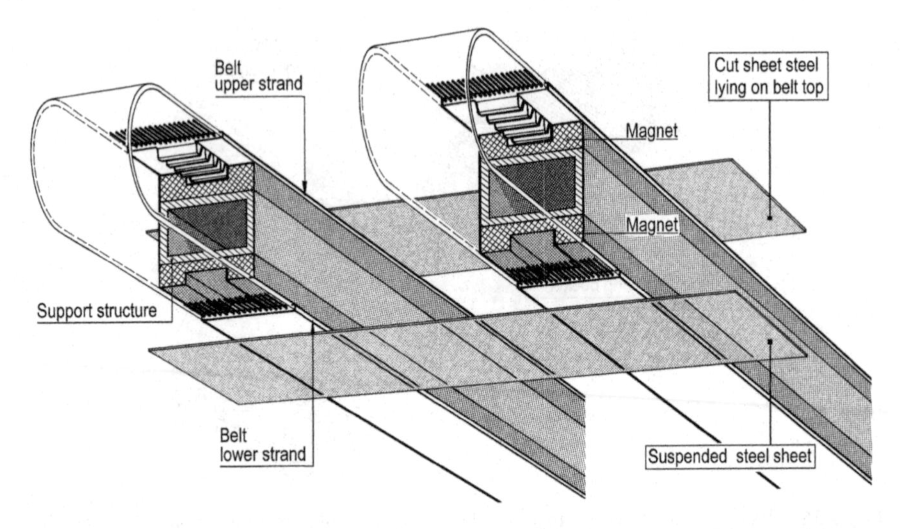

Fig. 5.37 Magnetic belts. A preferred solution for the suspended transport of ferrous sheets

The use of aluminium is increasing in the automotive industry and as described in reference [9] both steel (magnetic) and aluminium (vacuum technology) sheets are transported, handled and placed with high positional accuracy at a facility for VW-Shanghai, China. However this sort of system needs powerful vacuum technology with specially designed suction belts.

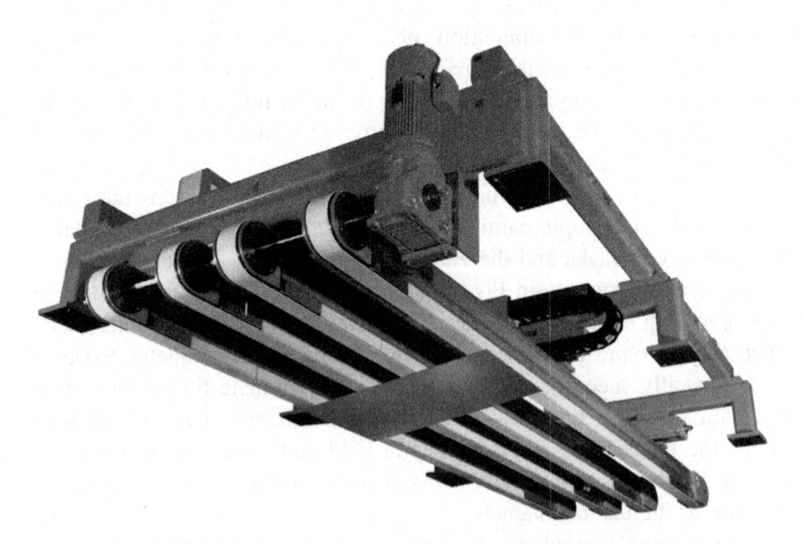

Fig. 5.38 Toothed belt conveyor for suspended transport from Neuhäuser [91]

Chapter 6
Timing Belt Failure

Abstract Timing belt failure is the loss of function before the end of the component's life, where the belt loses some or all of its properties in a predetermined manner and can no longer operate. The following analysis shows illustrations of the types of failure and helps to clarify the root cause of damage during failure. This section is recommended for troubleshooting, safety and achieving improved quality, safety and economy.

6.1 Sources of Failure

6.1.1 Design

Designing is a process to generate technical solutions for both production and use. The task consists of solving the given request from a set of requirements and secondary conditions. The desired result of a design is to measure and consider all appropriate influences and conditions which will affect the equipment over the entire life span and enable a correct drive deployment.

Typically, a belt driven assembly or machine has been designed in advance of its creation. Errors can occur if all operating conditions are not sufficiently known at the time of development. Further problems may occur through improper interpretation or misinterpretation of the drive calculation rules. The designer should always correctly implement the latest detail design techniques. For example, a flanged pulley drive (see Chapter 2.10) needs to be designed. The belt tension is not the only adjustment to be considered (see Chapter 3.7) as it is also necessary to know both the pre-tension and be able to give the final assembly department a plus/minus tolerance range for these values (see Chapter 2.17).

R. Perneder and I. Osborne, *Handbook Timing Belts*,
DOI: 10.1007/978-3-642-17755-2_6, © Springer-Verlag Berlin Heidelberg 2012

6.1.2 Manufacturing

Other possible causes of failure are manufacturing variations in the production of the belt and improper handling during storage and shipping. This also includes sources of error in the associated pulleys or other installation components that are directly related to the function of the drive.

6.1.3 Assembly

The installation of timing belts and pulleys into machinery requires a high degree of technical competence. For example, the running of long belts when their associated pulleys are far apart can be difficult. The assembly department therefore should have adequate measuring equipment available for testing the alignment and angularity of the installed components. Incorrect pre-tension is a frequently mentioned cause of drive problems. The provision of an electronic tension meter is a worthwhile quality assurance measure to prevent errors during assembly.

6.1.4 Operation

Drive system users must take into account maintenance and repair requirements as with increasing length of use the probability of drive damage rises. Slight deviations from the desired drive state will eventually lead to damage and the machine operator should pay attention to changes in the running of the drives (e.g. noise). After experiencing an initial problem he should try to estimate the remaining drive life and start preparations for changing the belt.

> After a problem incident, a failed timing belt should be kept and its installation position and direction of travel marked for further investigation.

6.2 Fault Analysis

In analyzing any case of belt/component failure it is important to establish the cause of failure amongst all the variables, then document any suspicious drive property deviations and distinguish between primary and secondary material damage. Table 6.1 lists possible causes of failure relating to visible modes of damage. This then provides a basis for subsequent drive optimization in order to increase reliability, safety and quality. Any aspect of failure analysis investigation is an important tool in all sectors of mechanical engineering. Failure analysis is also an integral part of FMEA in the field of quality management [43].

Table 6.1 Fault analysis

Fault type	• Possible causes ∘ Remedial measures
1. Tension members starting to break on one edge 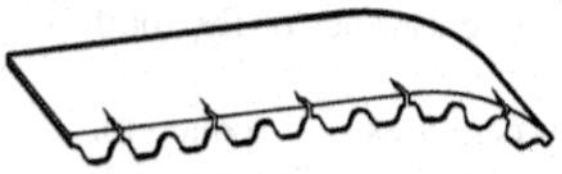 Tension member failure starts on one side of the belt	• Through parallel offset, angular error, or pulley eccentricity giving uneven load distribution across the belt section (see Chapter 2.10). This results in excessive tensile forces in the outer tension members ∘ Check and correct installation and pulley alignment
2. Broken belts  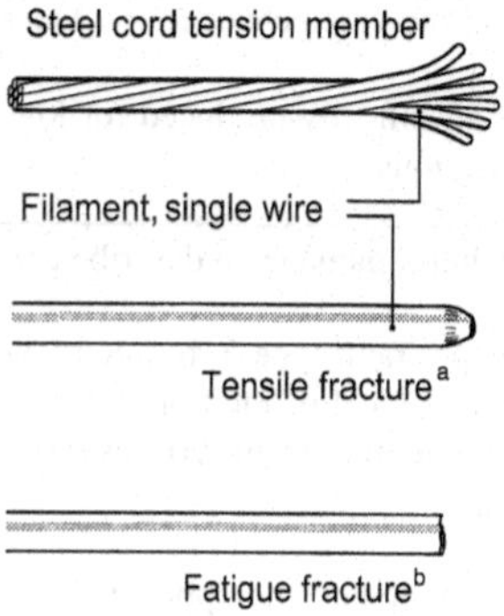 Steel cord tension member Filament, single wire Tensile fracture[a] Fatigue fracture[b]	• Initial damage with partially broken tension members as described in fault type 1 • Initial damage due to bent or crimped belts • Tensile damage due to pulleys being too small. By repeated bending and/or contraflexure developing fatigue breaks • Damage to the tension member through wear and tear, see fault type 6 • Catastrophic tension member failure due to a single, over-design shock load when starting, emergency braking or by accident • Belt pre-tension too low leading to tooth-jump on the driven pulley where the tension member cannot absorb the tooth-jump force which is followed immediately by total failure due to tension member failure ∘ Component examined for early damage. Check the drive dimensioning. Check minimum bending and flex diameters and compare with recommended values ∘ A tension member failure is normally unpredictable. Investigate proper pre-tension as per Chapter 2.7 and check the belt installation in accordance with Chapter 2.17. Compensate for a possible reduced pre-tension ∘ Distinguish whether the broken steel tension members have been subject to either a tensile[a] or fatigue[b] fracture ∘ Glass-fibre tension members show a frayed (fibrillar) mode and Aramid tension members go frizzy

(continued)

Table 6.1 (continued)

Fault type	• Possible causes ∘ Remedial measures
3. Cracks on the tooth base radius  Cracks are formed at the tooth base radius and continue along the boundary between the tension member and the elastomer body	• Pulley tooth-tip radius too small • Load carrying limit beyond belt tooth capability • Back tension idler is too small thus building high tensile stresses in the elastomer of the tooth base radius • Extremely low temperatures will reduce the resilience of the elastomer and increase the tendency of the elastomer to crack ∘ On a damaged belt cracking develops in the direction of rotation. Thus it can be determined which of the paired pulley flanks takes the main load and therefore where belt damage can be expected ∘ Replace pulleys where the tooth tip radius is incorrect ∘ Increase pulley sizes in drives with contraflexure ∘ Check drive design in respect to tooth shear strength For borderline cases change to a wider belt ∘ Use an elastomer specifically produced for low temperature applications
4. Individual belt tooth or tooth group shear 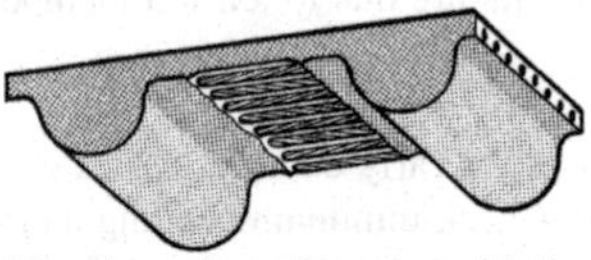 The shear line of tooth and belt carcass is formed along the boundary between the tension member and the elastomer carcass	• Initial damage includes damage as described in fault type 3 • Initial damage, as per fault type 3, has continued up to the loss of individual or multiple teeth • Individual teeth or multiple teeth torn off due to overload ∘ Make the same assessments and apply the same measures as described in fault type 3 ∘ If teeth are shearing due to overload first check the drive layout and where appropriate increase number of pulley teeth

(continued)

Table 6.1 (continued)

Fault type	• Possible causes ∘ Remedial measures
5. Wear on the belt tooth flanks 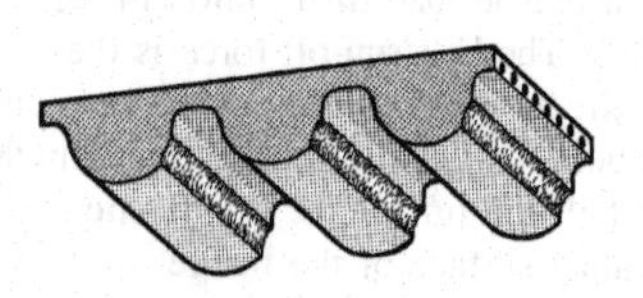 When flanks of a belt with polyamide tooth facing are abraded the material generates fluff and pellet formation however on a purely elastomer belt the tooth material is abraded away. Polyurethane belts will also show defects on the tooth flanks. Damage occurs over the entire length of the belt. For drives with a single direction of rotation material loss takes place on a single flank and on both flanks with a bi-directional drive	• Different pitching between the timing belt and the timing pulley caused by (a) Pre-tension too low (b) Pre-tension too high (c) Pulley outside diameter too large (d) Pulley outside diameter too small (e) Belt stretch • Pulley tooth tip radius too small (sharp edges) • Belt and pulley profiles do not match ∘ Flank wear and material loss can indicate a pitch difference between belt and pulleys. For a description see Chapter 2.7 (a) If wear is found on the working flank of the driven pulley then pre-tension is too low (b) If wear is found on the working flank of the driving pulley then pre-tension is too high ∘ Check drive dimensioning ∘ Measure the pulley outside diameter. Measure the belt pitch. Determine the pitch difference between the belt and pulleys ∘ Check the pulley tooth tip radius. Replace timing pulleys if necessary
6. Belt tooth base wear  The wear starts with visible fibres and fluff from the fabric tooth facing. It continues with loss of belt material and exposes the tension member. With the complete loss of fabric the tension members are abraded to destruction	• Tribological abrasive wear on the base line of the belt (in bottom of the tooth gap) where the belt tooth moves transversely across the pulley tooth gap, see Chapter 2.9. The pressure forces on the base line (the pulley outside diameter) results in friction and wear • Base wear is promoted by high frequency load changes such as in camshaft drives in automotive applications ∘ Use special toothforms with restricted backlash ∘ Use toothforms where the belt tooth rests in the pulley tooth gap (for example AT-series) ∘ Use belts with wear-resistant special fabrics on the toothed side of the belt

(continued)

Table 6.1 (continued)

Fault type	• Possible causes ∘ Remedial measures
7. Edge wear 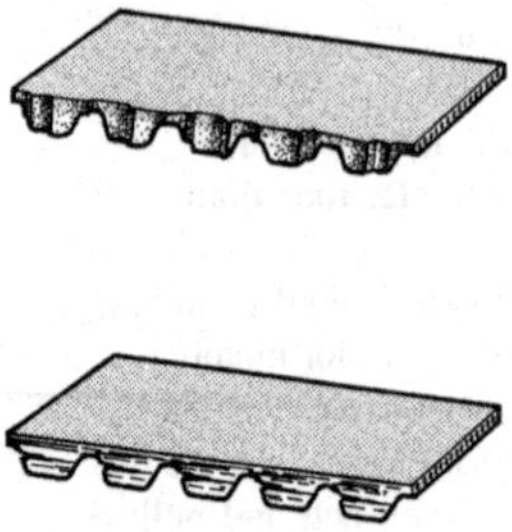 Belt edge vertically or longitudinally grooved	• Offset pulleys, angular errors or pulley run-out gives an uneven tensile load distribution in the tension members. The belt run-off force is the highest in the direction of the largest belt tension • The wider the belt and the shorter the span length the greater the belt run-off forces. The friction point is the contact surface of the flange • No flange lead-in, sharp edges ∘ The flanges should be supplied to the recommendations in Chapter 2.10. Correct build quality and pulley alignment problems. Check lead-in and edges on the flanges
8. Belt transverse bowing 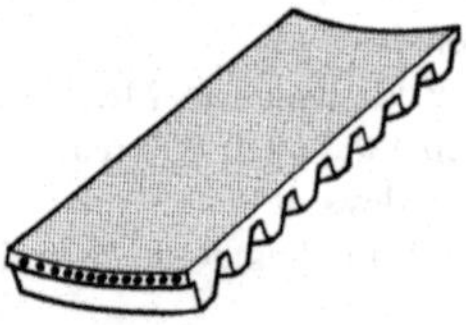	• Causes as in fault type 7 • Side forces cause a compression of the belt perpendicular to the direction of travel ∘ Remedial measures such as fault type 7
9. Deformed teeth 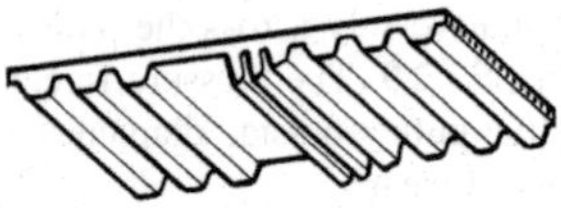 Displacement and deformation of individual teeth or groups of teeth on the tension members	• Bond between tension members and elastomer is insufficient. Lack of bonding agent or inappropriate application. Production process problems • The combined effect of high temperatures with high tooth flank stress can cause plastic deformation of individual teeth and tooth groups • The elastomer strength is reduced by aggressive media (oil, solvents, etc.). Flank forces may also cause deformities ∘ Quality of the adhesion agent to be checked by the manufacturer. Prevent high temperatures by changing environmental conditions. Use heat-resistant belt materials. Exclude contact with aggressive reagents ∘ Tension member embedment to be checked by the manufacturer and optimized where appropriate [118]. See description in Chapter 2.5

(continued)

Table 6.1 (continued)

Fault type	• Possible causes ○ Remedial measures
10. Corroded steel tension members Red/brown colouring, visible rust build-up	• Long-term effects of aggressive environmental factors such as acids, detergents or seawater • Frictional corrosion formation (rare) due to high speeds in conjunction with small idler pulley diameters ○ Use special stainless steel tension members ○ Frictional corrosion: Tension member embedment to be checked by the manufacturer and optimized where appropriate [118]. See description in Chapter 2.5
11. Loose flanges, Wear on guide elements Flanges damaged or loose	• Manufacturing variations in flange mounting. Improper attachment. Damage during handling, shipment or during installation • Drive alignment deviations can cause considerable side loads on the flanges • Protruding steel tension members can be abrasive on lateral guide elements ○ Apply featured flange designs and mounting as in Chapter 2.10 ○ Remove protruding tension members
12. Pulley tooth wear Tooth material removal on the loaded flank	• Environments containing dust, scale and gritty substances work together with the rotating belt to erode the pulley teeth ○ Cover the drive. Preferred materials for timing pulleys under abrasive conditions, if applicable, are PA, PE or POM plastics

[a] A *tensile fracture* is a failure in which a single, rapid and unidirectional stress, or peak load, which alone or together with a superimposed pre-tension exceeds the tensile strength of the tension member and brings about a complete material failure. Tension members at the fracture faces of the individual wires will exhibit typical necking

[b] A *permanent* or *stress fracture* occurs when the amplitude and the number of belt cycles lead to an overall load situation where the border area of statistical fatigue strength (according to *Wöhler*) is reached. A stress fracture includes time-dependent weakening in the form of either mechanical or chemical assaults which act by removing material or cracking the wire surface. With increasing crack growth, the effective tension rises in the remaining cross-section, until the eventual fracture occurs. It fractures *without* necking. The material break is characterized by a visible primary crack and final fracture

Chapter 7
Appendices

Abstract The design process consists of multiple repetitive steps of design, calculation and verification of technical feasibility and spatial implications reported by (Krause in Konstruktionselemente der Feinmechanik 2004). Solution-based approaches to machine design considerably reduce uncertainty and workload. With the use of timing belts, the drive space-efficiency and physical properties are iteratively modified to reach the desired target values. In order to efficiently accomplish the target solution, the necessary steps for drive design are summarized in the following chapter, which demonstrates easy ways to assess and determine the drive design and performance.

7.1 Overview of Drive Design

A timing belt drive is correctly designed when the total load, acting over the drive lifespan, is correctly calculated and used in conjunction with the chosen product. To dimension a drive, three main criteria must be considered:

- Tooth load capacity
- Tension member capacity
- Flex life

Tooth load capacity (see also Chapter 2.9)

The tangential force to be transmitted is distributed in the pulley angle of wrap through the number of engaged teeth. The designer must design around the pulley that has the highest single tooth loading in the selected drive geometry. In a two-shaft drive this is usually the small pulley, normally referred to as the driving pulley, which is used for the calculation of the tooth load capacity. In a multi-shaft drive it is not always clear which pulley has the largest tooth loading.

R. Perneder and I. Osborne, *Handbook Timing Belts*,
DOI: 10.1007/978-3-642-17755-2_7,

The assessment of the load bearing capacity of the maximum loaded individual tooth is the most important factor in the dimensioning of the required minimum width of a timing belt drive.

Tension member capacity (see also Chapter 2.6–2.8)

Knowledge of the real forces acting on the tension members is indispensable for the further understanding and qualitative assessment of the entire drive. This is the maximum occurring span load in relation to the F_{zul}-values of the product used. As a rule, safety margins are well above those of the tooth load capacity. Furthermore, the maximum tangential force is used in the determination of the pre-tension, shaft load and the natural frequency of vibration of the span. With linear, multi-shaft and positioning drives it is appropriate to operate with higher pre-tension forces. These are to be regarded as a pre-load and to run jointly with the operating load up to the maximum force in the tension member.

Flex life (see also Chapter 2.5)

The diameter, material and construction of the tension member essentially determine the timing belt flexibility. A distinction is made between the outside diameter of the tension member and the size of the filaments embedded therein. There are many ways to refer to the flexibility.

Belt manufacturers recommend a minimum number of teeth on the pulleys for the use of their products. With contraflexure, back idler pulleys should not fall below a predetermined diameter. Compliance with manufacturer recommendations will save the user extra calculation on the reference stress (the force built up from the tension and bending stress). The limit values in use are the parameterized tensile forces and associated bending diameters.

Variety of types and development stages

The belt profiles in the marketplace are often available with a variety of different modifications. Some manufacturers offer their range of products with differing tension members depending on application. A larger tension member diameter gives significantly higher values for load-carrying capacity and stiffness. However this measure is usually accompanied by a reduction in flex life. This may be addressed with a reduction in the diameter of the individual filaments. A sustained period of increase in the performance of timing belts has caused further developments in the elastomer materials and the processing of the composite structure. These include: sophisticated adhesive systems between tension member and elastomer body, direction-oriented fibre material additives to increase tooth load capacity, enhanced or multi-layer composite tooth fabric layers impregnated with low friction and heat-resistant elastomers based on peroxide-cross-linked rubber types. Carbon-fibre tension members were first used in Spring 2007 in timing belt manufacture. The trend towards further development progresses unabated as the belt manufacturers continue to set higher performance levels.

It is to be expected, with such a relatively young transmission element, that further reserves can be found and these potentials be used for performance improvements. For this reason drive calculations for new designs should always be checked against current manufacturer data. The provision of manufacturer-related data as well as its available updates cannot be fulfilled in a handbook such as this. In the long run, guarantees and warranties can only be made by the obligatory use of manufacturer data.

7.2 Balanced Drive Design

> Components and drive elements, which are functionally linked together, should exhibit meaningful dimensional relationships to each other in a balanced design. This also applies to timing belt drives. Optimal dimensioning has a certain relative importance to the surrounding structures and parts thereof.

Shafts for standard motors and drives are usually dimensioned for the torque to be transferred. The shaft diameter has thus a direct relationship to the drive torque. Since the effective geometry of belt drives is known, the comparable correlation of the shaft diameter can be used for the rough design of timing belt drives.

In the assembly below, the drive components of the shaft, bearing, timing pulley and belt are correctly dimensioned and have a balanced relationship with each other. The fundamental soundness of the sample design in Fig. 7.1 can be readily seen without the need for calculation. A first estimation of a toothed belt drive design is possible using the relationship of the shaft diameter d to size the surrounding components, see Table 7.1.

Table 7.1 Rough design ratios based on a shaft diameter of d

Timing belt pitch p	$(0.2–0.5) \times d$
Belt width b	$(1–2) \times d$
Timing pulley diameter d_w	$(2–5) \times d$

A timing belt has the task of transferring torque from the driving to the driven shaft. The drive designs in Fig. 7.2 show two comparable solutions for pulley diameter and belt width which transfer the same power in the ratio of $i = 1$.

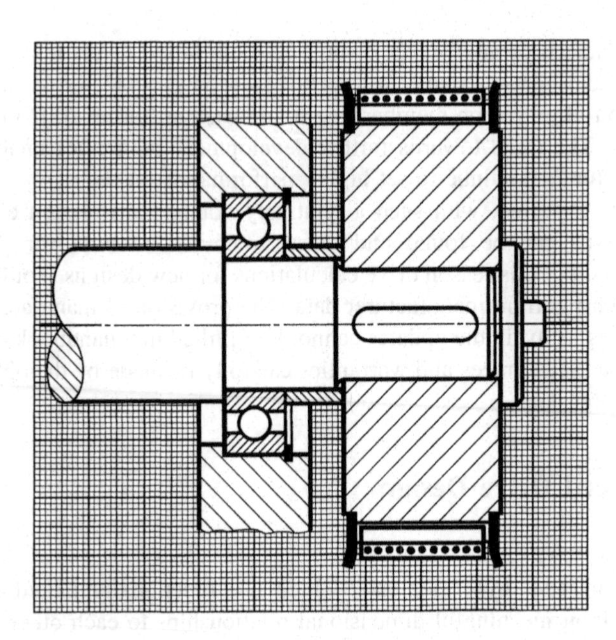

Fig. 7.1 A balanced design

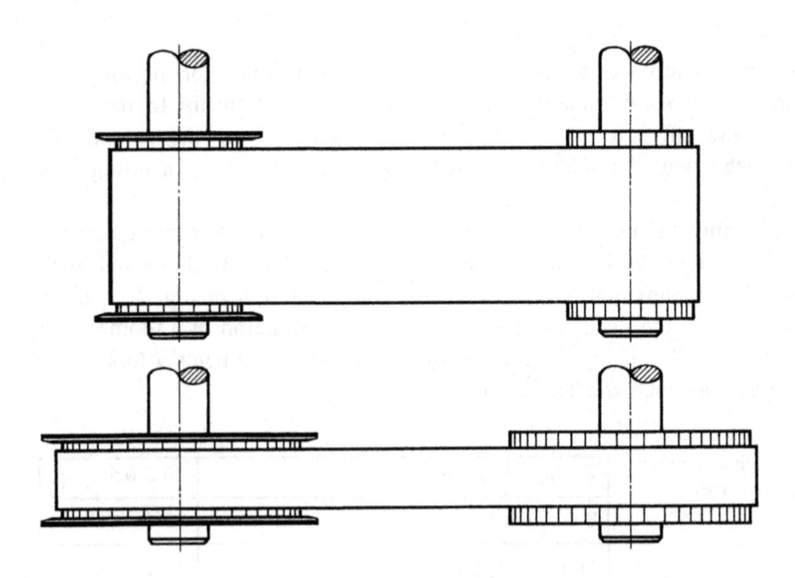

Fig. 7.2 Belt drives of the same power capacity

The drive with the narrow belt and large pulley diameters shows clear advantages in operating performance in comparison with the wider belt drive as it results in lower overhung shaft loads.

7.3 Tension Members and Tooth Stiffness

Knowledge of the overall drive stiffness is of particular importance in the qualitative assessment of timing belt drives. This is used in order to estimate the

Table 7.2 Allowable tension member loads as well as specific tension member and tooth-in-mesh stiffness for the ATL toothed belt in PUR 92 Shore A

Type	Character	Units	Belt width in mm					
			16	25	32	50	75	100
ATL5	F_{zul}	N	1,300	2,000	2,800	4,200		
	c_{Bspez}	10^3 N	330	500	650	1,050		
	c_{Pspez} Normal gap	10^3 N/m	144	225	288	450		
	c_{Pspez} SE gap	10^3 N/m	192	300	384	600		
	c_{Pspez} Zero gap	10^3 N/m	240	375	480	750		
ATL10	F_{zul}	N			7,200	11,200	16,800	22,400
	c_{Bspez}	10^3 N			1,800	2,800	4,200	5,600
	c_{Pspez} Normal gap	10^3 N/m			288	450	675	900
	c_{Pspez} SE gap	10^3 N/m			384	600	900	1,200
	c_{Pspez} Zero gap	10^3 N/m			480	750	1,125	1,500
ATL20	F_{zul}	N			9,800	15,400	23,800	31,500
	c_{Bspez}	10^3 N			2,450	3,850	5,950	7,800
	c_{Pspez} Normal gap	10^3 N/m			288	450	675	900
	c_{Pspez} SE gap	10^3 N/m			384	600	900	1,200
	c_{Pspez} Zero gap	10^3 N/m			480	750	1,125	1,500

The stiffness behaviour of other types of belt can be determined as follows
1. The value c_{Bspez} according to Eq. 2.36 is derived from F_{zul}, and is usually available as catalogue data from all manufacturers
2. The value c_{Pspez} in Table 7.2 refers solely to the ATL profile. Comparisons between the high power profiles AT, HTD, STD and RPP show they have a very similar fixed tooth base width s, see also Chapter. 2.4.9, Fig. 2.8. The specific stiffness of each individual tooth in mesh essentially determines the tooth base width. Thus, the table values are 1:1 transferable between these high performance profiles. In the case of dimensional variations or other belt body materials then an approximation by inter- or extrapolation can be attempted
F_{zul} allowable load in tension member
c_{Bspez} tension member stiffness based on a 1 metre belt length
c_{Pspez} specific stiffness of each single tooth in mesh

suitability of potential drive designs in regard to positioning ability and vibration response for the proposed application.

The high power AT tooth profile is used in belts whose preferred use is in accurate angular rotational drives and positionally accurate linear drives. Its associated geometrical dimensions are specified in Chapter 2.3.3. ATL belts have the same tooth profile and are geometrically exchangeable; however the "L" suffix refers to its preferred area of application in linear drives. Belts specifically used for linear drive tasks are manufactured with stronger tension members. Applications of this kind demand minimized elongation behaviour with ever increasing traverse distances. Furthermore, manufacturers produce short-pitch versions, which are offered with a length-minus tolerance range. Such a belt, when subjected to an appropriate pre-tension force, will achieve the nominal pitch when installed. The standard tolerance for ATL belts under a standard measuring load is –0.4 to –1.0% (measuring the pitch length is dealt with in Chapter 2.17). The technical data for ATL toothed belts, specified in the following Table 7.2, relies on documentation [85] from BRECO as well as the investigations in [25]. Since most product ranges are subject to changes, due to technical improvement, it is recommended that the current data from the chosen manufacturer is requested.

References

1. Bahr Modultechnik GmbH. www.bahr-modultechnik.de
2. Bartenschläger, J., Hebel, H., Schmidt, G.: Handhabungstechnik mit Robotertechnik. Friedrich Vieweg & Sohn Verlagsgesellschaft, Braunschweig/Wiesbaden (1998)
3. Bekeart N.V., S.A.: Firmenschrift Fine steel cord product catalogue, version 01 (12/02), Belgium-8790 Waregem
4. Bikon-Technik GmbH. www.bikon.com
5. BMW AG, München. www.bmwgroup.com
6. Böttger, A.: Lärmminderung von Polyurethanzahnriemengetrieben. Dissertation TU Dresden (1995)
7. Bosch Rexroth AG. www.boschrexroth.com
8. Brandt, G.: Verarbeitung von Stahl und Aluminium auf einer Anlage. ATZ Automobiltechnische Zeitschrift 102, Seite 612,2000.
9. Breco Antriebstechnik Breher GmbH and Co. www.breco.de
10. Breco: Firmenschrift Zahnriemen mit aufgeschweißten Nocken, Ausgabe 6/96
11. Breco: Firmenschrift ATN-System Zahnriemen, Zahnscheiben und Zubehör, Ausgabe 2004
12. Clarke, A.J.: The evolution of small pitch belts. Tagung Zahnriemengetriebe, TU Dresden, (Tagungsband) (2006)
13. Colt International GmbH. www.coltgroup.com
14. Compomac S.p.A.: www.compomac.it
15. ContiTech Antriebssysteme, GmbH.: www.contitech.de
16. ContiTech: Firmenschrift Hochleistungszahnriemen CXP III und CXA III, Ausgabe WT 533 D/E 08.04
17. ContiTech: Riemenantrieb auf den Kopf gestellt. Verlag für Technik und Wirtschaft, Der Konstrukteur, Heft 6/2005
18. Cybertron GmbH. www.cybertron.de
19. DIN 3051 Drahtseile aus Stahldrähten
20. DIN 7721 Synchronriementriebe, metrische Teilung, Teil 1: Synchronriemen, Teil 2: Zahnprofil für Scheiben (1979)
21. DIN EN 29283 Industrieroboter. Leistungskriterien und zugehörige Testmethoden (1993)
22. DIN ISO 5294 Synchronriemen, Scheiben (1989)
23. Dreckshage GmbH & Co. KG. www.dreckshage.de
24. DIN ISO 5296 Synchronriemen, Riemen (1989)
25. Dubbel, Taschenbuch für den Maschinenbau, 16. Auflage, Seite O10. Springer-Verlag, Berlin/Heidelberg (1987)
26. Dubbel, Taschenbuch für den Maschinenbau, 16. Auflage, Seite O10. Springer-Verlag, Berlin/Heidelberg (1987)

R. Perneder and I. Osborne, *Handbook Timing Belts*,
DOI: 10.1007/978-3-642-17755-2_6, © Springer-Verlag Berlin Heidelberg 2012

27. Elatech S.r.l. www.elatech.com
28. Ebert Kettenspanntechnik GmbH. www.roll-ring.com
29. Erxleben, S.: Untersuchungen zum Betriebsverhalten von Riemengetriebe unter Berücksichtigung des elastischen Materialverhaltens. Dissertation RWTH Aachen (1984)
30. ETP Transmission AB. www.etp.se
31. Farrenkopf, M.: Leistungssprung bei Synchronriemen durch Karbonfaser-Zugstrang. Verlag für Technik und Wirtschaft, Antriebstechnik, Heft 6/2007
32. Fenner Drives. www.fennerdrives.com
33. Flender-Gruppe (Walter Flender GmbH): Firmenschrift PowerGrip® HTD-Zahnriemen, Ausgabe HD 403. www.walther-flender.de
34. Flender-Gruppe (Walter Flender GmbH): Werkszeitung Sonderausgabe Juni 2001
35. Focke GmbH & Co. KG. www.focke.de
36. Gates: Konstruktionshandbuch Poly Chain® GT2-Zahnriemen, Ausgabe 01/05
37. Gates power transmission Europe BVBA. www.gates.com
38. Gates Mectrol GmbH. www.gatesmectrol.de
39. Gerstmann, U.: Robotergenauigkeit: Der Getriebeeinfluß auf die Arbeits- und Positioniergenauigkeit. VDI-Verlag, Düsseldorf, 1991.
40. Gesellschaft für Antriebtechnik mbH. www.gat-mbh.de
41. Goedecke, W.D.: Linearachsen im Vergleich. Tagung Zahnriemengetriebe, TU Dresden, Tagungsband (2004)
42. Goodyear, Engineer Products. www.goodyearep.com
43. Grosch, J.: Schadenskunde im Maschinenbau. Expert-Verlag, Renningen (2004)
44. Hahn, K.: Komponentenoptimierung im Steuertrieb. MTZ 11/20006, Jahrgang (1967)
45. Harmonic-Drive Antriebstechnik GmbH. www.harmonicdrive.de
46. Henkel AG & Co. KG aA. www.loctitesolutions.com
47. Hepcomotion GmbH. www.hepcomotion.com
48. Hesse, S.: Industrieroboterpraxis. Friedrich Vieweg & Sohn Verlagsgesellschaft, Braunschweig/Wiesbaden (1998)
49. Hopf, L.: Mühlentechnisches Praktikum, Band II, Mühlenbau, Hugo
50. Hung, T.: Kinematische Genauigkeit von Zahnriemengetrieben. Dissertation TU Dresden, (1989)
51. Isotec Automation and Technologie GmbH. www.isotec.at
52. INA, Schaeffler KG. www.ina.de
53. Intercorsa: Produktkatalog, Mühlhausen. www.cordus.de (2005)
54. ISO 13050 Krummlinige Zahnprofile für Riemen und Scheiben (1999)
55. ISO 5288 Synchronous belt drives vocabluary (2001)
56. Item Industrietechnik GmbH. www.item.info
57. ITI-SIM Simulations-Software. ITI GmbH, Dresden. www.iti.de
58. IEF Werner GmbH. www.ief-werner.de
59. Jansen, U.: Geräuschverhalten und Geräuschminderung von Zahnriemengetrieben. Dissertation RWTH Aachen (1990)
60. Kaden, H.: Tribologische Untersuchungen zum Einsatz von Trans portzahnriemen. Tagung Zahnriemengetriebe, TU Dresden (2006). (Tagungsband)
61. Kollmann, F.G.: Welle-Nabe-Verbindungen. Springer Verlag, Heidelberg/Berlin (1984)
62. Karolev, N.: Optimierung der Kräfteverhältnisse in Zahnriemengetrieben. Dissertation TU Dresden (1987)
63. Keiperband GmbH. www.belting-partner.de
64. Kiel, E.: Antriebslösungen, Mechatronik für Produktion und Logistik. Springer, Berlin-Heidelberg (2007)
65. Kollmann, F. G.: Welle-Nabe-Verbindungen. Springer Verlag, Heidelberg Berlin.
66. Korsch AG, Berlin. www.korsch.de
67. Köster, L.: Untersuchungen der Kräfteverhältnisse in Zahnriemenantrieben. Dissertation TU Hamburg (1981)

68. Koyama, T., Murakami, K., Nakai, H., Kagotani, M., Hoshiro, T.: A Study on Strength of Toothed Belt. Transactions of the JSME, 1. Re- port, Vol. 44 No. 377, 1978
69. Koyama, T., Kagotani, M., Shibata, T., Hoshiro, T.: A study on strength of toothed belt, Bull. of the JSME, 2. Report, 22(169) (1979)
70. Koyama, T., Kagotani, M., Shibata, T., Sato, S., Hoshiro, T.: A study on strength of toothed belt, Bull. of the JSME, 5. Report, 22(23) (1980)
71. Krause, W.; Metzner D.: Zahnriemengetriebe. Verlag Technik, Berlin und Dr. Alfred Hüthig Verlag, Heidelberg 1988
72. Krause, W.: Konstruktionselemente der Feinmechanik. 3. stark bearbeitete Auflage. Carl Hanser Verlag, München, Wien (2004)
73. KTR Kupplungstechnik GmbH. www.ktr.com
74. Kullen GmbH and Co. KG. www.kullen.de
75. Lenze SE. www.lenze.de
76. Librentz R.: Das Übertragungsverhalten von Synchronriemengetrieben. Dissertation TU Berlin (2006)
77. MAV S.p.A.: www.mav.it
78. Megadyne Group. www.megadyne.it
79. Metzger M, Bayer Materialscience AG, Leverkusen: Tagung Zahnriemengetriebe, TU-Dresden, (2007) (Tagungsband)
80. Metzner, D., Nagel, T., Heinrich, A.: Dimensionierung von Zahnriemengetrieben. Maschinenbautechnik 39 (1990)
81. mk Technology Group. www.mk-group.com
82. Motus Tech S.r.l. www.motus-tech.com
83. Müller GmbH, Wilhelm Herm. www.whm.net
84. Mulco-Gruppe: Belt-pilot, www.mulco.de
85. Mulco-Gruppe: Firmenschrift AT-/ATL-Linearantriebe, Hannover, Ausgabe 3.000 5/98
86. Mulco-Gruppe: Firmenschrift Breco ATN-System, Hannover, Ausgabe 40/04/7500 dt
87. Mulco-Gruppe: Gesamtkatalog Ausgabe Hannover (2003)
88. Nagel, T.: Vergleichende Untersuchungen zu Verschleißverhalten und Übertragungsgenauigkeit von Zahnriemengetrieben. Dissertation TU Dresden (1990)
89. Nagel, T.: 55 Jahre Zahnriemengetriebe. Verlag für Technik und Wirtschaft, Mainz, Antriebstechnik Heft 7/2001
90. Nagel, T.: Stand der Technik, Überblick über Zahnriemengetriebe. Verlag für Technik und Wirtschaft, Mainz, Antriebstechnik Heft 12/2005
91. Neuhäuser Magnet- und Fördertechnik GmbH. www.neuhaeuser.com
92. Norditec Antriebstechnik GmbH. www.norditec.de
93. NWT Haug GmbH. www.nwtgmbh.de
94. Optibelt: Firmenschrift power transmission. Höxter. www.optibelt.com
95. Paletti Profilsysteme GmbH & Co. KG. www.paletti-profilsysteme.de
96. Perneder, R.; Schmidt, G.: Optimierung von Antriebssystemen mit Synchronriementrieben. Verlag für Technik und Wirtschaft, Mainz, Antriebstechnik Heft 8/1997
97. Perneder, R.: Verdrehsteife Roboterantriebe dank Polyurethan-Zahnriemen. Verlag für Technik und Wirtschaft, Mainz, Antriebstechnik Heft 2/1993
98. Polygon & Rund Schleiftechnik GmbH. www.polygon-gmbh.de
99. Promess Montage- u. Prüfsysteme GmbH. www.promessmontage.de
100. Ratiotec GmbH. www.ratiotec.de
101. Ringspann GmbH. www.ringspann.de
102. Rodriguez GmbH. www.rodriguez.de
103. Rollon GmbH. www.rollon.com
104. Rose + Krieger GmbH. www.rk.rose-krieger.com
105. Schiffbau u. Entwicklungsgesellschaft Tangermünde mbH & Co. KG. www.set-schiffbau.de
106. SNR Wälzlager GmbH. www.ntn-snr.com
107. Spieht Maschinenelemente GmbH & Co. KG. www.spieht-maschinenelemente.de

108. Stemme, A.G.: www.stemme.de
109. Stüwe GmbH. www.stuewe.de
110. Sumitomo Cyclo Drive Germany GmbH. www.sumitomodriveeurope.de
111. Szonn, R.: Zahnriemen in beliebigen Längen. Konradin Verlag, Leinfelden-Echterdingen, KEM Heft 3/73 (1973)
112. TAS Schäfer GmbH. www.schäfer.de
113. Terschüren, W.: Erhöhung der Lebensdauer von Zahnriemen durch Verbesserung der Glascordeigenschaften. Tagung Zahnriemengetriebe, TU Dresden, (Tagungsband) (2005)
114. Thiemig, W.: Prüfstände für Fahrzeug-Getriebe, ATZ Jahrgang 1963, Heft 8, Seite 239 (1961)
116. Tilkorn, M.: Untersuchungen an einem Zahnriemen-Linearantrieb für die Fahrbewegung von Brückenkranen. Fortschr.-Ber. VDI Reihe 13 Nr. 46, VDI Verlag Düsseldorf (1997)
117. Tollok S.p.A.: www.tollok.com
118. Vanderbeken, B.: Trends and Developments in Steel Cord for Timing Belts. Tagung Zahnriemengetriebe, TU Dresden (2007; Tagungsband)
119. VDI/DGQ 3441 Statistische Prüfung der Arbeits- und Positioniergenauigkeit von Werkzeugmaschinen. Grundlagen. Beuth Verlag Berlin (1977)
120. VDI 2758 Riemengetriebe (1993)
121. VDI 2861 Blatt 2: Kenngrößen für Industrieroboter. Einsatzspezifische Kenngrößen. Beuth Verlag Berlin (1989)
122. VDI-Richtlinie 2860, Handhabungsfunktionen, Begriffe, Definitionen, Symbole. Beuth-Verlag Berlin
123. Vollbarth, J.: Übertragungsgenauigkeit von Zahnriemengetrieben in der Lineartechnik. Dissertation TU Dresden (1998)
124. Volmer, J.: Getriebetechnik-Grundlagen. Verlag Technik Berlin, Berlin (1995)
125. Weicon GmbH & Co. KG. www.weicon.de
126. Witt, R.: Modellierung und Simulation der Beanspruchung von Zugsträngen aus Stahllitze für Zahnriemen. Fortschr.-Ber. VDI-Reihe 13 Nr. 54. VDI Verlag Düsseldorf (2008)
127. Wolka, D.W.: Robotersysteme. Springer-Verlag, Heidelberg/Berlin (1992)

Glossary

Latin Uppercase

B	Pulley width in mm (m)
B_f	Clearance width between sides of the support guide rail in mm (m)
B_s	Clearance width of the vacuum guide in mm (m)
C	Centre distance in mm (m)
F	Force, span force in N
F_1	Force in the loaded span in N
F_2	Force in the unloaded span in N
F_A	Axial force in N
F_B	Acceleration force in N
F_H	Lifting force in N
F_R	Friction force in N
F_t	Tangential force in N
F_V	Pre-tension force in N
F_{max}	Maximum span force in N
F_{tspez}	Specific tangential force per belt tooth per cm belt width in N/cm
F_Z	Centrifugal force in N
F_{zul}	Allowable tensile force in the tension member in N
$\sum F$	Sum of forces in N
G	Modulus of rigidity in N/m^2
ΔL	Belt specific noise increase or decrease in dB (A)
$\Delta L_{(1m)}$	Actual tolerance of a pre-tensioned belt in mm per 1 m belt length
L_{WA}	Noise intensity level in dB (A)
M	Torque in N·m
M_B	Acceleration torque in N·m
M_{max}	Maximum torque in N·m
M_{spez}	Specific torque per belt tooth and per cm belt width in N·m/cm
P	Power in kW
P_a	Positional variance in mm (m)

R. Perneder and I. Osborne, *Handbook Timing Belts*,
DOI: 10.1007/978-3-642-17755-2_6,

P_{ab}	Power output in kW
P_{max}	Maximum power in kW
P_s	Positioning tolerance (repeatability) in mm (m)
P_{spez}	Specific power per belt tooth per cm belt width in kW/cm
P_u	Positional uncertainty in mm (m)
P_{verl}	Power losses in kW
P_{zu}	Power supplied in kW
U	Reversal error in mm (m)
ΔT	Temperature difference in K

Latin Lowercase

a	Acceleration in m/s^2
a_A	Length-temperature coefficient in 1/K
b	Belt width in mm (m)
b_f	Clearance width between the flanges in mm (m)
b_w	Timing pulley tooth base width in mm (m)
b_s	Width of vacuum groove in mm (m)
c_B	Belt tension member stiffness in N/m
c_{Bspez}	Belt tension member stiffness per metre of belt length in N
c_{ges}	Total stiffness of the belt in N/m
c_{m1}	Tangential backlash in mm (m)
c_{m2}	Radial play in mm (m)
c_P	Belt teeth stiffness in arc of wrap in N/m
c_{Pspez}	Specific stiffness per single tooth in mesh in N/m
c_t	Tooth tip backlash during tangential meshing in mm (m)
d	Central bore diameter of the timing pulley / tension roller in mm (m)
d_F	Pulley tooth base diameter in mm (m)
d_K	Outside diameter of the timing pulley in mm (m)
d'_K	Pulley outside diameter with profile alteration in mm (m)
d''_K	Pulley outside diameter with double profile alteration in mm (m)
d_S	Running diameter of the tension roller in mm (m)
d_W	Pulley pitch circle diameter in mm (m)
d_Z	Tension member diameter in mm (m)
f_e	Frequency of vibration in s^{-1}
f_R	Pulley radial run-out in mm (m)
g	Gravitational acceleration in m/s^2
h_a	Height of the vacuum track in mm (m)
h	Flange height in mm (m)
h_d	Double-sided belt thickness in mm (m)
h_g	Tooth gap depth of the pulley in mm (m)
h_r	Belt back height in mm (m)
h_s	Belt total thickness in mm (m)
h_t	Belt tooth height in mm (m)
i	Ratio

k	Torsional stiffness in N·m/rad
l	Shaft length in mm (m)
l_1	Belt part length pulling in mm (m)
l_2	Belt part length pulled in mm (m)
l_B	Belt length in mm (m)
l_T	Span length in mm (m)
Δl	Elongation length of the belt in mm (m)
$\Delta l_{(1m)}$	Elongation length of the belt per metre of belt length in mm (m)
m	Mass to be moved in kg
m_B	Mass of the timing belt in kg
m_L	Mass of the linear guide in kg
m_m	Mass per metre in kg/m
m_S	Mass of tension pulley in kg
m_{red}	Reduced mass in kg
m_{spez}	Specific belt mass in kg per mm belt width and per metre belt length
m_Z	Mass of timing pulley in kg
n	Number of revolutions (rpm) in min^{-1}
n_z	Number of revolutions (rpm) of the pulley in min^{-1}
p	Pitch in mm (m)
p_b	Pitch of timing belt in mm (m)
p_{fl}	Surface pressure in N/m^2
p_p	Pitch of timing pulley in mm (m)
r_a	Belt tooth top radius in mm (m)
r_b	Pulley tooth base radius in mm (m)
r_r	Belt tooth base radius in mm (m)
r_t	Pulley tooth top radius in mm (m)
s	Belt tooth base width in mm (m)
s_B	Acceleration distance / deceleration distance in mm (m)
s_{ges}	Total distance, total travel in mm (m)
s_N	Profile thickness in mm (m)
s_V	Traverse distance where v = constant in mm (m)
s_1	Compliance of teeth and tension member elongation in mm (m)
s_2	Belt clamping compliance in mm (m)
s_3	Belt length tolerance in mm (m)
s_4	Pitch run-out in the belt in mm (m)
s_5	Radial run-out in mm (m)
s_6	Polygonality in mm (m)
s_7	Elongation due to temperature effect in mm (m)
s_8	Tangential motion in mm (m)
t_B	Acceleration time / deceleration in time in s
t_{ges}	Total time in s
t_V	Time of motion where v = constant in s
u	Pitch circle distance in mm (m)
v	Speed in m/s
v_K	Profile correction in mm (m)

v_t Profile alteration for tangential meshing in mm (m)
z Number of teeth in the timing pulley
z_A Number of teeth of the drive motor pulley
z_B Number of teeth in the timing belt
z_e Number of teeth in mesh for tension load calculation
z_m Teeth in mesh

Greek Uppercase

Φ Flank angle of the pulley teeth in °
Θ Moment of inertia in kg·m^2

Greek Lowercase

α Span angle in °
α_1 Angle of wrap in °
β Angle of wrap of the small pulley in °
γ Flank angle of the belt teeth in °
δ Angle of rotation of motor in °
ε Slope angle of conveyor belt in °
η Efficiency
λ Belt length ratio
ρ Density in kg/m^3
φ Torsion angle in °
ω_e Natural angular frequency in s^{-1}
ω Angular speed in s^{-1}
$\dot{\omega}$ Angular acceleration in s^{-2}

Subject Index

R. Perneder and I. Osborne, *Handbook Timing Belts*,
DOI: 10.1007/978-3-642-17755-2_6, © Springer-Verlag Berlin Heidelberg 2012